URBAN PLANNING IN EUROPE

International competition, national systems and planning projects

Peter Newman and Andy Thornley

London and New York

First published 1996
by Routledge
11 New Fetter Lane, London EC4P 4EE

Simultaneously published in the USA and Canada
by Routledge
29 West 35th Street, New York, NY 10001

Routledge is an International Thomson Publishing company

Typeset in Garamond by
Florencetype Ltd, Stoodleigh, Devon

Printed and bound in Great Britain by
Clays Ltd, St. Ives PLC

British Library Cataloguing in Publication Data
A catalogue record for this book is available from the British Library

Library of Congress Cataloguing in Publication Data
Newman, Peter.
Urban planning in Europe: international competition, national systems and planning projects / Peter Newman and Andy Thornley.
p. cm.
Includes bibliographical references and index.
1. City planning – Europe. I. Thornley, Andy.
II. Title.
HT169.E8T48 1996
307.1'216'094–dc20 95–42836

ISBN 0–415–11178–1 (hbk)
ISBN 0–415–11179–X (pbk)

CONTENTS

List of figures viii
Acknowledgements ix
List of abbreviations xi

Part I International, national and urban influences

1 INTRODUCTION 3

2 THE INTERNATIONAL CONTEXT 9
Globalisation 9
Global cities 12
A European urban hierarchy 14
European urban and regional policy 17
Europe, nation-states and cities 21
Creating a market in eastern Europe 23
Conclusions 26

3 THE NATIONAL FRAMEWORK: PLANNING SYSTEMS 27
Introduction 27
The legal and administrative systems of Europe 28
The 'families' of Europe 28
Trends in national governmental approaches 38
National planning systems 42
Planning systems in the British family 42
Planning systems in the Napoleonic family 45
Planning systems in the Germanic family 60
Planning systems in the Scandinavian family 63
Planning systems in eastern Europe 69
Conclusions 71

4 URBAN GOVERNANCE AND PLANNING 76
Introduction 76
Partnership, entrepreneurial and 'successful' cities 78
Approaches to urban governance 80
Urban governance and planning 85

Frankfurt: governing the world city 85
Milan: political influence 87
Barcelona: building on consensus 91
Prague: reinventing urban governance 95
Berlin: world city versus unique character 98
Conclusions 104

Part II National planning approaches and development projects: Britain, France and Sweden

INTRODUCTION 109

5 GREAT BRITAIN: THE LEGACY OF THATCHERISM 111
Introduction 111
The Thatcherite project 112
Local government 112
Reorienting the planning system 114
Thatcherite urban policy 115
Boosterism in the cities 116
The planning system and urban policy in the 1990s 119
The greening of Thatcherism? 119
A new plan-led system? 120
New experiments in urban regeneration 122
Conclusions 125

6 ENGLISH CASE STUDIES 128
Birmingham: planning for a European city 128
The city centre 128
Birmingham Heartlands 131
City Pride 133
The Birmingham approach 134
London: planning projects, centralisation and partnership 135
The story of Canary Wharf 137
King's Cross railway lands 139
Greenwich Waterfront 142
The fragmentation of London planning 145
Public and private strategic planning 147
Subregional alliances 150
Conclusions 151

7 FRANCE: REORGANISING THE STATE 154
Decentralisation of planning 155
Planning in the regions 156
Urban planning in the communes 161
The 'communal public sector' 162
Styles of urban planning 165

Urban policy 167
Problems with post-decentralisation planning 169
The re-emergence of national planning 171
Issues in French urban planning 174

8 FRENCH CASE STUDIES 176
Introduction 176
Seine Rive Gauche: city regime and international competition 176
La Plaine Saint-Denis: intergovernmental rivalry 183
The Euralille project: institutional responses to European competition 189
Approaches to urban development and planning 194

9 SWEDEN: FROM 'MODEL' TO MARKET 200
Economic crisis 200
The consensus culture 201
The ideological shift to the right 203
Decentralisation of power? 204
The Swedish planning model 208
The 1987 Building and Planning Act 210
The 1980s boom and 'negotiation planning' 212
Enhancing the market ideology 216

10 SWEDISH CASE STUDIES 221
Introduction 221
Negotiation planning of the 1980s: the story of the Globe 221
The Dennis package: restructuring urban policy for the 1990s 225
City competition and the motorway revival 225
Private interests and Projekt Österleden 227
The move to the national stage 228
The contents of the Dennis package 230
Fast-track implementation 231
A restructuring of power? 233
The Öresund Bridge: developing a new European image 236
The Örestad project 238
Malmö's official plans 239
Euroc City 240
Issues raised 242
Conclusions 243

11 CONCLUSIONS 245
Fragmentation of government responsibility 246
Urban planning and economic growth 247
Social and environmental objectives 249
Legitimacy and accountability 253

References 255
Index 277

FIGURES

3.1 The legal and administrative 'families' of Europe 29
6.1 Birmingham 129
6.2 London 136
8.1 Paris 177
8.2 Lille 189
10.1 Stockholm 226
10.2 Malmö 237

ACKNOWLEDGEMENTS

In writing a book of this scope we are clearly indebted to many people for their ideas and comments. Our approach and detailed research has matured over a number of years during which time we have benefited from innumerable discussions and debates with colleagues throughout Europe. An odd remark here or a small suggestion there has often led to fruitful avenues of exploration. We cannot possibly mention all the people who have helped in such ways but we would like to thank a number of people who have provided particular support.

Lyn Davies of the University of Reading was a continual source of useful information on planning issues in the European Union. He also read an early draft of the book and made extremely valuable comments. Beverley Taylor read several chapters and helped us improve our style and clarify our presentation. A number of people provided information for Chapters 3 and 4, where we review planning in different countries and cities, or checked over our work. In particular we would like to thank Harry Coccossis, Kostas Lelenis, Vess Troeva, Andrzej Kurzawski, Piotr Wedrychowicz, Enrico Beltramini, Guido Borelli, Gabrielle Pasque, Tim Marshall and Rafel Llusa.

For our studies of the national planning systems and projects in France and Sweden we made many visits to each country, interviewing academics and planners. We are particularly grateful for the time and trouble people took to provide us with information and interpretations about local procedures and practices. Grants from the University of Westminster's European Awareness Fund supported a number of short visits to France and made it possible for Peter Newman to make an extended stay as a visiting lecturer in 1994. Peter also benefited from a brief period of study leave from the University of Westminster which helped to speed up the production of the manuscript. Alan Jago from the University of Westminster was a constant source of encouragement and Miffa Salter assisted in the early research for the French case studies. The chapters on France benefited from the help of a large number of people who agreed to be interviewed or commented on early drafts. Jean-Claude Boyer was a continual source of insights into French urban planning, and Gilles Verpraet, Michel Micheau, Rodrigo Acosta and

Jean Robert all provided invaluable assistance in pointing to sources of information and in discussing interpretations of planning practice, politics and theory. We are also grateful to Bertrand Rouzeau, Jean Moreau de Saint-Martin, Gilles de Mont-Martin and Christophe Demazière for providing information and giving time for interviews on the case studies.

We are extremely grateful for two grants that made our study of Sweden possible. One, from the Swedish Institute, enabled Andy Thornley to spend three months in Sweden while the second from the Swedish Council for Building Research provided the travel costs to enable the research to extend beyond the confines of the capital. We would especially like to thank Hans Mattsson who helped organise this finance, acted as host, and was the source of many contacts and information. The Department of Real Estate Planning at the Royal Institute of Technology was a very congenial home for the study. Many people contributed to the Swedish chapters through interviews or reading drafts. Henrik Lagergren of the Ministry of the Environment provided invaluable insights into the workings of central government and the detailed deliberations of the Committees looking at legislative changes. Ingemar Elander and Stig Montin from the University of Örebro contributed stimulating and informative discussions on Swedish local government. From the University of Stockholm, Thomas Hall was very generous in offering comments based on his great depth of understanding of Swedish planning practice and Lennart Tonell on his unique and detailed knowledge of the 'Dennis Package' case study. Other academics who provided their views and ideas included Göran Cars, Henrik Chambert, Vibeke Dalgas, Anders Gullberg, Barry Holström, Thomas Kalbro, Björn Larsson, Gerhard Larsson, Björn Malbert, Kerstin Sahlin-Anderson, Anders Törnqvist and Evert Vedung. We are also very grateful to the planning practitioners in the Stockholm area and Malmö municipality who were so hospitable and generous in giving their time; Magnus Björkman, Leif Blunqvist, Dag Boman, Suzanne Dufresne, Karl-Erik Eriksson, Larisa Freivalds, Christian Gauffin, Hans Hede, Peter Hellsten, Hans Henecke, Jan Inghe, Bjarki Johannesson, Eider Lindgren, Jan Linzie, Lars Magnussan, Thomas Miller, Olov Tyrstrup and Leif Wretblad.

Finally we would like to thank Mina Moshkeri and Jane Pugh of the LSE drawing office for turning our scrappy notes into clear diagrams and our editor Tristan Palmer for his encouragement and patience as several deadlines slipped by.

Peter Newman and Andy Thornley
London, November 1995

ABBREVIATIONS

APUR	Atelier Parisien d'Urbanisme
BDA	Barcelona Development Agency
CBI	Confederation of British Industry
CDC	Caisse des Dépôts et Consignations
CDU	Christian Democrat Party (Germany)
CIAT	Comité Interministériel pour l'Aménagement du Territoire
CLAQ	Coordination et Liaison des Associations de Quartier
CU	*Communautés Urbaines*
CUDL	Communauté Urbaine de Lille
DATAR	Délégation à l'Aménagement du Territoire et à l'Action Régionale
DDE	Direction Départementale de l'Equipement
DRE	Direction Régionale de l'Equipement
DSQ	*Développement Social des Quartiers*
EIA	environmental impact analysis
EU	European Union
EZ	Enterprise Zones
GOL	Government Office for London
GWDP	Greenwich Waterfront Development Partnership
IROs	Integrated Regional Offices
LDDC	London Docklands Development Corporation
LPAC	London Planning Advisory Committee
NIMBY	'not in my back yard'
NPPA	National Physical Planning Agency (Netherlands)
PAZ	Plan d'Aménagement de Zone
PFI	Private Finance Initiative
POS	Plan d'Occupation des Sols
PPG	planning policy guidance
SCET	Société Centrale pour l'Equipement du Territoire
SEM	*Société d'Economie Mixte*
SNCF	Société Nationale des Chemins de Fer
SPD	Socialist Party (Germany)
SPZ	Simplified Planning Zone

SRB	Single Regeneration Budget
TECs	Training and Enterprise Councils
UDC	Urban Development Corporation
UNCED	United Nations Conference on Environment and Development
ZACs	*Zones d'Aménagement Concerté*
ZUPs	*Zones à Urbaniser en Priorité*

Part I

INTERNATIONAL, NATIONAL AND URBAN INFLUENCES

1

INTRODUCTION

During the 1990s European issues have entered the forefront of debate. The international competitive climate has made it difficult for individual countries to operate in isolation, resulting in ever increasing collaboration through the European Union. Reappraisal of the traditional roles of the nation-state has been required and this has led to tensions and conflicts. This is evident at the political level, for example in the splits within the British Conservative Party, and within the European population at large, as witnessed by the differences of opinion expressed in recent national referendums. However, a more European perspective is inevitable. Many collaborative activities are already in place, discussions on further EU responsibilities are well under way and there is a queue of countries wanting to join. The likely future scenario is for further expansion in both activities and geographical area. Urban planning will clearly be affected by this.

The Single European Act of 1992 removed restrictions and trade barriers in many areas of European activity. Although primarily aimed at increasing economic interaction the more open and communicative climate has also generated greater co-operation between city governments with the establishment of many new networks (Goldsmith, 1993). This process of 'Europeanisation' has also had its effect on urban planning. For example at the professional and educational levels we have witnessed the formation of the European Council of Town Planners and the Association of European Planning Schools. There has been a greater interest in the exchange of ideas about urban planning theory and practice. The literature has begun to respond to this. There are now a number of books which outline the planning and property systems in particular European countries (e.g. Needham *et al.*, 1993; Dieterich *et al.*, 1993; Berry and McGreal, 1995) and the EU itself has commissioned a Compendium of European Planning Systems. This body of information is extremely valuable as the awareness and understanding of planning in Europe requires the basic facts about different approaches. However this literature is fundamentally descriptive. Other research is oriented towards the processes of change in European regions and cities (e.g. Dunford and Kafkalas, 1992; Parkinson *et al.*, 1992; Harding *et al.*, 1994) but does not focus on urban planning. The purpose of this book is to fill this gap. It explores in detail the many forces influencing urban planning in cities throughout Europe and provides a framework for understanding the similarities and differences in their responses.

The book focuses particularly on political and economic forces which create common trends in urban planning and explores the scope for national or urban governments to deviate from these trends and adopt their own approach. The dominance of conservative, market oriented, politics is a world-wide trend (e.g. Gourevitch, 1986; Piven, 1991) although the extent of this rightward direction varies. The collapse of communism in eastern Europe gives an added dimension to this trend. The role of planning is to intervene in market processes to achieve certain broader aims. Its position at the interface between market and public interest means that it is particularly influenced by political and ideological positions concerning the appropriate degree of market intervention. One of the common trends throughout Europe over the last decade has been the increased entrepreneurial attitude of governments at national and city level, reinforced by the increase in competition between cities to attract inward investment. A dominant theme is the increase in what is termed 'public–private partnership'. Does this competitive imperative mean that all cities have followed the same path? Other social objectives, local priorities or political agendas can influence the direction of urban planning. Our book explores the existing degree of variation in approaches to urban planning in Europe and the underlying reasons for this.

Urban planning is the focus of our attention and so other dimensions of planning are not covered, for example rural planning or regional planning which may be influenced by different factors. But what is meant by 'urban planning'? Each country has its own set of ideas about 'town and country planning', '*aménagement du territoire*', or '*raumordnung*'. The European Commission has attempted without success to capture the variety of national legal and cultural meanings in the terms 'spatial planning' or 'territorial planning'. This book provides a framework for examining the forces that affect decisions on physical development in European cities. The aim is to explore these forces from a number of different and overlapping perspectives. There is first the combination, and relative strength, of economic and political pressures. Then there is the interaction of the influences from different levels of government, from European through national and regional to local. The third perspective relates to the relative strength of the different interests involved including the private sector in its different guises, democratic government, appointed agencies and the local community. Urban planning changes in response to this range of forces operating on cities. Our concept of urban planning is therefore both broad and flexible (see also Hall's argument, 1994). It involves the exploration of the influences on development decisions incorporating both the regulatory planning system as an important dimension and the financial and administrative urban programmes of governments.

There is also much debate over what constitutes an urban area and how this might be defined for comparative purposes. Some (e.g. Cheshire and Hay, 1989) argue that comparisons should be based upon economic relationships and that there is a need to define 'functional urban regions' as city boundaries tend to be arbitrary. For many comparative purposes this is clearly a necessary definition. However this book is concerned with the way in which planning decisions are made and the influences on these decisions. In most cases functional urban regions do not have decision-making structures. The book therefore focuses on the political entities in which debates take place and policies are formulated (see for example Fainstein *et al.*, 1992, for a similar approach).

The actions of local political authorities will clearly have an effect on planning at the urban level. It is also necessary to explore the interaction between these authorities and other interest groups. Central to this is the role of business interests and the exact nature of power relations in the catch-all concept of 'partnership'. The stance adopted by local political parties can be crucial in determining the way in which the balance is struck between different interests. Another variable is the role of central government and the extent of its involvement in planning at the urban level. Urban planning is also set within a legislative framework which provides it with legitimacy and the book explores the degree to which countries vary in their legislative approaches and the implications of this.

Urban planning in a particular city will be responding to both global economic competition and local political forces within a framework of its national legal and institutional circumstances. Given these varying circumstances the book explores whether economic forces are generating similar urban planning responses in cities throughout Europe, and the scope for individual cities to be more innovative and undertake alternative strategies. The increasingly global nature of the economy and the expansion of competition within Europe creates a new context for urban planning. Economic change has variable impacts on places and social groups. Public policy is involved in both enhancing economic forces and averting or ameliorating undesirable social or environmental consequences. The policies emanating from the EU can be seen to reflect this dual role. At the urban level new alliances and forms of government are evolving to generate and implement a range of economic, social and environmental policies. Urban planning forms part of this new policy arena.

In order to explore the full range of interconnected forces outlined above an approach is needed that can deal with the complexity involved while also allowing for comparative analysis. The increase in activity within the EU has provided the need for more information exchange and the justification for descriptive studies of cities and urban planning in different countries (EC, 1994b). However the influences on urban planning are complex and varied, for example an understanding of the detail of local practice may require an interpretation of local cultural traditions (see Booth, 1993). In order to compare cities there must be an understanding of differing national political, institutional and cultural backgrounds, and studies of the interrelationship between cities and national context are emerging (see Logan and Swanstrom, 1990; Le Galès, 1993). However the national context does not necessarily account for the variation between cities within a country and so the dynamics of change at the urban level must also be examined. In his review of comparative studies in political science Rose (1991) sets out a series of objectives which include being able to identify similarity and difference, the reasons for difference and, in policy related work, to consider the consequences. These objectives shape the approach adopted in this book. Our analysis is informed by a variety of perspectives and draws on a range of generic concepts offered by contemporary social science.

Global forces have a major impact on the operation of local markets and planning responses. Thus there is a need to undertake a wide coverage in order to detect the full range of forces operating and to identify any patterns that might be developing. However, in order to do justice to the complexity of urban planning it is also necessary to examine issues at the detailed level. Any interpretation also needs to be grounded in reality in order to be convincing and to give the detail necessary to avoid superficiality. This need for both breadth and depth creates a methodological challenge. The book is structured to allow an appreciation of the vertical hierarchy of influences on planning – from international to national to urban – and also to draw out horizontal comparisons between countries and between cities and projects.

Part I takes a European-wide approach and explores the wide range of factors influencing urban planning. It covers all the countries of western Europe and outlines the main trends in the former eastern bloc but excludes the territory of the former Soviet Union. The discussion begins with an account of the changes in the international economy and the differential impact this has had on cities. The implications of the European Union, its institutions and policies are then covered. The relative autonomy of the EU, nation-state and city in determining urban policy is raised as a key issue. The collapse of communism in eastern Europe is clearly another major event affecting the geopolitics of Europe. The discussion then continues at the level of nation-state exploring the way in which national legal approaches and institutional arrangements provide the structure within which the planning systems of each country operate. A typology of five different groupings or 'families' of countries is constructed: British, Napoleonic, Germanic, Scandinavian and East European. Using this framework the planning systems of all countries are outlined. Within each 'family' a similarity in the planning systems can be identified with further convergence a possibility. However the legal and institutional differences between 'families' creates considerable variation in the planning systems across 'families'. It is concluded that a move to a common European planning system is impossible unless these contextual factors are also changed, which is very unlikely. However there are common trends which are discussed such as the shifts in the importance of plans and increased flexibility. An important distinguishing feature between different planning systems is their degree of decentralisation. A review is then undertaken of the pressure on urban planning at the level of the city. Throughout Europe there has been a move by city governments to take a more entrepreneurial approach and compete to attract inward investment. This has led to the formation of particular coalitions of interest which have dominated urban decision-making and there has been a tendency for urban planning to support these interests. This has commonly been associated with the establishment of special, often undemocratic, agencies within which urban planning operates. However the picture is complex and considerable variety can also be detected. Factors which create this diversity include intergovernmental relationships, party politics, political leadership and community responses. Part I concludes with reviews of the planning processes in Frankfurt, Milan, Barcelona, Prague and Berlin to illustrate the interplay of forces affecting urban planning at the city level.

Part II develops the analysis in greater detail through the investigation of urban planning in three countries, Britain, France and Sweden, which are drawn from different positions in our typology. These are countries which therefore demonstrate different approaches, legal, administrative and political. In each case we analyse recent trends in the operation of their planning systems and the influence of contextual forces. This analysis is then pursued through an even more detailed examination of particular cities and development projects. In Britain there has been a trend towards greater partnership between central government and local authorities and this has affected urban planning with a more collaborative approach to urban regeneration and more importance given to plans. Environmental issues have also become more important. However, the legacy of Thatcherism continues with the private sector still playing an important role in decision-making and central government retaining strong overall control. These trends are illustrated in the case studies. Planning in Birmingham has been affected by the competitive approach taken by the city and there has been a significant involvement from the private sector. However, the local growth coalition eventually had to accept greater central government involvement for financial reasons. The London studies of Canary Wharf, Kings Cross and Greenwich Waterfront show a shift from a closed property-led approach to one with greater involvement of local government and community interests. However, financial dependency again leads to greater influence of central government and, increasingly, the EU. Although a concern for social issues can be detected as a result of the more inclusionary approach this is severely restricted and the fragmentation of planning strengthens central government's guiding and monitoring role.

The decentralisation reforms of the 1980s created a highly fragmented planning system in France. Central government is attempting to retain influence over sub-national levels by means of new contracts with regions and communes and by encouraging intercommunal co-operation. Meanwhile strong entrepreneurial attitudes have developed in many cities. However, the public sector plays a much greater role in developing pro-growth strategies than in Britain. Two case studies in the Paris region, Seine Rive Gauche and Plaine Saint-Denis, illustrate the significant impact of local market conditions and the variation in the power of communes to intervene in urban development. In Lille the integration of levels of government and local politics through the pivotal role of the mayor contributed significantly to the speedy implementation of the Euralille project.

The two chapters on Sweden show the implications of the collapse of the Swedish welfare state model and the political move to the right. In common with many European countries Sweden has experienced a process of decentralisation of government responsibilities and this has given considerable planning power to the local authority. However, at the same time central government has increased its controls over finance and the private sector has a strong influence over large developments. During the 1980s greater negotiation was detected between local authorities and the private sector and this is illustrated in the first case study of the Globe development in Stockholm. The other case studies cover strategic transport planning in Stockholm and the schemes in Malmö associated with the bridge to Denmark. These both demonstrate the increasing influence of city competition and the corporate nature of decision-making. Growth coalitions strongly influenced events and the normal planning procedures were by-passed.

The main message of the book is that throughout Europe increasing competition and the priority given to economic objectives has led to a fragmentation of the planning process and a greater involvement of the private sector. However, there has been variation in both the degree to which this has occurred and the form it has taken. These variations are due to different national characteristics in culture, intergovernmental relations and politics. Variation can also be detected at the city level caused by particular local political pressures and leadership qualities. The competitive trend and the associated involvement of growth coalitions has generated certain problems. As a result the new agenda will encompass issues of accountability, strategic vision and social and environmental concerns.

2

THE INTERNATIONAL CONTEXT

This chapter examines the changing international context for urban planning. It reviews the literature that points to a 'globalisation' of economic relations and the consequences for cities. Both 'global' change and developments within Europe have created new relationships between European cities. Related to arguments about international economic forces is the new concern with international governance and the growing influence of European institutions on urban planning. The collapse of communism in eastern Europe has created another major dimension of change within Europe. This has generated a sudden and significant shift in the geopolitics of the continent.

GLOBALISATION

There is a growing literature on the 'globalisation' of the world economy and financial system (see Thrift, 1994; Moulaert and Demazière, 1994 for succinct reviews). It is argued that several interconnected changes have occurred in the international economy. Finance has become less dependent on national regulatory systems. Changes in communications and information technologies have meant that financial and other information is available world-wide, day and night, and that the markets which this information feeds have become incessant. The 'informational society' forms an essential part of the infrastructure for global capital (Castells, 1993). In addition, large companies have decentralised production and service delivery throughout the globe and it is argued that a new international economic elite has emerged to manage this global business.

Alongside these changes in the international economy new forms of agreement between business and government have developed. It is argued that national economic planning which had been a feature of the early post-war years has been abandoned in favour of deregulation. This is most dramatically demonstrated in eastern Europe. Deregulation at the national level has been accompanied by re-regulation of the economy at an international level (by means of, for example, the GATT rounds) and also at the level of regional trading areas (in north America, Europe and the Pacific Rim). In Europe the

Single Market and closer economic integration have served to remove national barriers and bring about a new competition between cities. Cumulatively these changes signal a fundamental difference with the past. It is argued that this global economy now controls events and not macroeconomic policy conducted by nation-states (Stopford and Strange, 1991).

However, there are theoretical and empirical objections to the arguments about 'globalisation' (see Thrift, 1994). These focus less on the overall direction of change and more on its extent. How far has the change from the national to the global been completed and is the behaviour of firms truly 'global' – that is, world-wide – as opposed to merely crossing some international frontiers? It is widely accepted that national economic regulation has weakened and that international markets have become dominant. However, the nation-state may not have disappeared completely as a locus of political organisation and economic regulation although supporters of the globalisation thesis see little future for it.

Just as the nation-state is weakened by global forces and international regulatory bodies so, it is argued, has it come under political challenge from sub-national groupings (see Harding and Le Galès, 1994). Global economic forces impact at local and regional level, replacing old economic activities or creating new ones. Deindustrialisation has had profound effects on northern European cities. In the 1970s the future of many cities seemed in doubt as economic decline and inner city crises combined to point to a bleak future. The recovery of urban and regional economies has however happened in the context of a global economy and not through national economic planning. Some western cities have recovered better than others. However, the service sector industries that replaced manufacturing jobs in the 1970s and 1980s have, also, entered recession and no longer guarantee a secure future. The new international division of labour and shifting locations for production create uncertainty about the economic future of most cities. This new concern with urban economies has been accompanied by a new assertiveness in urban politics. In a situation where global forces dominate, cities have to think about how they position themselves in the market. The leaders of economically successful cities claim to be able to manage their relationships to economic interests without the interference of national government, rather like the city-states of the past. Castells (1993) argues that some European cities, such as Amsterdam, have strong political traditions on which to draw in forging new political roles. In parts of Europe there has been a strong regional dimension to economic success and cross national interregional alliances threaten the coherence of national politics and policy. The nation-state is therefore under threat from the global economy, international regulation, and potentially powerful city-states and regions.

Arguments about changing economic forces are therefore linked to the changing nature of government. Nation-states are less important as economic regulators. Their role in welfare provision has also weakened (Bennett, 1993).

Hay and Jessop (1993) see common moves away from welfare states and a 'hollowing out' of the nation-state as the main level of intervention in economic and social life. There has been considerable theoretical interest in relating these changes in the form of the state to economic change. One of the dominant debates has developed around 'regulation theory' (Stoker, 1990). It is argued that change in the economy, in the mode of accumulation, will be reflected in changing social institutions and that the role of 'social regulation' is to attempt to stabilise inherently unstable economic relations. Thus if the world economy has moved on from Fordism (the car plants typifying a mass production economy with a skilled and disciplined workforce) as the dominant mode of economic regulation, new forms of social regulation should be expected to replace the large state bureaucracies which, it is argued, helped to maintain the industrial economy and its workforce. New local modes of social regulation may appear in advance of economic change (Peck and Tickell, 1992) as in the spread of western ways of working into central and eastern Europe. One of the objections to such analyses of changing social and economic organisation is their overabstraction and there is indeed some distance between identification of new forms of governmental organisation and the idea of a global shift out of Fordism. However, regulation theory has been developed to attempt to understand changes in local governance (see Stoker,1990; Goodwin *et al.*, 1993). These authors argue that new forms of central and local government may play a role in shaping a new mode of regulation. In addition this work identifies the complexity of relationships between the economy, government and policy and the absence of any one universal process of change. The search for a simple linkage between economic change and city government and urban policy has been unproductive. Preteceille (1990) argues forcefully that change in local institutions cannot be deduced from global economic change. There is a developing view that responses to global change vary and that they do so for a number of reasons. There may be a common direction to recent change but cities still demonstrate individual responses. At the end of their summary of the globalisation debate Dielman and Hamnett (1994) conclude that the outcomes of the processes of globalisation will differ both between countries and between cities. The new global economic system has had profound effects on cities, though how economic forces feed through into development on the ground, into social and political effects and into planning policy, is dependent on local circumstances. Logan and Swanstrom (1990) come to a similar conclusion in reviewing recent approaches to comparative urban studies. It may be that nations are being supplanted by international and sub-national forces, but national contexts still make a difference in shaping the competitive advantages of cities in the global market.

Economic change has therefore had different impacts in different parts of Europe. It is partly through the redevelopment and adaptation of cities that different countries have been able to win in international economic competition. The next part of this chapter looks at some of the major trends in the economic fortunes of European cities and their responses to change. It starts by examining those cities at the top of the urban economic hierarchy.

GLOBAL CITIES

The urban impacts of globalisation may be most obvious in those cities that have been 'successful' – those cities that have captured the high-level functions of the global economy. Some authors have argued that global economic trends have produced a limited number of cities which act as centres for the control of global finance, as concentrations of finance and business services, as places where new products are produced and, simultaneously, as the markets for new products (see Friedman, 1986; Sassen, 1991). The global economy has produced 'global cities'. The large city offers its traditional benefit of agglomeration and risk reduction to firms operating in the harsh world of global markets. The literature on the global or world cities is in part an academic product but also an important tool in the competition between big cities for core economic functions (see, for example, the Coopers and Lybrand Deloitte study of London in 1991). The definition of world cities is not just an academic issue.

In Europe the notion of *Weltstadt* historically denoted the cultural dominance of a few European cities, London, Paris, Vienna and Berlin in the 1920s (King, 1990). This imperial ranking has been replaced by criteria relating to new economic functions. On the basis of seven criteria Friedman (1986) identified Zurich, Frankfurt, Rotterdam, Paris and London as first-rank cities in Europe alongside Tokyo, Los Angeles, Chicago, and New York, world-wide. Meanwhile Brussels, Milan, Vienna and Madrid represented a second division. Sassen (1991) identified New York, Tokyo and London as the centres of global power in the 1980s. In the wide range of studies employing different factors that have been undertaken in recent years (see Shachar, 1994) Frankfurt, London and Paris vie for European dominance and other cities and regions such as the Randstad offer serious competition.

The global or world city is a centre of financial power and business. It will have a prestigious office centre where that power can be displayed. A substantial policy and financial effort is required to develop a Canary Wharf, La Défense or Shinguku Centre. In addition the global cities attempt to concentrate cultural investments in opera houses, education and the arts, for example. The dominant cities will also be located next to the major international airports and, more recently, in Europe they will have high-speed rail stations.

The new global cities have other characteristics. The success of the command and control centre of the global economy is counterbalanced by increased polarisation in terms of class, gender and race (see Smith and Feagin, 1987; Claval, 1994). The exact nature of polarisation is contested however

(Mollenkopf and Castells, 1991). Whilst there is undoubted evidence of growing inequalities the polarity caused by high immigration and the creation of low wage/low skill jobs is argued only to be the case in some world cities (Hamnett, 1994). Hamnett argues that in London growth at the top end of the occupational structure has not been matched by an increase in poverty. The complex structure of housing markets also makes it difficult to generalise about the segregation of housing areas. In most cases it is also difficult to separate city from national trends in inequality. However, most commentators agree that there are fundamental economic and social inequalities in the world cities. Frankfurt for example attracted large numbers of migrants from central and eastern Europe and by 1993 28 per cent of its population were foreigners (EC, 1994b).

A further cost of world city status is the environmental problem created by increased commuting and traffic congestion. Returning to the earlier theoretical debates, how far are cities able to manage the advantages and costs of accommodating the command and control functions of the world economy? Dielman and Hamnett (1994) argue that planning policies and government structures do make a difference to the effects of globalisation on cities. In the case of New York and London, Harloe and Fainstein (1992) argue that despite the strong similarity in the ways in which governments have adapted to economic change, local policies had the opportunity to modify the urban impacts of change. Some of the key factors in shaping economic trends are identified as the social base of politics, institutional arrangements and national ideological responses to economic trends (Fainstein, 1994).

It is not just the organisation of government that mediates global economic forces. Different countries have different traditions and practices in the organisation and financing of property markets. For example, the behaviour of real estate interests differs in significant ways between London and New York (Fainstein, 1994). Ambrose (1994) examines the unique combination of elements behind British property development – developers, forms of finance, nature of the construction industry – which produces particular development outcomes including the rebuilding of London Docklands. Fainstein (1994) identifies the reasons behind the property crash in London and New York as the behaviour and belief systems of developers and investors. Property markets whether for housing (see Barlow and Duncan, 1992; Ambrose, 1994) or offices (see Lizieri, 1991; Fainstein, 1994) or industry (see Wood and Williams, 1992) operate in specific ways according to national fiscal and legal structures.

The property development industry has itself responded to the challenge of globalisation. There have been two significant international trends in Europe. The first is the internationalisation of utilities companies and their interest, not just in service provision but in development and providing planning expertise to local governments (see Drouet, 1994, for example).

The second is the lead taken by many banks (see Aczel, 1993), and property companies (see Hudson, 1993) in transforming urban decision-making in eastern Europe. These trends in property markets and the development industry have impacts on all cities, not just those at the top of the hierarchy.

A EUROPEAN URBAN HIERARCHY

European cities have been affected by global economic change. As a result of the opening up of markets, competition has increased between cities both for command and control functions and other economic development. Within Europe a process of economic integration has removed national trade barriers. The relative position of cities has also been affected by changes in communications and the opening up of eastern Europe. Improvements to systems of communication, the upgrading of airports and construction of new high-speed rail networks have had urban impacts. Increased airport capacity allows some cities, Frankfurt and Dusseldorf, for example, to compete with London, Amsterdam and Paris for international traffic. Large development projects such as that at Roissy, next to Charles de Gaulle airport, and the development poles attached to Madrid airport feed off the concentration of transport investment. The Randstad can expect to benefit from becoming the meeting point of the French TGV and German ICE high-speed train networks. High-speed trains have proved financially successful, for example the TGV between Paris and Lyons has almost recouped its development costs (Lambert, 1992). The stations on these new high-speed lines attract development such as the large shopping and commercial centre proposed for the high-speed link station at Ebbsfleet in north Kent.

Recently the European economy has been in recession. The strong markets of the 1980s in commercial property and housing have been substantially weakened. It has been argued that the single European market will bring few benefits to peripheral regions but will rather increase competition between already dynamic areas, in particular the metropolitan centres (Bremm and Ache 1993; Dunford, 1994). Recession may have accentuated this trend (Dunford, 1994). There is also new competition between the western and southern periphery and demands from the proposed integration of east European states. The opening up of eastern Europe has provided a new set of influences on the hierarchy of cities, and the capital cities of eastern Europe with their traditional national dominance are likely to benefit. Their city centres enjoyed a brief property boom in the early 1990s on the expectation of western investment, and cheap housing areas adjacent to these centres were targeted for conversion to more expensive uses. In most cases the property boom was short lived. In addition to the impacts within the dominant cities the functions of cities linking east and west have been affected by the changing relationships. As trade with the east grows so new 'gateway' cities will be defined. For example, Vienna has attracted US and Japanese headquarter offices and new commercial and hotel space has been allocated in the Donau-City project to accommodate the growth expected from enhanced east–west links. Another gateway is Helsinki, close to St Petersburg. The city is planning enhanced international facilities both in office accommodation and cultural attractions in order to perform this function as an entrance point to the east (Schulman and Verwijnen, 1993). It is the western cities that can expect to benefit most from their gateway position. In St Petersburg for example numerous projects for international hotels, business centres and technopoles have been discussed, but uncertainties about the administrative and financial systems prevent speedy progress (Limonov, 1993). The pace at which the opening up of eastern Europe feeds through to urban impacts is difficult to predict. There are political obstacles. Increasing co-operation between the European Union and the east will necessitate reducing subsidies to western agriculture and giving priority to investments in the east in order to boost demand. Peripheral regions in the west and the agricultural lobby may resist such a potential loss of EU support.

The dramatic changes in economic fortunes of cities in recent years have, not surprisingly, aroused substantial academic interest in the locational decisions of companies. There has been much analysis of the characteristics that make a place successful in a global economy and discussion of an emerging 'new urban hierarchy'. Of course, global economic change does not only affect the few cities at the top of the hierarchy. As Bonneville argues, 'internationalisation must be viewed as a process applicable to all urban spaces and which renews the urban hierarchy by excluding some spaces' (1994, p.271). His study goes on to identify a number of types of international cities. There are those with a strong technological base such as Stuttgart and Turin, those which are regional gateways to the world economy such as Milan and Lyons, and those with international regulatory roles such as Frankfurt and Geneva. There are numerous other perspectives on new city roles and positions in a new hierarchy. For example, the Netherlands National Physical Planning Agency (NPPA, 1991) proposed a fourfold typology of cities – Metropoles (London, Paris), Europoles (Amsterdam, Berlin, Milan), Eurocities (Antwerp, Florence, Birmingham) and smaller cities. Bremm and Ache (1993) suggest a hierarchy extending from 'international finance spaces' to rural backwaters. Other attempts to redefine the urban geography of Europe have focused on broad geographical areas rather than listing individual cities. Keil and Lieser (1992) suggest three types of urban region created by the new competitive world economy. Their three types of post-Fordist urban region are the internationally competitive cities, locations for new production, and those areas marginalised by economic change. Parkinson *et al.* (1992) consider three parts of Europe, 'old' and 'new' cores – the traditional and more recent concentrations of production – and a periphery. Economic diversification was seen as one of the key factors leading to the success of particular cities such as Hamburg, Rotterdam, Dortmund,

Montpellier and Seville. Elsewhere economic weakness and poor infrastructure and communications contributed to the failure of cities to adapt to the new competitive economy. Examples suggested were Marseilles, Dublin and Naples. The work of DATAR (Brunet *et al.*, 1989) suggested a *dorsale* of core regions from Milan to Birmingham and a 'sunbelt' of cities stretching from northern Italy into Spain. Additionally commentators are concerned with the future success of cities. Hall (1993), for example, speculates that Brussels, Amsterdam, and Frankfurt at the heart of the 'golden triangle' of Europe will benefit from current trends. This may mean that London and Paris do less well. Nijkamp (1993) reviews other conceptualisations of European space that are based on economic trends and changes in communications networks. There have also been a number of 'league tables' produced based upon statistical analysis of selected variables (e.g. Cheshire, 1990; Lever, 1993). These are oriented towards showing the relative attractiveness of different European cities, particularly for inward commercial investors. Such tables clearly depend upon the variables chosen and the emphasis given to the various economic performance measures.

There have been, then, a multitude of attempts to reclassify the relationships between European cities whose relative fortunes are widely considered to depend on successful adaptation to global economic trends. Political as well as economic reasons will influence the fortunes of cities. The decision to relocate some federal government functions in Berlin destabilised the hierarchy of German cities (Bremm and Ache, 1993). According to Hall (1993), Berlin is likely to benefit from this decision and from its new position as a 'gateway' to eastern Europe.

Cities are not just passive places in which international capital or prestigious functions locate but, in the new global competition for economic growth, have themselves become important actors in creating opportunities for economic development and influencing the new urban hierarchy. Cities have been repositioning themselves through marketing strategies and the creation of new images and by preparing the spaces to accommodate new economic activities (Ashworth and Voogd, 1990; Gold and Ward, 1994; Kearns and Philo, 1993). Economic development has often been accompanied by the desire for new cultural symbols – an opera house or symphony orchestra – and by competition for international events such as the Olympic Games and cultural festivals that can enhance the European and world ranking of cities. Cultural displays also serve to reinforce the assertiveness of city governments and highlight the relative weakness of national planning. Cultural industries have become an important part of regeneration strategies (see Lim, 1993 for a review). Since the example of Glasgow in 1990 other cities have used the European City of Culture annual arts festival designation as part of a city-wide economic strategy and marketing drive.

Information on European rankings has become a vital part of the analysis that informs city strategy making and marketing (e.g. Birmingham City

Council, 1994). In a Europe of competitive cities there is a tendency to ignore the fact that there are only so many international business travellers, or potential stations on high-speed lines, or opera houses to go round. Competition has losers as well as winners. The less dynamic cities and regions will, according to Bremm and Ache (1993), depend on international aid through the European Structural Funds for their futures. Economic change and competition has, as we saw in the case of the global city, not only impacted on the hierarchy of cities but had spatial impacts within the city. Here too governmental support is needed. The global cities are not alone in demonstrating social and racial tensions and urban unrest has occurred in many medium-sized and small cities (e.g. Oxford and Dreux).

The relationships between European cities in the global economy are not solely competitive. Many groupings of cities have been established to allow exchanges of information and expertise (Goldsmith, 1993). Some networks relate to economic sectors and the locational consequences of industrial change while others, for example Eurocities, relate to size. The Eurocities network was established in 1986 and meets every year to exchange experience and organise ways of promoting their common interest in the European Commission (Davies, 1993). In the next section we turn to the question of how urban and regional problems have been addressed at this European level.

EUROPEAN URBAN AND REGIONAL POLICY

Concern about the impacts of global economic change can be argued to lie at the heart of international policy-making in Europe. The establishment of a 'common market' involves removing border obstacles, promoting economic integration and providing the necessary infrastructure to enable all of the member states to compete for economic development. Alongside this economic objective the European Union has shown increasing concern for the social and environmental effects of economic change. Environmental Impact Assessment for major projects has been required under EU law since 1985 and this has been extended to plans and programmes. The Single European Act in 1986, which amended the original Treaty of Rome, legitimised this environmental interest and gave the Union a firm mandate to intervene in environmental and regional affairs (Davies and Gosling, 1994). The Union has powers to produce binding legislation and the Directorate General which deals with environmental matters acts like an international regulatory agency setting a range of environmental standards and objectives. The Commission's current expenditure programme on environmental issues – the Fifth Environment Action Programme (EC, 1992) – places a strong emphasis on spatial planning as an instrument for achieving environmental objectives. This report was given the title *Towards Sustainability*, illustrating its priorities which include integrated action and sustainable mobility. This area of policy has become increasingly accepted in the member states.

International concern is not confined to the European Commission. The Agenda 21 programme arising out of the 1992 United Nations Conference on Environment and Development (UNCED, 1992) has encouraged sub-national governments to adopt environmental policies. This has had a considerable impact; for example, according to a Swedish Ministry report 30 per cent of Swedish municipalities have passed formal council decisions to adopt Agenda 21 programmes and most others are in the process of doing so (Ministry of the Environment and Natural Resources, 1994). The environment has become an accepted policy issue at international, national and sub-national levels.

The EU has always been concerned about the distributional effects of economic change and has increasingly developed regional policies. These have been oriented around the funds available to overcome regional disparities. There was a review of these Structural Funds in 1993. The programme for the period 1994 to 1999 aims to be more ambitious and co-ordinated (Chapman, 1994), to support innovation and networks between cities. Member states make submissions for Community Support Frameworks and the Structural Funds are allocated in response. These funds are focused on three kinds of areas; the 'lagging regions requiring structural adjustments to their economy, smaller areas affected by declining industry and high unemployment', and thirdly the more remote rural areas. The extension of eligibility was marked with the inclusion of parts of the capital cities of Berlin, Madrid and London in the Objective 2 category in the recent review of the Structural Funds (EC, 1993a).

The Maastricht Treaty of 1993 further modified the Treaty of Rome. It reinforced the regional dimension with the establishment of the 'cohesion fund' to supplement the Structural Funds through assisting the poorer member states, Ireland, Greece, Spain and Portugal. A Committee of the Regions was also established to represent the interests of regional and local governments within the Union. Meanwhile the Director-General for Regional Development carried out a programme of studies to explore the emerging pattern of development in Europe. The subsequent report, *Europe 2000* (EC, 1991), focused on economic integration and social cohesion and for the first time attempted to set out a common framework for regional and urban planning. *Europe 2000* was accompanied by an urban research programme including twenty-four city case studies of responses to economic change (Parkinson *et al.*, 1992).

The Maastricht Treaty also revealed a growing urban dimension to Union policy. This Treaty for the first time gave the Union the ability to undertake measures concerning town and country planning. However this ability was made subject to both the unanimity rule requiring agreement from all member states and also the subsidiarity rule stating that the Union will only intervene when action cannot be undertaken by member states themselves. The likelihood of any Union intervention in town and country planning is therefore extremely unlikely. However, many of its other actions will have a significant bearing on planning within nation-states, such as the environmental and regional programmes mentioned above. Whilst the European Union has no specific remit on urban policy its interest in this area has increased significantly in recent years. *The Green Paper on the Urban Environment* (EC, 1990) indicated the Commission's desire to tackle the problems of European cities. This report identified the common problems in the urban environment of European cities and suggested a greater emphasis on mixed uses and higher densities. The follow-up to the report has involved pilot projects on environmental planning, urban public spaces and urban mobility. The Union is also giving priority to developing trans-European networks, including both road and rail transport and information highways.

Another illustration of the Union's concern for urban policy is the Delors White Paper on employment of 1993 (EC, 1993b). In this document sustainability, environmental and transport objectives are included to support the main task of reducing unemployment. The White Paper sought a level of homogeneity across Europe so that everywhere business could take advantage of the Single Market and each part of the Union could be competitive. The consequence of this aim is to prioritise investment in communications to reduce geographical barriers and to prioritise training to reduce disparities in skill levels. Completing the necessary transport network includes such links as those from London to the Channel tunnel, the TGV from Lyons to Turin, and a Dresden–Prague motorway. In the medium term transport investment is likely to enhance the accessibility of already prosperous cities in the *dorsale*. There are also great differentials in access to information across Europe. For example, investment in telecommunications for the citizens of Luxembourg runs at about twice the level of investment in Portugal (EC, 1994b).

The Union also allocates funds to special measures through Community Initiatives, the funding for which is top-sliced from the structural funds. The 1994 URBAN initiative (EC, 1994a) was introduced to target urban problems. As a result, 600m ecus were allocated to projects in deprived areas of cities, the bulk of which will go to those in peripheral regions. About 200m ecus will be concentrated on innovative and experimental projects in big cities, targeting areas of dense population, high unemployment and bad housing. There are other related initiatives such as Anti Poverty programmes that tackle urban problems and specific funds to support city networks. Various experimental projects, for example on economic development, environmental initiatives and the protection of historic centres bring funding into a large number of cities. Finally, the INTERREG programme provides support for cross border co-operation within the Union and other programmes of co-operation with neighbours.

Following the report *Europe 2000*, a review of spatial policy since Maastricht was put to Ministers in September 1994 (EC, 1994b). This took into account the proposed accession of new members to the Union – Sweden,

Austria, and Finland. The report, *Europe 2000+*, makes a substantial case for intervention in regional and urban policy. Mindful of the political arguments that dogged the Maastricht debates, the narrow majorities in national referendums and the continuing hostility of some member states to greater European intervention, the Commission reaffirms that spatial policy is a national, regional or local responsibility. It goes on however to also argue for higher level intervention, 'the complexity, diversity, and the growing interdependence of areas in the Union together with the increasing importance of transnational issues mean that policies also need to be undertaken at the European level to influence the development of the territory as a whole' (EC, 1994b, p.23). The case for the European level is reinforced by the focus in the report's detailed studies on cross national 'living areas'. If Europe's problems are cross national then some supranational policies are justified. A number of cross national areas are identified by the report. There is the 'Centre Capitals' region comprising the large metropolitan systems of London, Paris, Rhine-Ruhr, the Randstad, and Rhine-Main. The 'Alpine Arc' includes Baden-Württemburg, Milan, Switzerland and Lyons, while the 'Continental Diagonal' comprises the central parts of Spain and France. Then there are the New German *Länder*, the Mediterranean area, the 'Atlantic Arc' stretching from Scotland to Portugal, and the 'North Sea Regions'. The graphically named 'Ultra Peripheral Regions' includes the French *Départements d'Outre-Mer* and gives a truly Eurocentric view of the world! The discussion of urban issues in these regions includes policy prescription at a very detailed level. In the Centre Capitals region for example, *Europe 2000+* argues for green belts, traffic restrictions and actions on social exclusion. In the New *Länder* the report is concerned about the risk of development concentrating around Berlin. The Commission's interest in urban policy has become more specific.

The argument in favour of smaller towns is a strong theme of the report. It is argued that the big conurbations that have been the main beneficiaries of economic growth also impose disproportionate environmental and social costs, for example through congestion and pollution and crime. Growth in the conurbations has encouraged more commuting which carries environmental costs. It has also been big cities rather than neighbouring smaller towns that have benefited from economic integration. The Commission identifies substantial investment flows from northern to southern Europe as a positive outcome of the Single Market. However, investment has flowed into the major cities of the peripheral regions. The problem of the concentration of growth in the main conurbations may also be exacerbated by the Trans-European road and rail networks. Links between the main centres will encourage further concentration and the Commission stresses the importance of secondary networks to bring medium-sized and small towns the benefits of improved communication. The Commission encourages networks of towns to come together to tackle their problems and sponsors cross-border co-operation, for example in the integration of a system of towns around Maastricht and Liege, and another stretching from Lille into Belgium.

Europe 2000+ can be seen as an attempt to consolidate the Commission's role in urban and regional policy. The document required approval by the member states and, given the political tensions in some states, it is not surprising that drafts of the document were not available for consultation. The policy process at Commission level is therefore inevitably criticised for being closed (Kunzman, 1995). A further criticism raised is that the treatment of regional study areas overconcentrates on internal linkages; for example, the Atlantic Arc covers a diversity of circumstances. Areas within this broad region may be more interested in their link to the capitals than to other peripheral areas (Moylan, 1995). However, despite the shortcomings of international planning the Commission would like to go further than *2000+,* and at a series of ministerial meetings between 1993 and 1995 proposals for a spatial development plan for Europe were approved and a 'Trend Scenario' published under the French presidency in March 1995.

The spatial policy in *Europe 2000+* reflects an increasing concern with the borders of the Union. The Commission's analysis therefore also covers groups of potential Union members, those covered by association agreements, Poland, Hungary, Czech and Slovak republics, Romania, Bulgaria, Albania, Slovenia, Baltic States, and the countries of the southern Mediterranean. It should be remembered that the European Commission is not the only international planner in Europe. During 1994 Finland, Estonia, Russia, Latvia, Lithuania, Belarussia, Poland, Germany, Denmark, Sweden and Norway prepared a Baltic Region Plan outlining the economic development potential of the area incorporating environmental and social objectives.

The scope and detail of international urban policy in Europe have expanded and environmental policy is well developed. Since the late 1980s the Commission has been increasing its interest in urban affairs through its studies and inputs to the Treaty of European Union. In addition to this policy development work, the Commission has a substantial interest in urban policy through its expenditure programmes.

EUROPE, NATION-STATES AND CITIES

In principle expenditure programmes such as the Structural Funds are run as a partnership between the Commission, national, regional and local authorities. In practice the national governments are the main interlocutors with the Commission. Relationships vary according to the structure of nation-states. Only Germany and Belgium have developed regional structures where regional level negotiation replaces much national intervention. The funds are also increasingly geared to large projects and national governments not surprisingly want to control the negotiation. However, the relationship of sub-national government with the Commission has changed in recent years.

Between 1989 and 1994 seventeen new offices of UK sub-national government opened in Brussels, making twenty in all. The city of Birmingham has been there since 1984 (John, 1994). The UK, France and Germany have the biggest number of representations in Brussels, though German and Spanish offices typically have many more staff. The UK representative offices include some sub-national groupings of elected and non-elected bodies who come together for European lobbying (John, 1994). The Catalan office has a similar mix of public and private interests.

It is European policy-making, but more importantly European money, which has brought this ambassadorial influx to Brussels. Some cities have won large-scale European funding to rebuild infrastructure and contribute to a new style of civic boosterism enhancing local rather than national politics (see Bennington, 1994). Thus despite the strong role of national governments in relation to the Commission, sub-national governments have shown strong support for the 'Europeanisation' of government. The increasing role of the Commission with its developing relationship with sub-national government has led some to suggest that there has been a fundamental change in the relationship between levels of government. Bennington and Harvey (1994) claim that contemporary European politics is less about hierarchical co-ordination than about overlapping spheres of influence – local, regional, national, and international. Others suggest a form of developing clientelism between the Commission and sub-national governments. On the other hand, despite its significant influence, Keating (1993a) argues that the Brussels bureaucracy remains small and has failed to by-pass national governments. However, the evidence on the growth of representation of sub-national government in Brussels and the enthusiasm of sub-national governments for European urban and regional programmes indicates that this international level of policy-making has grown in importance. The Treaty of European Union established a permanent Committee of the Regions to represent sub-national governments and act as a forum for consultation by the Commission. The Committee gives further scope for sub-national governments to interact directly with the Commission and for the Commission to by-pass national governments in debates on urban issues. The Committee has limited powers, but nevertheless this forum and other informal links between sub-national government and the Commission continue to reinforce a new institutional politics. It remains to be seen how far national control over urban policy will diminish. Indeed it can be argued that as macroeconomic planning has become less important within nation-states, national urban policy has become more important. *Europe 2000+*, for example, draws attention to urban policy initiatives in France (*contrats de ville*) and Britain (City Challenge) as examples of new national policies that offer lessons for other countries.

It is undoubtedly the case that in Europe international policy has developed to respond to global economic change and its economic and social consequences. The objective of the Single Market can be seen as an expression of economic liberalism in European policy, as advocated by Mrs Thatcher. Deregulation and removal of barriers clearly fit the imperatives of a global economy. Economic liberalism dominated political discourse throughout the 1980s and political parties of the right were successful in many countries. The collapse of communism and the subsequent enthusiasm in the east for western market values reinforced the ideological dominance of liberalism. The northern European countries with developed welfare systems shifted towards 'post welfare' states, breaking down state bureaucracy and privatising services (Bennett, 1993). The ideological shift was international (Piven, 1991). In the industrialised countries and cities support for left wing parties declined although there were of course differences between countries. Piven suggests that some of the reasons for this political shift were the long-term deindustrialisation of cities, the decline of nation-states and the rise of local as opposed to class politics. Similar reasons are given by Harloe and Fainstein (1992) to explain the absence of any coherent opposition to economic policy in London in the 1980s. However, this broad shift in political ideology informs only part of the response to globalisation from the EU. The Single Market and its completion post-Maastricht was not defined by the Commission in such terms. The project pursued by Jacques Delors was one of 'harmonisation', a technical rather than political adaptation to the global economy (Barry, 1993). Delors did not espouse the old welfarist form of the state but developed the idea of *état-animateur* in which a range of institutions, public, private and voluntary, would help shape the direction of the Union. Harmonisation and the subsequent pursuit of cohesion across the Union required the technical reorientation of policies and spending programmes.

Dynamic sub-national governments, as we have seen, have set up European offices and oriented themselves to the Union's programmes. The Structural Funds and other initiatives are allocated through processes of detailed discussion with national and selected sub-national participants. As urban policy has become an increasingly important aspect of international regulation so many decisions about the future of cities have been drawn into this technical, depoliticised process of negotiation. There is therefore a specific form of politics associated with the international response to globalisation in Europe. The identification of a 'democratic deficit' in Europe should not be confined to the transfer of powers from elected parliaments to officials and Ministers, but must also include the technical and procedural rules and norms that allow selected parties to influence policy and determine the scope and beneficiaries of spending programmes. Policy-making, for example in *Europe 2000+*, is closed and decisions on funding restricted to a few knowledgeable partners.

CREATING A MARKET IN EASTERN EUROPE

The concentration of power within the EU is illustrated by the eagerness with which the countries of eastern Europe have sought to gain access. In this final section we will explore a little further the conditions in this part of Europe.

The events of 1989 in the Soviet Union and eastern Europe were revolutionary in nature leading to the collapse of the whole economic and political structures in these countries. The removal of communism was immediate – its replacement has been a much slower and tortuous process. New structures have only evolved slowly while ordinary life has had to continue, including the economic and social life of cities and demands for new physical infrastructure. Thus the planning and implementation of land use and development has had to take place in contextual uncertainty – one might even say vacuum. In this situation many *ad hoc* approaches have developed, often incorporating elements from before 1989. This is clearly a transitional period and only when the broader framework has settled down can the planning system adopt any firm shape. The discussion about urban planning in eastern Europe at the present time can therefore only cover these transitional arrangements and the trends that can be detected. The major debate is over the preconditions needed for a market-oriented planning system.

That these countries are moving to a more market-oriented approach at least seems clear. However there is no one market model. As Hutton (1995) shows, there are a number of models each of which has its own institutional support structures and set of values. The post-1989 years saw advocates such as Mrs Thatcher and the British lobby group Aims of Industry campaigning for the virtues of their particular brand of market economy. Many US and UK agencies are active in these countries linking their particular approach to financial support. Sýkora (1994) argues that in the immediate period following the velvet revolution Czechoslovakia was used as a 'playground' for liberal economics. The proximity of countries such as Germany and Austria with their social market system also has an influence, and Scandinavia has established strong relations with the Baltic states. The route taken by a particular country will eventually affect the kind of planning system adopted.

There is therefore uncertainty over the particular form of market economy each east European country will adopt, or even whether a new form might evolve (Dahrendorf, 1990). There are a number of specific issues that these countries will have to address in developing their new economic systems and supportive welfare systems (Deacon, 1992; Deacon *et al.*, 1992). There is a legacy of social problems left by the old regimes such as the amount and quality of housing and the needs of those neglected by the communist welfare system such as widows, young unemployed people and gypsies. A further difficulty arises in coping with the popular expectation that the market will solve all problems. In reality it will create new ones as inequalities and unemployment rise. Conditions attached to western credit will force the abandonment of many universal subsidies such as that supporting the low level of housing rents. Deacon (1992) has speculated on the likely direction of the east European countries based upon the strength of their working classes, the role of the Church, their economic position and the culture of individualism or authoritarianism.

However, it is too early to assess the new directions as the political systems are taking a long time to settle down. The sudden collapse of communism led to a release of political energy oriented towards a liberal and market-oriented system. However, decades of underground opposition did not necessarily create the skills and experience necessary for actual government. There has been a period of adjustment while the political systems set up in the euphoria of freedom are adjusted to the process of generating some degree of cohesion from the political fragmentation. There has also been evidence in many countries of a backlash against the limitations of the new political forces and a return to favour of some of the former communist parties, although under new labels. Political stability is needed as a precondition for changes in legislation and the establishment of new institutional structures. We can illustrate the difficulties in creating such political stability with the typical example of Poland.

In Poland numerous parties were set up to contest the 1991 elections. These represented all shades of opinion including the Polish Beer Lovers Party which won sixteen seats. However, in line with the general trend to fragmentation, this party later split into squabbling factions – Big Beer and Little Beer (*Guardian*, 1 April 1992). Twenty-nine different parties won seats and forming a government was a long process of negotiation leading to a five-party centre-right coalition. The former communists, now called the Democratic Left Alliance, were still a major force with sixty seats. The minority government found it difficult to pass legislation and the parliamentary process was described as one of personality battles. A new coalition was formed after a year but problems of generating consensus continued against a background of industrial unrest and negotiations with the IMF over debts. In order to deal with the political impasse a new electoral constitution was agreed requiring parties to achieve a minimum of 5 per cent of the vote. The government was defeated again in 1993 and new elections held under the revised rules. New alliances and parties were formed for this election. The Liberal Democratic Congress which had been the party at the forefront of the market reforms failed to pass the 5 per cent barrier and right wing supporters took to the streets to demonstrate. Meanwhile the old communist party, now professing to support capitalism with a human face, substantially increased its vote to become the largest party, although not securing a majority of seats. The Polish experience shows how the process of political experimentation with the new pluralist system has led to a blockage to decision-making. In this climate it has been difficult to change the administrative and legal framework to that required for a new planning system. The struggle between market-oriented politics and the reformed communists continues and so the exact role of the state and degree of public intervention, crucial for determining the scope of planning, has yet to be resolved.

CONCLUSIONS

The European Commission has an expanding role in urban policy. Nation-states have not lost their powers but relations with the Commission differ from country to country. Sub-national governments are developing new relationships with the international level, and dynamic cities are asserting their independence from national regulation. These are developing trends and their outcomes are unclear. Keating (1993a), for example, argues against the simplistic notion of a Europe of regions and expects to see greater institutional differentiation between states as they respond to current trends. Preteceille (1991) argues that the power of the capital cities of Europe has not been displaced by the growth of supranational government. This is more likely to hold true for the global cities where economic power is concentrated than for capitals further down the economic hierarchy.

The role of the Commission in relation to urban and regional policy will be discussed during the 1996 Intergovernmental Conference. Some member states may want to restrict the Commission's interest. Sub-national government will also want to renegotiate the role of the Committee of the Regions.

The international context for urban planning is uncertain in both western and eastern Europe. The national level seems to be reducing in significance while cities and international institutions are asserting themselves in the context of global markets. Some commentators (Castells, 1993) see a future in which city-states relate to a European state; most are less clear about directions of change and recognise the wide range of political and institutional factors involved in shaping national and city responses to global forces. In the next chapter we explore some of these factors that operate at a national level and are likely to influence urban planning.

3

THE NATIONAL FRAMEWORK

Planning systems

INTRODUCTION

Urban planning takes place within a particular national framework. This setting will create certain opportunities and constraints for the planning of a city. This chapter explores the variation in national planning systems across Europe, identifying their similarities and differences. This is particularly interesting given the expectation in some quarters that national planning systems will move towards greater uniformity. There are however severe constraints on any such move as this chapter will seek to demonstrate. Whereas in the previous chapter we noted some global economic and European-wide institutional factors which pressurise urban planning into following a common path, in this chapter we highlight the continued and ingrained differences between European national systems.

Planning gains its power through its embodiment in the legislation and regulations which form part of the legal apparatus of a particular country. The nature and style of this legal apparatus can vary from country to country and therefore have an effect on generating different approaches to planning. Secondly, the implementation of planning occurs through the administrative system which again varies considerably across the countries of Europe. Thus as Healey and Williams put it, planning systems can be differentiated by 'variations in national legal and constitutional structures and administrative and professional cultures' (Healey and Williams, 1993, p.701).

In seeking to identify the factors influencing the variation in planning systems the chapter starts with an investigation of different legal styles, generating a typology of what have been called legal 'families'. The different administrative structures will also be explored and integrated into this typology. Here the emphasis will be upon the way these structures affect the relationship between a national planning system and planning at the urban level. Other administrative trends and issues which have been taking place within cities themselves will be taken up in the next chapter. The legal and administrative review will then be used to create a division of countries into legal/administrative types. Within each type one would expect considerable similarity of planning approach and, vice versa, considerable contrasts between countries of different types. The second part of the chapter will explore the planning systems in each country utilising the typology. Clearly it is not possible to explore each system in detail – there is a growing body of literature which provides a description of the planning system in each country. Our concern is to make linkages between the various factors influencing urban planning.

THE LEGAL AND ADMINISTRATIVE SYSTEMS OF EUROPE

The 'families' of Europe

We begin with a discussion of the variation in legal and administrative structures. Zweigert and Kötz (1987) have carried out a review of the literature on the different legal families of the world (see also David and Brierley, 1985). They point out that there is a danger of basing the division on a single factor or dimension because the reality is more complex. Thus two systems may be similar in one branch of law, say private law, but different in another, say constitutional law. However, they overcome this by using the concept of legal style. By style they mean the distinctive elements which give the system its particular form. They suggest that the factors which contribute to this style are its historical development, its legal mode of thought, its legal sources and its ideology. They identify five different European legal styles or families.

The way in which local government is administered also varies throughout Europe and this variety is likely to continue. As Marcou puts it:

> despite the fact that common values of local government are nowadays accepted by all European countries, there is no tendency to harmonisation, no hint of a move towards a common standard as regards the structures, the distribution of responsibilities, the forms of local democracy, local finance or local development.
>
> (Marcou 1993, p.53)

Nevertheless within this variety certain groupings of countries can be identified. Leaving aside eastern Europe, the most common distinction made is that based upon a North–South divide (Page and Goldsmith, 1987), but this still leaves much to be explored. To carry out a more detailed analysis it is necessary to specify the factors one is using to test similarity and difference. Here we will be concentrating on the current structures of administration, although these often display their historical roots. The balance between central and local government is a central issue as the locus of power will have a significant effect on the autonomy and strength of urban planning. An analysis of the constitutions of each country is a good starting point for the

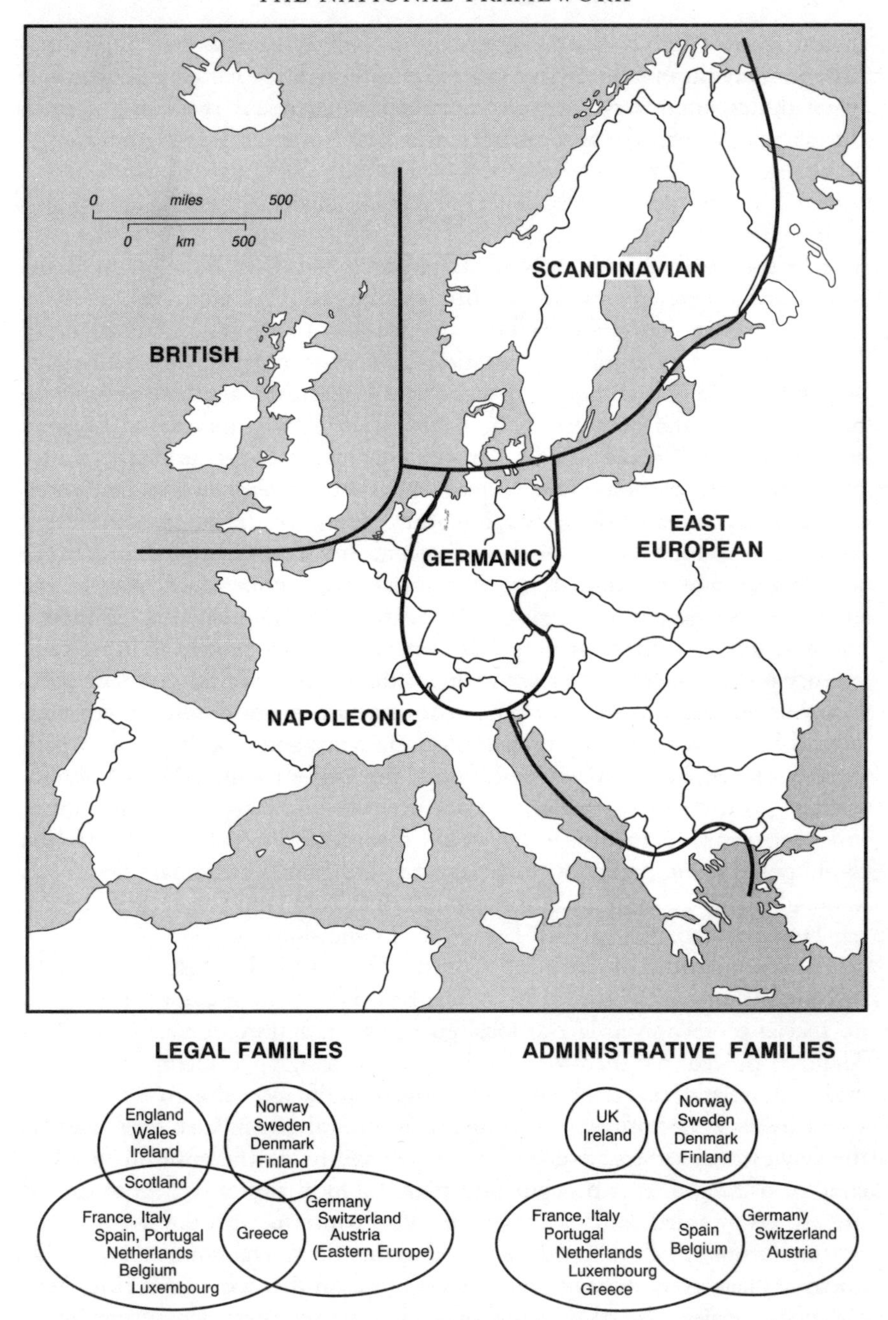

Figure 3.1 The legal and administrative 'families' of Europe

investigation. There is general agreement in the literature on these topics that European countries fall into five categories, although the names given to these five families often vary. For our purposes we will use the labels British, Napoleonic, Germanic, Scandinavian and East European (see Figure 3.1).

The British family

The British legal style is easily identified as it stands in isolation from the others. It has evolved from the tradition of English Common Law; a system of case law that has gradually built up decision by decision. The mode of legal thinking is to consider the relationships between parties and their rights and duties. There is an empirical slant to this approach and an emphasis on past experience and precedent. The British style of law originated in England and spread through colonisation to influence many countries of the world. However, nearer to home the Scottish legal system maintained its own identity because it was an independent kingdom until the early eighteenth century. In order to counter the threat from England the Scots developed an alliance with France and established strong cultural links. Thus in contrast to the isolated development of English law, Scottish law developed a distinctive combination of local customary law and Roman law. From the nineteenth century onwards Scottish law came under the influence of English Common Law, but elements of the past remain to produce many distinct differences. Ireland has a longer history of English dominance going back to the twelfth century. Seven hundred years of using the English common law system, with its principle of precedent, has not been dislodged by twentieth-century independence (Moran, 1960). In Ireland there was never any attempt to fuse the imposed common law with the existing indigenous customary law. Thus although there are some small differences, such as in land law, recent practice has been to adapt the English law to Irish conditions.

The comparison of national Constitutions also demonstrates a basic division in Europe (Norton, 1991). The British 'unwritten constitution' gives no special protection in law to local government, and experience under Mrs Thatcher proved that there were no barriers to changing a system based upon tradition. Ireland, due to its history of British dominance, also gives no special status to local authorities. Traditionally, local authorities have been seen as the deliverers of services and their scope defined by central government. The concept of *ultra vires* is brought into play if a local authority tries to exceed the powers awarded to it by the centre. This contrasts with the norm in the rest of Europe which is based on the 'doctrine of general competence'. This doctrine takes as its starting point the assumption that local authorities have a general power over the affairs of their communities. The principle of subsidiarity applies in that higher levels of government only become involved if the lower levels are unable to perform their function. This is particularly strong in the countries with a federal constitution.

The administrative system in Britain and Ireland has been described as a dual system (Leemans, 1970, quoted in Bennett, 1993). This implies that central government departments set legal and financial constraints for local authorities and act in a supervisory role. Stoker (1991b), drawing on Clarke and Stewart (1989), identifies three models for the relationship between central and local government, one of which is called the agency model. In this model local authorities are seen as agents carrying out central government policies and so central government regulations, laws and controls are formulated to allow this to happen. In this model there is little need for local taxation and finances are largely obtained from central grants. Stoker suggests that Britain is moving very close to this agency model. In the last decade the autonomy of local government has been consistently eroded as central government has increased its financial controls (Duncan and Goodwin, 1988).

Traditionally the political spheres of the two levels of government have been regarded as very distinct. This separation has been described through the concept of the dual polity (e.g. Bulpitt, 1989; Batley, 1991) indicating that the two levels tend to operate in their own separate worlds. There is little movement of professionals or politicians between central and local government. This differs from many other European countries where politicians often hold office at more than one level of government. A further distinguishing feature of the British approach is that local government is administered through political committees with the mayor playing only a symbolic role. In most countries in Europe the mayor is a powerful political figure operating with a political executive. As local government in Britain has been based on the ethic of efficient delivery of services the units of government are fairly large. As we shall see this is in contrast to some other countries in Europe where the identity of local administration is closely linked to local communities.

The Napoleonic family

This family, originating in France, is the largest in Europe in terms of the number of countries it contains. As we shall see, one of the results of this extensive coverage is a considerable amount of internal variation.

The Napoleonic legal style has a tendency to use abstract legal norms and enjoy greater theoretical debate than the British style. The great jurists of the Continent have been professors while those of England have been judges. The aim has been to think about matters in advance and prepare a complete system of rules based upon the codification of the abstract principles. The most influential period for the Napoleonic style was during the eradication of feudal institutions through the French Revolution and the establishment of the Civil Code in 1804 which provided the model for all codes of private law within this legal family. The revolutionary nature of the changes to society resulted in a very rapid and comprehensive statement of the new legal principles. As Zweigert and Kötz put it, these principles were 'founded on the creed of the Enlightenment and the law of reason that social life can be put into a rational order if only the rules of law are restructured according to a comprehensive plan' (1987, p.88). It can be seen that this is a far cry from the cautious evolutionary approach of English case law.

The influence of the French Code spread largely as a result of military expansion under Napoleon. However, when these countries regained their independence they retained the basic elements of the French Code. This was due to another factor that led to its widespread influence, namely its inherent quality. Belgium, Luxembourg and the Netherlands were very quickly encompassed into the French Empire and had the Code imposed upon them and, even though before this time the Dutch, for example, had developed an independent and detailed legal tradition, it is the Napoleonic style which remains today. Napoleon's armies also took the Code to Italy and even though it was repealed after liberation the new legal system was built upon the French one, although with some variation as the Church had considerable influence. In Spain the laws which developed in the Middle Ages and were particular to different localities have retained much of their importance. The history of legal evolution has been a gradual one of attempting to get a national code in each area of law. Mortgage and land law arrived first and a Civil Code was established in 1889. As these national codes were devised they were modelled on the French ones; however, complete legal unity still does not exist. In contrast Portugal has been fairly unified since the fifteenth century and modern codification has also largely followed the Napoleonic style.

The structure of local government and the degree of power attributed to the lower tier of government is strongly affected by historical developments. Bennett (1993) traces these influences across Europe from the early establishment of city autonomy in the feudal period. He points out the enduring nature of the commune as the basic building block of local administration which still has considerable importance in France, Belgium and Switzerland. Administrative systems which place importance on the local commune are likely to have numerous authorities at the lowest level rather than large authorities based upon efficiency of service delivery. The commune originally derived from the administrative structure of the Catholic Church, and the Reformation led to the Protestant north and west devising a new administrative system oriented round the nation-state, with a more corporatist and professional orientation. In Catholic southern Europe the process of nation building occurred later and in some countries the imposition of dictatorship delayed the establishment of democracy. These historical roots and the different paths to democracy led to different administrative structures. Ashford (1989) illustrates this through his analysis of France and Britain which he sees as polar opposites. In France the commune was an integral part of the revolutionary democratisation process and government had to work hard to impose its central authority – as a result a strong intergovernmental network was essential. In Britain the local landed aristocracy maintained their local patronage in the counties and the issue was how to ensure that local level remained subordinate to parliamentary sovereignty.

Leemans (1970) has categorised authorities which have strong links between central and local levels as 'fused' systems. The purest example occurred in France before the 1980s where, as we have just noted, strong linkages were needed between government levels. Then, as with the legal code, the military expansion under Napoleon ensured that this system had wider application. Central government established a uniform system that ensured central control over lower tiers. The department prefect is a career civil servant from central government who, in the past, appointed the local mayors. Now the latter are elected but still have a strong association with national policies through the prefect (Bennett, 1993). So within this system local government is not simply the local agency of central government but contains local representation, albeit with strong central controls. This system was extended to Italy, Belgium, the Netherlands, Portugal, Spain and Greece. In the last three cases dictatorial regimes ensured an even more centralised system but recent reforms have brought these countries more in line with the typical fused system. The degree of centralisation is also affected by the economic prosperity of a country. If it is poor with a scarcity of skills and finance then a more centralised approach is likely. During the 1980s in Spain and Belgium there has been a move away from the typical fused system with attempts to create more regional autonomy. However, in Spain the central state has retained certain important functions such as control of finance and the influence of the regions over national decisions is limited (Norton, 1991).

The Germanic family

The Germanic legal family can in some ways be regarded as a distinctive branch of the Napoleonic one. It includes Germany, Austria and Switzerland. Both families share the legal approach of codification but their historical influences differ. In Germany there was no central power to impose a unified legal system as there was in England and France, and the existing laws became more and more obsolete. The lack of central power meant that there was no authority to rationalise the various existing laws into a new order. However, there was a system to hand which could provide the necessary overall framework and this was the ancient Roman Law. Other countries had moulded the concepts of Roman Law on to their local legal traditions but because of the German political situation Roman legal ideas and institutions were adopted in a more comprehensive way and developed through the fifteenth and sixteenth centuries. Then the Enlightenment brought the view of systematic order and codification which was taken up with great rigour by German jurists but without the ideology of change associated with the French Code. The Germanic approach is more abstract and intellectual.

The German Code was widely admired for its scholarship and technical merits although its abstract conceptual language caused difficulties. Most continental countries had already adopted their codes by the time the German one was formulated, but it had considerable influence in eastern Europe. Greece was also influenced. In the early nineteenth century at the time of Greek independence there was much debate about a legal code. The spirit of revolution made the French code attractive but others wanted to base the law on that of Roman Byzantium. The latter view predominated and this led to a natural affinity to the German system, and German lawyers had considerable influence on the development of ideas. Eventually the Greek Civil Code came into force in 1946 and adopted the Germanic style, utilising German and Swiss approaches although its administrative structures were clearly modelled on the Napoleonic system.

A feature of the Germanic family is the importance given to the written Constitution (Basic Laws). This sets out very clearly the powers of different levels of government and it requires a constitutional amendment to alter the balance of responsibilities. The approach taken in the Constitution is a Federal one. The central state shares much of its powers with the regions (*Länder*) which have their own constitutions and representatives taking part in national decision-making. In Federal systems the organisation of administrative responsibilities can become quite complex. There is variation between regions (*Länder*) each having their own arrangements for dealing with the counties (*Kreise*) and communes (*Gemeinden*) within them. In Germany, for historical reasons, there are also a number of free-standing cities (e.g. Hamburg and Bremen) which possess the combined powers of the different levels. The *Länder* have less power in Austria because of the country's legacy as part of the Austro-Hungarian Empire. Within the Empire a regional level of administration had been set up as an agency of the Empire and there was no general autonomy at the local level. The same system was applied throughout the Empire and the later nation-states such as Czechoslovakia, Yugoslavia and Austria have suffered from this democratic deficit and have had to struggle with the tension between the national system and the *Länder* divisions.

The Scandinavian family

This includes Denmark, Sweden, Norway, and Finland. The family is clearly different from the British one but less distinct from the other two. The links between the Scandinavian countries stem from the history of conquests over the years by the Danish and Swedish Empires. In medieval times the Nordic laws were based upon old Germanic law and these were then centralised and codified in the seventeenth and eighteenth centuries. At this time Swedes often obtained their training in German universities. However, Scandinavia was later to be influenced by the ideas of the French Revolution which affected the modernisation of the legal codes in the early nineteenth century. Towards the end of that century increased co-operation developed between lawyers throughout Scandinavia which was linked by trade and similar language. Thus the Scandinavian legal style developed its own path and avoided the 'scientification' of the codes as happened in Germany – in fact a complete legal code has never been formulated and Scandinavian lawyers have taken a more pragmatic attitude. One of the features of the Scandinavian style is its accessibility and clear written style. It has been influenced by Germanic and Napoleonic styles and even to some extent by the British, but has also its own local characteristics.

The administrative approach of the Scandinavian family can also be regarded as a hybrid. Following the Napoleonic approach there is a strong relationship between central government and regions. Central government usually has its own agency operating at the regional level to implement national policy and staffed by personnel appointed by the centre. At the same time, although local authorities have been reorganised into larger units for efficiency, local self-government has a long history stemming from the strength of peasant politics and in some cases the far-flung nature of the countries. Local self-government is seen as one of the cornerstones of the Scandinavian constitution.

The East European family

It is difficult to say anything definitive about the countries of the former eastern bloc as their post-1989 systems are in such an early stage of development. In some cases legal and administrative reforms have occurred but they often have the appearance of being transitional. In other countries the unstable nature of post-communist politics has delayed the process of reforms and the new laws and administrative structures are still to be devised. Each country will work out its own path and, although exploring the approaches available from other countries, will want to establish a system that is relevant to its particular situation. However, given the common past under communist rule there are likely to be some similarities. Another question is the extent to which a country draws upon the system that existed before the change to communism, and common historical roots with Austria or Germany may re-emerge in the future arrangements.

As we saw in Chapter 2 the economic approach and political structures in the countries of eastern Europe are still in a state of flux. This means that the framework for market operations and regulatory processes are developing slowly and the conditions necessary for a market-oriented planning system have still to be put in place. In particular a market has to be created in land and property. Political stability is required to pass the complicated new legislation and there is also a lack of skills in market valuation and market responsive planning. At the moment even the word 'planning' is impossible to use because of its connotation with the centralised approach of the command economy; thus, other phraseology has to be adopted such as strategic or urban management.

A first step in establishing a market in land and property has been through the process of restitution, giving the land back to former owners. However this has also been a slow and fraught process. First there is the problem of information in evaluating people's ownership claims and long legal battles when there are any disputes. A second issue concerns how far back one goes in the process. Should one go beyond the communist period to include the action by the Nazis? The Polish Church is claiming that restitution should go as far back as the expropriation by Alexander the First. Having established the claims to ownership there are then alternative routes possible. One removes the current tenants and restores the previous owners – this is particularly difficult, for example, if huge blocks of flats have been built on previous farmland. There is the alternative, particularly in relation to farmland, of giving the owners an equivalent piece of land elsewhere. A common approach is to give compensation instead of actual possession. However, this raises economic problems as the government often does not have the finances to back up the compensation vouchers that it might issue. The degree of the problem varies from the simpler cases of agriculture and single family homes to the complexities of multiple occupancy flats and urban areas. In Łódź, the second largest Polish town, the ownership of about 70 per cent of the real estate units identified are being contested in the courts (Ciechocinska, 1994). This naturally puts off potential investors.

During the communist period administrative systems were clearly highly centralised. The principle of uniformity of state authority gave no room for local policies (Regulski and Kocan, 1994). The state was controlled by the Communist Party who selected the candidates for local elections, and councillors, once elected, had to implement the Party's national programme. There was usually a regional level which acted as an agent of central government. Each country has been developing its new local administrative and electoral systems and although there are different approaches there is a common tendency to react against this highly centralised past and adopt very decentralised approaches. The aim is to increase participation in local decisions including planning. However, the plans of the old regimes also claimed to involve participation and the cynicism of the public has to be overcome. Once again the reforms tend to be emerging rather slowly. It is not possible to cover all the countries of eastern Europe but some illustrations will be given to show the main trends in establishing new administrative structures.

In Poland there had been pressures for decentralisation before 1989 (Regulski, 1989). However, the concessions that were made were then followed by subsequent centralisation as the Communist Party felt a challenge to its control, the powerful centralised bureaucracy re-established its position and economic problems demanded control over local finances. A local self-government Act was passed in 1983 which accepted that local authorities would take decisions unless they were made the explicit responsibility of a higher level of government (Elander and Gustafsson, 1993). Greater discretion was given in the field of local economic planning, the appointment of officers and the local budget. However, politically the central state did not loosen its control and so the new legal openings were constrained by the Communist Party structures. During 1989 the Communist government entered into discussions with the Solidarity opposition and local government was a major theme. The aim was to give local authorities some legitimacy and this could only be achieved by freeing them from central political control. Eight parliamentary Acts were rushed through and amendments made to nearly 200 other Acts (Regulski and Kocan, 1994), and in the hurry some internal incoherence resulted (Swianiewicz, 1992). New democratic elections took place in 1989 even before the roles and responsibilities had been sorted out. The detailed regulations were left to the post-election period. Many were prepared in 1990 and were criticised for controlling local authorities too much. It has been suggested that this was a result of the monetarist national economic policy and the influence of the International Monetary Fund and World Bank (Swianiewicz, 1992). One of the architects of the Polish privatisation programme of the time spoke at a Conservative Party Conference in Britain saying that they took their guidance from Mrs Thatcher. Sorting out the powers and responsibilities has taken much longer than expected, each national political change generating new ideas. As Regulski, a major figure in the 1989 changes, says, 'it is easy to change laws, it is less easy to change institutions, and it is most difficult to change people – their mentality and habits' (Regulski and Kocan, 1994, p.64). Discussions were still underway at the time of writing.

Thus local power elites have been replaced by a new political grouping; however, the legal system is still not in place, many parliamentary provisions have proved unworkable, and there is insufficient finance (Ciechocinska, 1994). New local politicians are inexperienced and good-quality people may not be attracted. One of the threats that this poses is a continued disillusionment with local government as hopes do not materialise (Elander and Gustafsson, 1993). Another aspect of the communist legacy should be mentioned here. Residential location was not the locus of involvement or social provision. This happened through workplace and professional organisations. Thus these bodies provided such services as child care facilities, cultural programmes, holidays, sports facilities, and often health care and housing. The sectoral approach to the economy also meant that urban facilities were subordinated to industrial needs. The reform of local authorities therefore is greater than might at first appear as new services have to be covered, the expertise established and the funds found.

Administrative reforms in Poland in the mid-1970s had reduced the three-tier local government system to two. This consisted of forty-nine regions or voivodships and about 2,500 communes. The post-1989 reforms have concentrated on the commune level. However, there is agreement that at some point something needs to be done about the upper tiers. At the moment the voivodship remains a central state agency although this is supplemented by a consultative body of people elected by the communes (Marcou and Verebelyi, 1993). However this is probably a transitional arrangement and many believe that the country needs about twelve regions with an economic policy emphasis. If so then another level will be needed between these regions and the local authorities.

It has been said that in Czechoslovakia the political and legal aspects of local government reform were co-ordinated with greater harmony (Erlander and Gustafsson, 1993). The new local government system was established in 1990 in which local communes would exercise self-government and take on some of the previous responsibilities of the state. The outstanding feature of the reforms was the fragmentation of communes as, in reaction to past centralisation, every settlement demanded its local autonomy. The number of communes was thus increased by 40 per cent to reach 5,769 units. In Prague there is also a lower tier. This fragmentation makes it particularly difficult to implement any strategic policy. At the regional level the Regional National Committees covering the ten provinces of Czechoslovakia have been abolished. At the time of the 1990 reforms there was a debate over the nature of a second tier authority. It was felt that a second tier would complicate the role of communes regarding property and financial resources (Zarecky, 1994). The only intermediate level is the district which consists of a state administrative office, of which there are seventy-five in the Czech Republic.

According to Dostál and Kára (1992), in the existing economic climate the Czech administrative system is near collapse. This is because of the costs of improvements needed to infrastructure and housing, the lack of relevant experience, and the small size (average 1,800 inhabitants) of the communes. As in Poland many social services were provided by enterprises in the past. Many are closing down or are privatised. In a climate of cost consciousness these social budgets will be the first to suffer cuts. There is a danger that the small and financially weak local authorities will not be able to pick up this welfare function (Illner, 1992). This increases the pressure to create larger units of government.

TRENDS IN NATIONAL GOVERNMENTAL APPROACHES

This review of legal and administrative approaches has been very summarised and clearly does not do justice to the richness of the subject (for further details see, for example, Batley and Stoker, 1991; Bennett, 1993; Marcou and Verebelyi, 1993). However, it has been sufficient to identify some of the main differences across Europe. A caveat needs to be added at this point. So far an assumption has been made that if a country has a set of legal and administrative regulations that relate to planning then these will shape and control the way in which planning is carried out. This implies an acceptance of the legitimacy of laws and the use of formal political machinery to carry out decisions. The appropriateness of this assumption, though, varies throughout Europe. In crude terms it is possible to distinguish a North–South divide with countries in the North more likely to conform closely to legal and formal arrangements. In the South there is a greater tradition of alternative informal mechanisms and greater flexibility in conforming to the law. Thus in certain parts of Europe, such as Italy or Greece, it is quite common to find a disparity between the formal laws and regulations and implementation.

The clearest conclusion that can be drawn from the above review is that the British family is very distinct and has very little overlap with the other families. It has developed historically in isolation from the rest of Europe. In administrative terms the British family is also significantly different from all others, having a much sharper division between central and local government and a high degree of centralised monitoring and control. Whereas most other countries comply with the principle of subsidiarity and general powers for local government, in Britain and Ireland the powers of local government are defined by central government and are highly constrained. A second family that is fairly distinct is that of Scandinavia. Here the distinguishing characteristic is the mixed and flexible nature of its structures. It draws on the other families while also having its own Nordic characteristics. Its administrative structures combine elements of local self-sufficiency with national supervision.

The picture becomes a little more complex when looking at the remaining countries. It is possible, however, to distinguish a broad division revolving around the Napoleonic and Germanic styles. The first centres on France and the Civil Code and an administrative structure with a strong interplay between central and local government. The centre has strong power which it extends down through the prefectural approach, thus having a presence at the more local level. Meanwhile the local commune has strong claims to autonomy, based upon the old links with the Catholic Church. One feature of the resulting 'fusion' is the exchange of politicians between the different levels and multiple political office holding. Although each country has its own particularities it is possible to place a number of countries in the same group as France, largely as a result of the Napoleonic influence; these would be Italy, the Netherlands, Belgium, Luxembourg, Portugal and Spain. However, development towards stronger autonomy for the regional level in both Spain and Belgium means that they cannot be described as having a 'fused' relationship between central and local government and that they could develop into a more federal structure. Italy also presents its complications as the unification of the country has not fully taken hold and differences are still strong between North and South. The extreme variation in the level of economic development across the country and the power of informal networks also add to the fragmented picture. Although Greece draws upon the Germanic tradition in its legal style the strong Napoleonic administrative approach probably gives it greater affinity with the other countries in this family. The Germanic legal style has an approach based upon a written code which is even more rigorous and has different historical roots. The administrative approach is based upon a strong constitution and a federal system with important powers allocated to the regional level of government. Although exhibiting some differences, Germany, Austria and Switzerland form a clear grouping.

Although it is possible to distinguish this typology of different families there are also similar trends being exhibited across Europe in the reorganisation of the national structures of government. A common shift can be detected towards greater decentralisation and regionalism, encouraged – as we saw in Chapter 2 – by the initiatives of the EU. As Batley (1991) has pointed out, the language of reform is almost universal and focuses on deregulation, improved efficiency and responsiveness. The aim is to increase flexibility, reduce central supervision and increase user satisfaction. However, this decentralisation trend has not been experienced in Britain (Duncan and Goodwin, 1988; Crouch and Marquand, 1989). Throughout the 1980s the power of local government was systematically eroded and some metropolitan authorities even abolished. It might be said that some of the aims behind decentralisation were met in the British case by devolving activities to market process, and John Major's Citizen's Charter initiative fits into the trend for greater responsiveness. However, the British approach is highly orchestrated from the centre and lacks the development of a local political element. It has been suggested that the long tradition of a dual polity mentioned earlier has led to an absence of channels for conflict resolution (Batley, 1991).

Elsewhere in Europe one of the features of recent trends has been the growth and expansion of such conflict resolution procedures involving an increase in intergovernmental activity and complexity (Marcou and Verebelyi, 1993). In relation to the Netherlands, Toonen (1987, 1993) has stressed this feature. He says that the Netherlands should not be regarded, as it has often been in the past, as a decentralised unitary state where the balance between the centre and local autonomy is maintained through a hierarchical system of procedures; rather, it should be regarded as a consensus state. In the latter approach unity is achieved through consensus-building and mutual adjustment between government levels. The German federal system has always had its procedures for negotiation between the various levels but this has become increasingly important in recent years (Hesse, 1987; Balme *et al.*, 1994).

In most cases the trend to decentralisation has involved the devolution of powers to local or regional levels. This has often been a long-term process, such as in Italy where over the last forty years there has been the gradual move from a centralised state to a mixed model (Leonardi *et al.*, 1987) in which the region has gained in political importance, as witnessed for example in the rise of the Northern League political party. In recent years legislation has reinforced the role of the province. In the formulation of a democratic state in the post-Franco era, Spain has drawn strongly on the German federal model and created Autonomous Communities. However, the implementation of this system is subject to political tension between the centre and the regions. It has been said that the 1985 Local Government Act reversed the regional trend as it gave local authorities greater ability to work directly with the central state and hence bypass the regions (Valles and Foix, 1988). The Belgian case perhaps demonstrates the breakdown of the central state in most dramatic fashion as powers are transferred to the three regions; here also, though, the exact relationship is still evolving (Delmartino, 1988).

Perhaps enough has been said to show that the issue of decentralisation is not a simple one (Goldsmith and Newton, 1988; Wolman, 1990). It can be viewed in many different ways and can vary depending on the aspect of government one selects for analysis. Bennett (1993) has identified three major dimensions: the extent of delegation of powers, formal legal responsibilities and the political and financial resources to provide services. It is possible to have decentralisation in one aspect but not in another. For example the changes in Sweden have been described as a decentralisation of responsibilities but a centralisation of finances (Erlander and Montin, 1990). One of the reasons for the complexity in the trends regarding central/local relations is the diverse objectives which can be pursued. Pretecielle (1988) shows how the decentralisation process in France, which included greater powers for the lower levels of government and an elected regional authority, was initiated by the Socialist Party drawing upon the ideology of self-government and democracy. The changes were also supported by the right who saw opportunities for improving efficiency and financial control. According to Mény (1988) the French decentralisation reforms have not changed the government process itself but simply the composition of the decision-making elite. The same mix of financial efficiency and democratic accountability has been noted in the Swedish movement towards decentralisation (Montin, 1993). The use of decentralisation to create financial savings is part of the general fiscal crisis of the state. This is another common trend across Europe over the last decade or so, although the timing of the difficulties has varied between countries (Mouritzen, 1992). It has often been noted that the centre has tried to export some of its financial problems on to lower levels of government. In some cases, however, this strategy has met with opposition. Germany provides one example where the demand for greater democracy has been important, with strong grass-roots pressure, highlighted by the activities of the green movement, for a 'revival of politics from below' (Hesse, 1987). Hesse cites this as evidence of growing opposition to the off-loading by the central state of its economic and social problems.

We have now established a framework for exploring the range of planning systems in Europe. Some common trends in governmental arrangements have also been noted, particularly the move to decentralisation and regionalism underpinned by the desire for both greater efficiency and local accountability. The rest of this chapter will explore the planning systems in each country. As these systems are one element in a nation's legal and administrative arrangements it will be expected that the features identified above will have an important influence on the nature of planning systems.

NATIONAL PLANNING SYSTEMS

Planning systems in the British family

Britain

The first comprehensive planning legislation was passed in 1947 and, although this was largely replaced by subsequent Acts, the basic principles set out in this post-war period remain relevant. These principles include the division of planning into three broad functions – development control, development plans and central government supervision. Development control involves a local authority receiving applications for development and making a decision, taking into account the policy framework and detailed local circumstances. This decision is made by local politicians based upon advice from local planners, who have a considerable amount of administrative discretion in formulating their view. The local authority also prepares a development plan which sets out the land-use policies for its area. This plan is an important consideration in making the decision on an application but it is not legally binding and 'other material considerations' can over-ride it in some situations. Originally these plans had to be approved by central government but this has now changed. Central government is responsible for enacting the legislation and issuing policy guidelines, which are also important considerations in making a development control decision. Another important feature of the system is that applicants have the right to appeal against a development control decision and the appeal is decided by central government. This is particularly important when central and local government have different views on planning. In making their decision, local planners will always be thinking about the implications of the applicant going to appeal and this can therefore influence their judgement. This broad system is very flexible and involves a lot of interpretation. Over time there has also been considerable fluctuation in the degree of involvement of central government and also the strength given to the development plan. Let us look in more detail at the current situation.

Central government policy has a strong influence on the rest of the planning system through the instrument called Planning Policy Guidance Notes. At the time of writing there were twenty-two of these in operation.

There are two kinds, one topic-based and the other area-based. Examples of topics covered are Green Belts, Housing, Archaeology and Control of Advertisements. These often have a big impact on local policy such as whether to allow out of town shopping centres. The second kind is Regional and Strategic Guidance for specific areas which sets out in very broad terms the planning policies to which lower tier plans have to conform.

Development plans are at two levels, structure plans prepared by counties and local plans by district authorities. In metropolitan areas these two tiers are combined into a Unitary Development Plan. Since the 1991 Planning and Compensation Act these plans have been given more importance after a decade during which they were downgraded. A public inquiry is held before the adoption of the plan, which does not have to have central government approval, although the government has reserve powers to step in if it wishes. The principles of the development control part of the system have not changed much over the years, although there has been an increase in the practice of negotiation over planning permissions, called 'planning gain' (renamed 'planning obligations' in more recent legislation). These negotiations, which are conducted in secret, cover such matters as financial responsibilities, provision of infrastructure and whether the developer will provide certain community-oriented uses in the scheme. There is no right for third parties to challenge planning decisions.

As we have already noted Scotland has its own legal and administrative arrangements and this causes variation in the planning system there. The key central government department in Scotland is the Scottish Office which has considerable autonomy over planning matters. One of the main differences is the development of a stronger strategic approach (Begg and Pollock, 1991) with more comprehensive statements of national planning policies. This has included the preparation of National Planning Guidelines setting out policies for issues of national importance, providing a framework for local authority plans. These have recently been replaced by a new instrument, the National Planning Policy Guidelines, which attempts to give greater clarity to strategic planning policy (Lloyd and Black, 1995). Reflecting its varied geography and settlement pattern the administrative structure of Scotland used to contain a mixture of single-tier and two-tier arrangements. The regional level used to produce regional reports which were another distinctive feature of the Scottish system. However, the British government has recently conducted an administrative reform creating a new system of single-tier, all-purpose authorities of very variable size causing problems for strategic planning. Northern Ireland is also a special case with the curtailment of local authority powers in 1973. Planning matters are concentrated in the Northern Ireland Office which combines most local planning responsibilities and central government powers.

In a comparative context the distinctive features of the British system are therefore its centralised nature, with central government having the ability to influence local policy through its Planning Policy Guidance, and its role in deciding appeals, which can be over matters of substance not just administrative procedures. A second feature is the amount of discretion in making decisions on planning applications. These decisions have to take into account central government policy guidance, development plan policy, local issues such as access, traffic generation, effect on neighbours and any other 'material consideration'. The balancing of these factors, in which the development plan is not binding, gives scope for flexibility. Variation can also be generated over time by central government through the adjustment of the priorities it gives to development plans and/or central guidance. The Planning Policy Guidance approach allows central government to change planning policies very quickly. Compared with many other countries there is a noticeable lack of significant plans at the national and regional levels.

Ireland

Central government passes legislation and formulates planning regulations which set out the procedures for development plans and planning permissions. The legal basis for planning is the 1963 Local Government (Planning and Development) Act with amendments in 1976, 1982, 1983 and 1990. Central government publishes policy guidance statements and can require amendments to development plans. It has also produced the National Development Plan (1994–9) which is a vehicle for attracting EU funds. This plan is therefore strongly related to the opportunities provided by the EU financial programmes and does not attempt a comprehensive assessment of national priorities. It has implications for physical planning because of the impact of major infrastructure projects. The Irish approach is highly centralised.

The whole country is designated as a single region in EU terms. There is no interregional policy as this might undermine the strategy of attracting EU funds for the whole country. However, a regional tier of eight authorities was created under the Local Government Act of 1991 and these came into effect on 1 January 1994. The function of these authorities is to co-ordinate the provision of public services in the region, bringing together policies of central government ministries and the various local authorities in the region. The regional council is made up of members appointed by the constituent local authorities. Part of the co-ordination process involves reviewing the local authorities' development plans for consistency and the implementation of European funded programmes. The region has to produce a report at least every five years.

The local authorities are responsible for producing development plans, which are obligatory, indicating the development objectives for their areas. Counties and urban authorities (eighty-eight in all) are given the same powers and functions. The plans do not have to be approved by a higher authority. The local planning authority also decides on planning permission which has to be sought for all development. In making the decision the authority will take into account the development plan and appropriate local planning requirements. Applicants can appeal against this decision to the Planning Appeals Board which is an independent tribunal. This Board was created in 1976 – previously appeals were heard by the Minister as in the UK. The aim was to remove the appeals from political influence by having an expert Board to adjudicate (Fehily and Grist, 1992). The decision of the Board is final and can only be challenged on a point of Law by the High Court. Third parties also have a legal right of appeal.

Planning systems in the Napoleonic family

France

A comprehensive planning system is also a post-war creation in France. Until the early 1980s the system was highly centralised but since then planning power has been shared with local government. The French system is characterised by a national codified law – the *Code de l'Urbanisme et de l'Habitat* which dates from the 1950s but is continually revised. The Code specifies types of planning activity and regulations for the development of land. Most of the local government system dates from the French Revolution. However, the regional level is a new addition and the distribution of competencies between levels, and relations between local and central government, were radically reviewed in the 1980s. There are four levels of government with an interest in planning – state, region, *département* and commune. The three sub-national levels are run by councils of elected representatives.

The state produces national rules and guidelines. It also periodically reviews the roles of other levels, and major infrastructure development is determined by the state. In the 1960s and 1970s the planning of large-scale housing estates through ZUPs – *Zones à Urbaniser en Priorité* – was controlled from Paris. More recently the influence of central infrastructure planning can be seen in the expanding TGV network. Since the mid-1980s the President himself took an active role in deciding the location and form of a range of *grands projets*, including cultural and office developments, in the Paris region. The national Ministry covering local government and the environment incorporates regional and urban planning and has local offices in most big towns. The government's *préfets* representing the state services still have considerable influence over local government. In addition the government is advised by DATAR – Délégation à l'Aménagement du Territoire et à l'Action Régionale – on regional and urban issues.

Periodic national plans set overall economic policies. Both state and region have specific interests in economic planning and co-ordinating transport, education and other public investments. The twenty-two regions (except Ile de France) can prepare development plans, but few have done so and there is a current debate about regrouping the regions into larger units. The *départements* have no specific land-use planning powers but have a wide range of functions which impact on urban issues and planning decisions.

The Loi d'Orientation Foncière in 1967 established a two-tier system of plans. Strategic plan making is not obligatory and by 1990 there were about 200 *Schémas Directeurs* (Acosta and Renard, 1991). A *Schéma Directeur* can be initiated by one commune but must carry a majority of the communes in the area concerned. The fragmented structure of local government means that strategic level plans must be an intercommunal responsibility usually produced in conjunction with the state. The content of plans varies but will be concerned with general objectives and major infrastructure and must fit with the programmes of the state and other public bodies.

The second level of plans – *Plan d'Occupation des Sols* (POS) – is the responsibility of the communes. All communes with a population of over 50,000 have a plan. Plan production is not obligatory but brings with it the power to determine applications to develop and, for some communes, the right to buy key development sites. The POS is a strict zoning plan. U zones are already urbanised and development will generally be allowed. NC zones are mostly agricultural, ND zones conservation areas. NA represents future growth areas. The POS will also indicate the socio-economic characteristics of the area. The state and other public bodies, the chamber of commerce for example, will normally be involved in plan production and there will usually be other public consultation and public hearings for objectors. Where there is no POS the state has comprehensive planning powers, through national regulations, governing the location and appearance of buildings. Special provisions cover sensitive areas such as mountains, coasts and historic centres. Detailed redevelopment plans – *Zones d'Aménagement Concerté* – follow similar plan production procedures.

Planning law also provides for considerable implementation powers, including compulsory purchase of land for public purposes. Communes with a POS can require vendors to notify them of land sales and have the right to buy at market value. This procedure enables action to be taken against land speculation and also allows land banks to be built up. Developers may be required to contribute to infrastructure costs through a range of mechanisms, and development rights above a standard level may be sold by the commune. Mayors of the communes issue *permis de construire* to applications which conform with a POS. Development which does not conform to the POS is illegal. Illegal development is more of a problem in the south of France. The state's *préfets* have the power to intervene if communes follow illegal procedures and the regional Cour des Comptes checks financial probity.

Luxembourg

Planning in Luxembourg operates at only two levels: the nation-state and the commune (EC, 1994b). The Council of State is responsible for drawing up a Master Plan (*programme directeur*) setting out the most important strategic objectives and the means of implementation over a period of 10–20 years. This plan is not binding on third parties but is an instrument of co-ordination. A plan was produced in 1988 with a time horizon to the year 2000, and the Ministry of Planning began to prepare a new plan in 1995. The Ministry also prepares development plans which are vehicles for implementing the Master Plan, are legally binding, and contain participation procedures. These cover such areas as the surroundings of the airports and national industrial zones. In principle communes are responsible for their own planning if they have plans which cover their entire area. The plans are drawn up by the commune and submitted to the central government body, the Planning Commission. The plans must conform to the higher level development plans and there is a local participation process.

Netherlands

There are three layers of government – the national level, the provincial and the municipal – each with clear roles set out in the Constitution. The Netherlands is described as a decentralised unitary state. This means that each level has independent legislative and administrative powers under the overall supervision of the central state. Each level can formulate its own regulations as long as it does not conflict with a higher level. All levels are involved in planning (see Brussard, 1986; Davies, 1989, for details). As Davies (1989) points out the planning system is formed by its broader legal context based upon a Napoleonic code and written constitution. The state guarantees citizens rights through a framework of rules and the concept of legal certainty (*rechtszeterheid*). In planning terms this translates itself into an emphasis on plans and regulations which have statutory force. Thus in principle if a proposal for development conforms to the plan and building regulations it cannot be refused.

The national government prepares the Physical Planning Act which sets out the framework for planning throughout the system. It also prepares statements of national planning policy called National Physical Planning Key Decisions. These co-ordinate the various sectoral bodies and are open to public discussion. They are consolidated in occasional Reports on Physical Planning in the Netherlands which set out national physical planning priorities (Faludi and Valk, 1994). The most recent was adopted in 1993.

Provinces, which have elected governments, vary in size and their budgets are small as most service delivery is carried out by municipalities. Toonen (1987) has described their role as one of 'intermediary cum co-ordinator' and as such they have an important role in physical and regional economic planning and environmental matters. At this level the general directives of the different ministries are integrated and translated into orders for implementation by the municipalities. The key instrument is the regional plan

(*Streekplan*) which is administratively binding on municipalities and other public sector agencies (Davies, 1989) and outlines the future development for the whole province. It is the basis for the approval of lower tier plans. The regional plan can be supplemented by reports on specific topics (Brussard, 1986).

The responsibilities of municipalities are characterised by autonomy of action within their own areas, but this is only loosely defined in the Constitution. In taking any initiative their actions are subject to supervision – municipal plan proposals and budgets need the approval of provinces. However, provinces do not command the municipalities on what they should do; the relationship is more one of checking and negative control – a 'blocking power' (Hupe, 1990). The arrangements for municipalities to implement national legislation are referred to as *medebewind* – translated as 'co-government' (Hupe, 1990). Thus the interaction between the three levels of government is a complex one of relative autonomy and supervision which is typical of the 'fused' nature of this family in the typology. In relation to planning, Davies (1989) has described it as very much 'a matter of vertical coordination between, and horizontal coordination within, the different levels of government' (p.340). This reference to horizontal co-ordination reflects the numerous departments and agencies involved at each level of government. The system is built around a high degree of consultation, either obligatory or voluntary (Brussard, 1986), and hence there is a lot of interaction and opportunities to influence policy at other levels. The consensual nature of Dutch society is often mentioned in which an attempt is made to reach a compromise between different views. It is said that this is based upon its social composition which comprises Catholic, Protestant and secular groups and results in the need to find a 'politics of accommodation' (Hamnett, 1982).

There are two kinds of plan that operate at the municipal level: the structure plan (*struktuurplan*) and the detailed *Bestemminsplan*. The first, which is not obligatory, can be for all or part of the municipality. It is intended to provide a context for the *Bestemminsplan* but is not binding and does not require provincial or national approval. The *Bestemminsplan* is the most important planning instrument at this level and is the only plan which is legally binding on citizens, organisations and public bodies. This contrasts with the previous plans mentioned which are all indicative, showing the wishes of the various government bodies but which cannot actually be used to prohibit development. The municipality is obliged to produce a plan for all of its area that is not already built up and can also if it wishes produce plans for built-up areas and single topics. According to Davies (1989) virtually the whole country is now covered by this type of plan. They show the intended use of every piece of land and determine whether a building permit will be given.

In recent years there have been moves to decentralise policy, but there has also been dissatisfaction with the pace of this (Hupe, 1990). This dissatisfaction is based on the perceived problem of increased centralisation through financial control and more detailed supervision of local policy. In recent years some housing matters and urban renewal have been transferred to the municipalities. However, the small size of many municipalities is a brake on the ability to decentralise powers. This links to the concern in the late 1980s that the administrative structure was not suited to dealing with metropolitan areas. A government advisory commission – 'The Montijin Commission' – reported in 1989 that there was a need for some kind of metropolitan government. A law was passed by national government in March 1994 (the Kaderlaw) which stated that the aim of central government was to develop metropolitan governments for seven urban areas. At the time of writing this law has still to be implemented; it is the municipalities in the seven areas that are required to set up the new bodies and this is subject to local political negotiations. Its future is still uncertain, but the first new metropolitan government is expected for Rotterdam in 1996. In many of the new metropolitan areas the new governments will take over the responsibilities of the provinces. The aim is that they will be able to produce more rational planning as they encompass both inner city areas and suburbs. It is hoped that they will be more innovative. Current plans in Amsterdam are for departments to cover city planning, public housing, economic affairs, transportation and the environment. The new authority will produce a structure plan for its whole area.

The Dutch system has been described as plan-led (Davies, 1989), giving considerable certainty for developers and citizens. There are also participation opportunities at all levels and a high degree of organisational interaction. Discretion at the level of planning permission is only possible if specified in the plan. With such a comprehensive and strong system of plans it is not surprising that one of the issues that has been much discussed in recent years is the way in which flexibility is introduced into this system (Davies, 1989; Thomas *et al.*, 1983). Needham *et al.* (1993) have clearly summarised the opportunities for flexibility that exist. First there are opportunities to revise or withdraw the plan. A partial revision can be made if the municipality wants to grant permission for development that does not conform with the plan, but this takes several years. If the municipality is in the process of preparing a new plan it can grant exemptions from the conditions in the existing plan if these exemptions are consistent with the proposed plan. This clause has been widely abused by municipalities and was tightened up in new legislation in 1985. This Act also created the opportunity to draw up the plan in more global terms with the details delegated to the municipal executive. The plan can also specify areas for flexible treatment by defining certain aspects that can be amended or over which the municipality can grant slight exemptions or impose more onerous conditions.

A final characteristic of the Dutch system should be mentioned. It has been said that it is the 'most planned country' in Europe (Dutt and Costa,

1985), and the public acceptance of this probably relates to its highly urban nature and its history of organising sea defences. One legacy of this background is that many municipalities have a very active land policy. They acquire the land, service it and sell or lease it to private developers or housing associations. Amsterdam for example owns 75 per cent of its territory. This affects the relationship between the state and the developer. The relationship is rarely just one of regulation but also involves partnership and negotiation and the municipality is oriented towards creating development opportunities (Davies, 1989). This attitude will also influence the way in which the *Bestemmingsplan* is drawn up – having invested much effort and finance in preparing land for development the municipality will want to ensure that developers will be attracted and will therefore take their wishes into account in preparing the plan.

Italy

The Italian National Constitution sets out the responsibilities of different government levels; however, it has been said that national integration has never really taken hold and thus regional fragmentation remains a practical reality (Balme *et al.*, 1994). This is of course most clearly evident in the North–South division and is expressed politically in the regional nature of the political parties to emerge in the 1990s – the Northern League and the neo-fascist MSI.

National government is responsible for producing planning legislation, but since 1972 this power has been shared with the regions and this has created considerable complexity. In theory there can be a two-level legislation in which national law sets out guidelines for regional law; however, these guidelines have not been produced and the relationship is much more convoluted (see Cattaneo, 1986). There is no national physical planning as such although clearly national sectoral policies such as transportation or development policies aimed at the South have an impact on planning. The national state co-ordinates the administrative activity of the regions, formulates policies on interregional balance, identifies planning guidelines on aspects of national interest and great environmental impact, and formulates planning standards (Ave, 1991).

Although there have been more recent modifications, the basis of the planning system is the 1942 Act which sets out a hierarchy of plans. These can be at three levels: regions, communes and sub-communes. The principal planning authority is the municipality or commune of which there are 8,091 varying considerably in size. The principal planning instrument is the master plan (*piano regolatore generale*) which zones the land use of the commune and is implemented through a more detailed plan (*piano particolareggiato*). The master plan, with a strong emphasis on physical and design issues, is the key instrument and sets out objectives and guidelines concerning new development and conservation throughout the commune area. The plan has an indefinite lifespan and this leads to changes being made over time, case by case, in a rather random fashion. The rewrite of the master plan is a lengthy and costly process.

The detailed plan puts the master plan into practice, but only a small number of these have been prepared, approved and implemented (Ave, 1991). However, several other kinds of plan have been introduced which play a similar role. These include planning instruments to cover low-cost housing, land subdivision, urban renewal and plans for productive activities such as industrial estates, commercial areas and tourist activities. As an example of the fragmentation of the planning system that results we can look at retailing. A law was passed in 1971 requiring communes to devise a specific plan for retailing to be used in the determination of applications for trading licences. The applicant must also seek registration on the traders' register from the local chamber of commerce (Veraldo, 1979). All communes must also prepare Building By-laws (*Regolamenti edilizi*). At the sub-commune level there are executive or implementation plans which cover small areas and, in conformity with the master plan, set out the physical changes allowed and the implementation standards (Ave, 1991). A proposal which conforms to the plans and by-laws must be given permission.

The region prepares general context or structure plans (*piani d'inquadramento* or *piani-quadro*) which provide the guidelines for the lower tier plans. However these tend to be very general relating to economic and social issues and therefore not giving a clear framework for physical planning. At the same time many communes are too small to deal with planning matters and the province which lies between the regional and the commune level will prepare a structure plan instead of the *piano regolatore*. In 1990 a new law, the Ordinamento delle Autonomie Locali, was passed to try to deal with the local government boundary problem. One result is that a subregional authority has the task of preparing a structure plan (*piano territoriale di coordinamento*) to act as a better framework. Of particular interest is the formulation in this Act of eleven metropolitan authorities for the biggest cities, analogous to provinces, with their own directly elected bodies and responsibilities for producing structure plans. The *piani regolatori* remain the responsibility of the communes. It has been said that the aim of this reform is to bypass the regional tier and enable cities to implement urban policies with the help of central government (Mazza, 1991). It could also signal a shift from the fragmented approach to a more comprehensive strategic framework. However, in common with most laws in Italy it is not clear how far this initiative will be implemented. There has also been an increase in concern recently for the environment and the introduction of EU legislation on environmental impact has been grafted on to the Italian system. Plans for the protection of the environment (*piani paesistici*) have been made compulsory and act as guidance to the master plans.

In the Italian case it has to be pointed out that the formal system of laws does not necessarily guide actual practice. Although the system was set up by the 1942 Act, actual development during the post-war period paid little attention to it and speculative renewal of city centres and urban sprawl developed unchecked (Calabi, 1984). Few southern municipalities have produced effective master plans. A number of attempts were made to pass new legislation in the 1960s and 1970s that would be more effective but they did not make a big difference. These tended to cover particular aspects of urban policy such as rehabilitation and conservation. It has been argued that it was impossible to create an effective planning system because of the patronage and clientelism that pervaded the political structures (Calavita, 1983). An illustration can be given of the way illegal development has become a regular feature of the Italian system. In November 1994 storms deluged the north-west of the country creating billions of pounds worth of damage and over sixty deaths. In the investigations that followed the damage was blamed not only on the way rivers had been canalised but also on the scale of development which had been carried out without permission – not just the odd extension but whole industrial areas and housing estates (Hooper, 1994). The judicial system cannot cope with the scale of irregularity, which is accentuated by the deals between politicians and construction companies, and in any case only small fines are imposed. The acceptance of such illegality is illustrated by the regular amnesties which are given to such developments and which therefore discourage conformity. Shortly after the floods described above Berlusconi announced, as part of his attempt to boost revenue in his crisis budget proposals, that people who had breached planning laws could buy freedom from prosecution by paying a fee. The scandals over certain development projects in the late 1980s have generated a conservative attitude amongst administrators. There is low visibility in the decision-making process and a scarcity of information. The legal vacuum left by the scandals led firms and public bodies to a negotiation approach to overcome the 'development freeze'. However, the procedures adopted were all different and so compounded the confusion and low visibility (Ave, 1991).

The formal system itself has its own problems and has been described as a legislative jungle (Ave, 1991), with approvals needed from various government levels and departments and fragmentation due to regional laws and different kinds of planning systems. 'Problems of effectiveness, accountability and co-ordination occur not just with particular proposals but lie at the heart of the whole machinery of planning' (Malusardi and Talia, 1992, p.135). Such complexity of procedures leads to long delays and difficulties in implementing projects. During the 1980s special 'extraordinary' procedures were adopted to deal with urgent problems such as city traffic or the housing crisis or special events such as the World Cup football. These procedures simplified the process, perhaps by accepting more limited documentation for proposals, bypassing the numerous approvals needed from democratic bodies, ignoring the regional level, central government managing finances directly, or automatic acceptance of changes to the master plan.

Portugal

The planning system in Portugal is underdeveloped and new planning legislation is only slowly being implemented. Urban and regional planning developed as a formal system from the 1930s and culminated in the 1944 Act. However, much illegal development took place during the 1960s and 1970s and is still continuing though to a lesser extent (Costa Lobo, 1992). The change to a democratic government in 1974 brought new laws and a hierarchy of plans is now established.

Central government passes laws, is responsible for national plans for socio-economic development and for organising regional level administration which acts as a decentralised arm of central government. Central government prepares a plan for the national road network which has considerable implications for local development. Two further national instruments impose important restrictions on development; these cover the most productive agricultural areas and those which are environmentally sensitive. One of the problems for planning at this level is the way responsibilities are scattered amongst national bodies (Vasconcelos and Reis, 1994). Central government also approves the municipal plans. Where no plans have yet been approved central government considers each development.

At the regional level Portugal used to be divided into eighteen districts each with a governor representing central government in Napoleonic style. Reforms have now taken place to change this system to one of five administrative regions. Regional Co-ordination Commissions, branches of central government, have been set up to co-ordinate and implement action at the regional level. Regional physical land-use plans (*Planos Regionais de Ordenamento do Território*) are prepared to provide a broad framework for investment and the plans of municipalities. Under the responsibility of the Commissions technical planning help is provided to the municipalities – offices (called GAT) have been organised for this purpose for subregional groupings of municipalities. The GAT thus provide an important integrating function, although this has not succeeded in Lisbon and Oporto (Williams, 1984).

The municipalities, of which there are 305, are the centres of regulatory planning power and once they have an approved municipal plan (*Plano Director Municipal*), which is a comprehensive plan for physical and socio-economic development and covers their whole area, they can prepare more detailed plans. There are the more detailed urban plans (*Planos de Urbanizaoção*) for parts of the area and also detailed layout plans (*Planos de Pormenor*) which involve the parishes (*freguesias*) of which there are over 4,000. To encourage the production of these municipal plans, and by 1993 only thirty-three had been prepared (EC, 1994b), the approval of a plan has been linked to grants and the powers of expropriation. The municipal plan has to be submitted for approval to the Technical Commission which is made up of various state bodies including the Regional Commission. Then it is considered at a public inquiry before approval by the Municipal Assembly and the Minister of Planning. Some municipalities have also started to prepare strategic plans as a framework for the municipal plan.

A more direct approach to deal with illegal housing was attempted. Central government established *comissariados* in areas that had intense problems of illegal dwellings and these were to record the extent of the problem and prepare plans to legalise them. They also provided technical help to the municipalities which had to implement these plans (Williams, 1984) and, according to Williams, the most important plans have been in Lisbon while elsewhere the results have been piecemeal. The planning process has been slow and centralised, and the initiative for development has been through privately sponsored schemes, including illegal ones. Since 1974 improvements to the situation have occurred, more autonomy being given to the municipalities; but there is a lack of expertise. It has also been suggested that the new municipal plans are too regulatory and do not allow enough flexibility (Vasconcelos and Reis, 1994).

Belgium

As already noted, Belgium falls generally into the Napoleonic family; however, recently it has adopted a federal structure which has been superimposed on the existing system. The planning system follows the general pattern of others in this group, but the federal division means that there are differences between the three regions. The Constitution was amended in 1970 to introduce the federal approach and through the 1980s the country has been shifting from a tightly organised central state structure to a federal state. This approach was adopted in two dimensions – the cultural and the geographical. At the cultural level three communities were defined, the French community, the Flemish community and the German-speaking community. These communities do not always follow geographical boundaries – for example, Brussels contained both French and Flemish communities. Each community has its own legislative and executive organisation. However, in planning terms these are only important in relation to cultural monuments. It is the geographical division that is important for planning. Here there are the three regions covering the Flemish, Walloon and Brussels areas. These regions formulate legislation on planning and prepare regional and subregional plans. The national level is no longer involved in planning except when essential for matters covering the whole territory, such as major roads.

Until recently the principles of the planning system were based on the Zoning and Planning Act of 1962. This Act set out a hierarchy of plans with the lower one subordinate to the higher one with some allowance for flexibility (Suetens, 1986). The plans were regional, subregional, general municipal plan and special municipal plan. Planning permissions were given if they accorded with the municipal plan, which was legally binding. However, many of the regional and subregional plans were not prepared. The first kind of plan possible at the municipal level was the General Zoning Plan (*Plan Général d'Aménagement – Algemeen Plan van Aanleg*) covering the whole municipality. This was considered too inflexible and in the evolving regional legislation broader 'structure' plans have become possible (Van Wunnick, 1992). More frequently used was the Detailed Zoning Plan (*Plan Particulier d'Aménagement – Bijzonder Plan van Aanlag*). Most plans were made for the municipalities by consultants and they had to be approved by the Regional Minister responsible for planning. Implementation of the plans has always been a weak factor.

However this system is now undergoing complete reform in the light of the constitutional changes. Each region is now adopting its own autonomous planning powers and following different routes (EU, 1994b). In the Flemish area three tiers of region, province and local are being proposed with two kinds of plan – the spatial structure plan providing a framework for the spatial implementation plan. Only two tiers, the region and the municipality, are being developed in the Walloon area. All acts and regulations are being codified in a new planning Code for Walloon. Each level will have to prepare a structure plan and a legally binding zoning plan. The Brussels Capital Region is also adopting a two-tier system, with a development plan for the Brussels region and separate plans for its nineteen municipalities which have to be approved by the region. Again both levels have to prepare a general framework plan and a zoning plan. One issue for the future will be how strategic planning issues which cross the regional boundaries will be co-ordinated.

Spain

Although its legal and administrative traditions demonstrate its Napoleonic legacy, Spain has also been developing in a federalist direction. This has resulted in a complex legal basis for planning. Although planning in Spain has a long tradition going back to the eighteenth century (García-Bellido, 1991), the basis for modern planning in Spain was an Act of 1956. However this was not suited to the rapid migration to urban areas that took place in the 1960s and early 1970s, and legal loopholes, illegal development, local political rivalry, corruption and local authority negligence contributed to the weaknesses of the Act and the rather chaotic urban development that occurred (Wynn, 1984). In an attempt to address this problem the 1956 Act was amended in 1975 and consolidated in a new Act of 1976.

However the death of Franco in November of that year brought a sudden change in political context. A new Constitution was formulated in 1978 and reflected the change from a highly centralised system to a democratic one. An important aspect of the Constitution was the autonomy given to the seventeen regions of the country. Each has its own legislative assembly and elected regional government. However, the degree of delegated power varies from region to region. Further variety results from the opportunities for each region to produce its own laws and regulations. Each region therefore has the power, either on its own or in conjunction with central government, to cover planning, economic, environmental, infrastructure and urban development. The regions are starting to produce legislation on these matters, providing guidelines and regional plans to which the plans of the municipalities must conform. National government retains responsibility for issues of national concern such as major infrastructure and national parks and for formulating guiding legislation. In recent years there have been conflicts between central government and certain regions over the autonomy of each body in formulating planning legislation. The central reforming legislation of the 1990s sought to limit the scope of the regions and this has led to challenges in the Constitutional Court which have yet to be resolved (Keyes *et al.*, 1993).

Municipalities, of which there are almost 9,000, are responsible for the preparation of local plans and their implementation. The 1956 Act, amended in 1975 and 1990 and consolidated in a new Act in 1992, provides the basic legislation for local planning. A general urban plan is prepared for the whole area of the municipality and this is compulsory for those with a population of more than 5,000. This plan, which is legally binding, designates land into three categories: land already developed, land available for development and land excluded from development. Each category is accompanied by differing degrees of detail in plan formulation (see Keyes *et al.*, 1991 for details). The important middle category, called 'reserve urban land', is further divided into programmed land and non-programmed land, indicating whether the land is expected to be needed in the short or longer term. The plans are programmed in two stages of four and eight years. Programmed land has to be developed in accordance with the general plan while non-programmed land will be developed later in urban programmes drawn up after the general plan (Bassols, 1986). The general plan provides a brief description of the use of each zone, designates communication networks, conservation areas, open spaces, community facilities and public buildings, and sets out the means for protecting the environment and heritage. The general plan has to be checked by the region for its legality and conformity with higher order plans and regulations. The aim of the 1975 Act was that every municipality would have its general plan, but this has not been realistic. However, provision was made for 'subsidiary guidelines' to be produced for municipalities as yet without a plan, or for rural areas. These cover in less detail the same kind of issues as the general plan but do not distinguish between programmed and non-programmed land.

Within the general plan or subsidiary guidelines there are a number of different instruments for working out the details at a smaller scale. These may cover parts of the municipality such as historic town centres, countryside with special character, new development areas or infrastructure projects. The *plan parcial*, which is used for control purposes, is one important example. This has to be produced in order for programmed urbanised land to be developed. It has to accord with the general plan and may be approved by the larger municipalities themselves. All plans have to be accompanied by a 'memorandum' which shows the analysis and alternatives and provides the justification for the plan. This also includes an account of the participation process and the authorities' response to it. Another instrument, 'Detailed studies', can be used as a supplement to the general plan or *plan parcial* and provides very detailed guidance where projects are complex – these can be used to control aesthetic aspects. They do not require higher order approval. Flexibility can be introduced into the system through a 'Programme of Urban Action' which can bring land forward for development in advance of the original intentions in the general plan.

In recent years control of development has been tightened through regulations passed in 1978. Licences have to be obtained for all development, demolition and land subdivision. Applications are considered by the municipality who check for conformity with the plans and building regulations. For major developments the applicant also has to get the approval of the college of architects and this has to accompany the application. The college will verify that the procedures have been correctly complied with and that the architect submitting the proposal is registered.

Although the law specifies a hierarchy of plans in which the lower level plans must conform to the higher one, current reality is more complex. The law mentions a national plan but this has never been produced and in the climate of regional autonomy is unlikely ever to materialise. Regional plans have been slow in preparation and so municipalities have little higher level guidance (Keyes *et al.*, 1993).

Greece

The Greek administrative approach is characterised by a very high degree of centrality – a French prefectural system was introduced in the 1830s. A major feature of the planning system is its constant state of flux. Laws and policies undergo regular change as there is no political consensus over planning and each change of government brings shifts in approach and laws (Coccossis and Pyrgiotis, forthcoming; Getimis, 1992). The legal context that has built up over time is therefore piecemeal and complex, involving a labyrinth of amendments, exemptions and special laws, and has not been properly codified (EU, 1994b). There is also a problem of enforcement, and the reality of development does not necessarily relate to the legal framework. The importance of land and property to individuals and the prevailing cultural values militate against planning intervention. It has been said that there is a long-standing coalition acting against planning involving the state, landowners, developers and certain professions (Delladetsima and Leontidou, 1995).

A Constitution forms the basic legal framework of the country and laws are passed by Parliamentary vote and ratified and issued by the President. The President also issues Presidential Decrees which are countersigned by responsible Ministers. The country has been divided into various administrative levels. There are thirteen recently created regions, fifty-four prefectures and 147 counties. There are also provinces, but these are more geographic entities than administrative. Each ministry is represented in the prefecture and the head administrator is the prefect who is appointed by the Minister of the Interior. A recent law has changed this structure and proposed that the prefect should be elected, that the Prefecture should have some economic autonomy and that it will be responsible for making staffing appointments. These reforms should reduce the influence of central government. The elected local authorities are cities (264) and municipalities (5,759). Generally cities and municipalities take the initiative for formulating plans while central government approves them (Makridou-Papadaki, 1992). The prefectural level can approve plans for towns of minor importance.

Although reference is made in Greek legislation to a national spatial plan, no such plan exists. Five-year national economic development plans are produced fairly regularly but since the 1988-92 draft plan was abandoned no new plan has emerged. Regional development plans have been produced since 1980, but these have tended to be resource allocation plans and are closely linked to negotiations on European Union funding – for example, the 1989-93 Regional Development Plan has been drawn up in the context of EU policy (Andrikopoulou, 1992). These are non-statutory guiding documents but the public works that emanate from these plans have obvious spatial planning effects.

The basis of the planning system goes back to 1923 and although there have been modernisation attempts these have always been frustrated by the actions of vested interests (Wassenhoven, 1984). In the context of corruption and political patronage it has proved impossible to find a way of generating enforceable legislation that commands consensus approval. The consequence is that the 1923 Law is still operative, with partial additions resulting from attempted reforms, particularly in 1979 and 1983. The result has been described as three legislative frameworks (Lalenis, 1993) covering differing situations. For example, the 1923 Law applies in revisions of urban plans which are not considered of major importance, the 1979 Law in revisions to urban plans of major importance not covered by the 1983 Law. This last law covers problematic housing redevelopment, extensions to existing urban plans and general development plans for housing areas. Meanwhile land development in rural areas is practically unconstrained and regulated by Presidential Decree (Coccossis and Pyrgiotis, forthcoming).

The 1923 Act allowed various agencies, including municipalities, to draw up plans which could be for entire cities or just one block. Plans then had to be submitted to the central Ministry for approval, modification or rejection (Wassenhoven, 1984). Although still in use in some areas this law is considered problematic; for example, it does not take into account the regional context, it is cautious in dealing with private property and there is no participation (Lalenis, 1993). The fall of military dictatorship in 1974 brought the first serious effort to upgrade the planning system, with new central government institutions and a Law of 1979. This involved the introduction of development plans under state responsibility to replace the previous system of a codification of norms with the responsibility placed upon private developers to conform (Getimis, 1992). The 1979 Urban Development Act was modelled on the French *zone d'aménagement concerté* concept. An attempt was made to ensure that all future development took place in designated 'urban development areas' to try to deal with the lack of funds and land for communal facilities and infrastructure. A general planning study had to be prepared as a context for the urban development area. However, it involved the contribution of land and finance from private landowners who succeeded in stopping the full implementation of the concept.

Significant reforms took place in the 1980s, set in the context of the first socialist government in post-war Greece, with an emphasis on decentralisation and participation. There was an attempt to decentralise some power to local authorities. Until then central government supervised all local government appointments and it was proposed to give mayors some control of this and also to delegate development control powers (Wassenhoven, 1984). The 1983 planning law set out two levels of plans: the general development plan and the 'implementation' plan which has to conform with it. It also included a programme for the planning of coastal resort areas, boundaries, and building regulations for 10,000 rural settlements and plans for 1,800 developing rural settlements (Makridou-Papadaki, 1992). A key policy was the 1983 Operation for Urban Reconstruction aiming to check the anarchistic urban sprawl. However, the economic crisis in 1986 led to the collapse of this programme with the playing down of participation and planning and more emphasis on the free market. This trend was stepped up by the 1990 neo-conservative government. Throughout this period, notwithstanding some of the rhetoric in the 1980s, the system continued to feature a highly centralised approach (Andrikopoulou, 1992). Thus central government continues to be involved in all levels of planning policy through the regional and prefectural administrations, and even development control matters remain largely a central government matter (Delladetsima and Leontidou, 1995) and building permits and regulation are administered through local offices of the central administration. However, very gradually local authorities are gaining some opportunities to formulate plans and certain approval powers. Special strategic master plans are being prepared for Athens and Thessaloniki. In 1986 a law was passed on the protection of the environment which included procedures for Environmental Impact Assessment. This has been described as one of the most complete environmental laws in Europe: however, it is not co-ordinated with physical planning and lacks the means of implementation (Beriatos, 1995).

The revision of the entire legal framework of town and regional spatial planning is currently under consideration by the Ministry for the Environment, Spatial Planning and Public Works. A second tier of local authorities with planning responsibilities is also under discussion and a new local government law is expected. The major problems that have to be faced include the overcentralised administration, the lack of control over urban growth, problems of enforcement and the lack of planning skills. This is against a background political culture of which it has been said 'Town Planning is but one more governmental function which is undermined through the network of family, extended kinship and political ties, and generally exploited for personal and political gains' (Wassenhoven, 1984, p.7). The gap between planning intention and reality is very great. This is likely to continue as long as the enforcement powers are limited and the 'rules of the game' continually change (Coccossis and Pyrgiotis, forthcoming).

Planning systems in the Germanic family

Germany

Since 1990 Germany has been a re-unified country and the planning system of Western Germany has been applied in the former eastern *Länder* with certain small modifications to allow for flexibility (Dieterich *et al.*, 1993). As mentioned earlier there are two key features of the Germanic approach: a strong legal framework and a decentralised decision-making structure. A constitutional division of power between the levels of government is set out in the Basic Law (*Grundgesetz*) of 1949. These aspects have a strong influence on the planning system with strong legal backing to plans and permissions and considerable variation in planning practice across the different states (*Länder*), which produce their own planning laws. Reflecting the general legal approach with its strong emphasis on codification, the planning laws are set out very precisely and there is considerable attention to legal interpretation (Hooper, 1989).

The planning system generally operates at the level of state (*Land*) or below. The Federal level just sets out a framework of regulations to ensure basic consistency in the planning legislation of each *Land*. A principal objective of the Federal legislation is to try to achieve equality of living conditions throughout the country. Much of the Federal planning activity is spent advising the lower tiers on the interpretation of these regulations. It is a basic principle of the system that plans and policies have to conform to those of the higher level but within the overall concept of subsidiarity which provides the lower authorities with autonomy in relation to the details of policy.

All sixteen *Länder* are obliged to set up state-wide comprehensive plans. The types of plan and rules differ between *Länder* but normally contain broad statements of development intentions covering such issues as population projections, settlement hierarchies and priority areas. Each *Länder* also has to set up its regulations for the protection of the environment. More detailed comprehensive plans are usually also prepared for subregions within the *Land*. Regional plans have to conform to the Federal guidelines set out in the Federal Comprehensive Regional Planning Law (*Bundesraumordnungsgesetz* – BROG).

The principal regulations regarding land use and development control are contained in the *Baugesetzbuch* which incorporates the Federal Building Law (*Bundesbaugesetz* – BBauG). These regulations assign the responsibility of implementing control to the local level and make provision for two further types of plan. First there is the preparatory land-use plan (*Flächennutzungsplan*) which is essentially a zoning plan for the whole local authority area (there are about 16,000 communes or *Gemeinden*). This plan is binding on public authorities but has no legal effects on the rights of private landowners. The second type of plan (*Bebauungsplan*) is however, legally binding and determines the accepted land use of plots and contains an environmental assessment. It has to conform to the *Flächennutzungsplan*. There is public participation in the initial stages of preparing this plan and people can also make objections when the draft is open for public inspection. Negotiation between the local authority and the developer takes place in the preparation of the *Bebauungsplan* which can specify such matters as height of buildings, floor space, mode of building construction, car parking and infrastructure. Approval of these plans has to be obtained from the *Land*, but this can only be withheld on legal grounds, not planning substance. The planning permission process then becomes an administrative task of checking conformity with the plan. Flexibility has been introduced into the system through certain sections (34 and 35) of the BBauG Law (Hooper, 1989). These sections come into effect when no binding plan has been prepared. Where this is the case the development can be allowed if it is in conformity with the surrounding area and infrastructure can be provided. This allows greater interpretation and flexibility and local authorities may choose this route in some locations.

A new instrument has been in operation in recent years called the *Vorhaben-und-Erschliessungsplan*. This was originally devised for use in the new *Länder* but since 1993 has been available throughout the country (Dieterich and Dransfield, 1995). Planning permission can be given without a local plan if a developer guarantees to prepare a local plan and finance and implement the servicing of the development. This speeds up the planning process in such situations. Another development of recent years has been the increasing overlap of policies and government levels. Some *Länder* have created administrative districts within their area to improve the implementation of *Land* legislation and these have often co-ordinated land-use planning activities. In response local authorities have then often grouped together and elected an advisory regional council to influence the work of the *Land* districts. In some *Länder* there have been reorganisations of local authorities into larger units but others have kept the autonomy of smaller authorities. Thus an intermediate tier between *Land* and local government has often been formed through the grouping of towns and communities and certain responsibilities have been decentralised to them (Balme *et al.*, 1994). There is also the county (*Kreis*) level, which performs some of the broader functions of local government which smaller authorities cannot afford, such as hospitals or major roads.

Austria

Austria is a federal state comprising nine independent regions (*Länder*). The division of responsibilities between the central state and the regions is set out in the federal Constitution which exhaustively itemises the role of central government, leaving the residual responsibilities to the regions. The central state retains control over national aspects e.g. national road system, railways, flood control, military matters, education and overall control of finances. The Constitution also identifies municipalities as administrative areas with their own self-government. Each region devises its planning laws which are similar. At the regional level, planning information is collected, regional plans are prepared, co-ordination is sought with central government and neighbouring regions, advice is provided for municipalities and municipal plans are supervised. The regional plans can be for all or parts of a region and can be subject plans. Each region has carried out its plans in different ways, although within the same general legal restrictions (Kühne and Weber, 1986). There are 2,300 municipalities which are therefore quite small and they have to prepare local physical plans which are legally binding on individuals. These have to set out development principles, means of implementation and a scheme for the use of all land in the territory. Each plot of land is given a classification such as green belt or building land. The latter is then divided further into use categories. The municipal plan has to be authorised by the region which checks that it conforms to regional plans and regulations and is financially sound.

Switzerland

The country is a federal state divided into three levels of government: the national or confederation level, the twenty-six cantons and over 3,000 communes. Each is involved in planning. At the national level a general law on planning aims and principles has been formulated – the Federal Law on Spatial Planning 1980. National government is also expected to co-ordinate its own activities and the planning at the canton level. However this has not been very successful (Ringli, 1992, 1994) and has led the government to develop a national spatial strategy called 'Guidelines for Swiss Spatial Development', which went to parliament for approval in 1995. As well as providing a framework for the plans of the cantons it is a reaction to European competition. The strategy aims to improve public transport between cities, each with their particular function, creating an integrated 'Swiss City' of 3 million inhabitants. The aim is to concentrate new commercial development on public transport nodes, to restrict the peripheral growth of each city thus preventing congestion and to enhance the attractiveness of the Swiss countryside. The Swiss City aims to compete with other cities, such as Munich, Frankfurt and Lyons, with a highly efficient urban structure together with an environmental quality that will attract 'top' professionals and their families (Ringli, 1994).

The cantons formulate their own Planning and Building Laws and produce canton regional plans. Within the federal guidelines each canton has pursued its own legislation according to its particular circumstance (Lendi, 1986). Its plans, called guiding plans, are binding on authorities at all levels and are approved by the Federal Council. They are said to be the central instrument of spatial co-ordination in Switzerland (Ringli, 1994). The canton gives directives to the communes and approves their plans. The communes regulate development through the instruments of a zoning plan which has to cover their whole area, and through building regulations. The zoning plan sets out the permitted use of all land and is binding on landowners.

Planning systems in the Scandinavian family

Denmark

There are three levels of government in Denmark – national, county and municipality. Local self-government and decentralisation of decision-making are considered key principles. The most recent planning Act is that of 1992 which builds upon the reforms of the 1970s, and in this legislation a hierarchy of plans is specified in which the lower level has to conform to the upper.

The national government controls issues of national significance and has a number of means of doing this (Östergård, 1994). It formulates regulations over the way in which the planning legislation is implemented and can also issue binding directives on lower tiers. These directives cover issues of national importance such as infrastructure and landscape protection. National government has a power of veto over regional plans, and state authorities can veto local plans in matters over which they are responsible. Central government can also call in plans for its own consideration. It also stimulates interest through its information role and through financial support for pilot planning projects. Central government has to prepare a national planning report after each four-yearly national election. In 1992 this was issued in the form of the report 'Denmark towards the year 2018' the content of which illustrates the trend towards marketing Denmark in the European context. It sets out a network of cities with the Greater Copenhagen or Öresund Region at the top of the hierarchy as an area of European significance. The report also stresses environmental quality. The report is not binding on lower authorities but intended as inspiration for them and the private sector and an instrument for negotiation with the European Commission (Östergård, 1994).

A basic element of the planning law is the division of the whole country into three zones: rural, urban and recreational. In rural zones only development relating to the commercial operation of agriculture, forestry and fishing is allowed. In urban and recreation zones development has to conform to the local regulations and plans. The change of rural areas into urban zones requires a local plan and also involves the landowner in paying a betterment tax.

The fourteen county councils and the cities of Copenhagen and Frederiksberg prepare regional plans which are also revised after each four-year electoral round. Until 1992 these had to be approved by the Minister for Environment but are now controlled by the means mentioned above. These plans, which cover the whole country, present policies on the broad structure of the urban pattern, environmental protection, the designation of urban zones and the location of large institutions and communication facilities. They are binding on municipal and local plans and also incorporate the Environmental Assessment regulations. The county is also responsible for planning decisions in rural zones.

There are 275 municipalities with an average population of about 20,000 – larger than in many other countries. The municipality is the prime authority for planning and it approves its own plans. The municipality prepares a structure plan (*kommuneplan*) within the regional guidelines. This is a comprehensive plan covering all the municipality and setting out the broad pattern of uses, specifying any transference of land between zones and providing a framework for local plans. As with other plans it has to be reviewed every four years. There is some degree of variation possible in the form of these plans – they can stress principles or be more regulatory. In recent years there has been a trend to the former with more emphasis on political priorities and strategic approaches to urban renewal, the environment or commercial attraction.

Whenever an important or large development is proposed, or when the municipality wants to promote an area, a local plan (*lokalplan*) has to be prepared which is legally binding on property owners. Local plans fulfil a number of roles and therefore vary somewhat in form. They can be theme oriented, as in the conservation of a historic centre, provide a neighbourhood plan, or be simply a scheme for the control of the extension to an individual property. Usually they will set out regulations for individual properties, specifying land use, size, location and appearance of buildings. The content of the plan can be negotiated between the municipality and development interests. There is no additional system of planning permission – if a development is in accordance with local regulations and the local plan then it is automatically approved, although building permits are needed. Appeals are possible on legislative procedure but not on content. The participatory requirements on structure and local plans in which comments are invited at both the formulation and proposal stages, are considered sufficient democratic safeguards.

Flexibility is built into the system by allowing structure plans to be broadly conceived and also allowing amendments, although, unless these are minor, they have to go through the normal procedures, including participation. A feature of the Danish planning system is the close link with the political electoral process which helps to give it both legitimacy and flexibility (Kjaersdam, 1992; Edwards, 1989). Participatory democracy is catered for through the participation opportunities at each stage in the plan formulation, and representative democracy through the requirement for a new plan at each level after an election.

Sweden

The basis of the post-war Swedish planning system was an Act of 1947. However, from the 1970s onwards discussions took place on reforming this legislation as it became increasingly irrelevant to current concerns (Holm and Fredlund, 1991). New legislation aimed at simplifying the processes, responding to the societal demands for increased participation and taking on board the general trend towards decentralisation in decision-making. A reform package was passed by government in 1986 and included the two new Acts: The Natural Resources Act 1987 and the Planning and Building Act 1987 (for details see Kalbro and Mattsson, 1995). The aim was to give as much responsibility as possible for planning to the municipality. The first of the two Acts sets out those aspects which are still the responsibility of central government based upon the principle of national interest. This means that central government can impose guidelines on such matters as ecologically sensitive areas, national recreation areas, water supply and reindeer raising grounds. It also approves the location of hazardous industry such as power stations and chemical plants. Central government can only intervene in municipal plans if such national interests are affected or there is a danger to health and safety.

The regional level of planning is very weak. In the large metropolitan areas the counties prepare regional plans but these are only advisory, providing an overview, research and co-ordinating function. With the increased interest in regional development in the European Union there is a move to develop regional co-operation between local authorities. However, the new legislation placed the emphasis for planning firmly with the municipality, which is described as having a 'planning monopoly'. This was initially used to describe the fact that developers could not formulate detailed plans – this could only be done by the municipality. However the phrase is also used to indicate the degree of municipal autonomy from central government. The Planning and Building Act of 1987 requires all municipalities to make a comprehensive plan (*översiktsplan*) for their whole area showing the basic features of land use and development. This plan is also intended as an instrument to show how the municipality has taken into account the matters of national interest. As these plans can be rather generalised some municipalities have produced more detailed versions for parts of their areas. The comprehensive plan is not binding on individuals but provides a framework for detailed plans and for the co-ordination of public sector activities. In practice most of these comprehensive plans have been vague so as to allow maximum flexibility, but there is a current debate about ways in which the plan can be made more meaningful. The comprehensive plan has to be submitted to the County Administrative Board, an arm of central government, which checks that it has adequately taken national interests into account. There is also a period of consultation with interested bodies and the public, and the municipality has to show how it has responded to the views expressed. Any appeal can only be on procedural matters.

The most important planning instrument is the detailed plan (*detaljplan*), which is legally binding. It is prepared when change is expected. Negotiation and discussion takes place with the developer during the formulation of the plan which can only be ratified by the municipality. Detailed plans can vary in form but all must specify the intended land uses, public spaces, building lots and an implementation period (Kalbro and Mattsson, 1995). It can also cover design, construction materials, lot sizes, floor areas, landscaping, parking, conservation and similar matters. The municipality must consult other public bodies and interested parties. The County Administrative Board again checks for conformity with national interests. There is also a programme of public participation, and appeals can be made to the County Board and thereafter to the government. The developer will then seek a building permit which is automatically given if the proposal conforms to the plan. Finally Special Area Regulations were a new form of control introduced in the 1987 Act. They are legally binding and are used to ensure that national interests and comprehensive plan objectives are secured for areas not covered by detailed plans, although, unlike detailed plans, they do not confer development rights. They are intended for simple cases which only encompass a limited number of issues such as areas for holiday homes.

Norway

There are nineteen counties and 454 municipalities in Norway and the municipalities have been given considerable power – it is said that this is partly due to the large and sparsely populated nature of the country (Falkanger, 1986) and the long history of independent freeholding (Holt-Jensen, 1994). Throughout the 1970s there was a move to decentralise decision-making from the state to the counties and sometimes also to the municipalities (Lorange and Myhre, 1991). The 'free local government' scheme which was introduced in 1986 continued this trend (Lodden, 1991). The aim of this scheme was to experiment with giving certain local authorities greater autonomy and an ability to adopt their own regulations in tune with local circumstances with a view to formulating new decentralising general legislation.

As far as the planning system is concerned the key act was passed in 1965 with revisions in the 1986 Planning and Building Act. A hierarchy of different kinds of plans was set out from regional plan, county plan, to general municipal plan and local plans. At the national level there are plans for such issues as highways and telecommunications and, as the peripheral areas are politically strong, there are also regional strategies to protect their interest. The 1965 Act laid out rules for regional plans to deal with issues that affect two or more municipalities (such as water supply or sewage disposal) and although not obligatory it was hoped that they would be prepared by all adjoining municipalities. However these plans were not successful (Falkanger, 1986) and ceased in the 1970s. Regional co-operation continues to be a problem; for example, in the Oslo region over thirty municipalities operate their independent planning policies (Lorange and Myhre, 1991) although large infrastructure decisions have led to some co-operation. The county level has the advantage of representing a political entity, and county plans (*fylkesplan*), which became obligatory in 1973, are comprehensive and act as guidelines for other plans, covering such items as population distribution, economic development, and use of natural resources. However, they tend to concentrate on the topics over which the county has a financial responsibility, such as hospitals and county roads.

The level which has the primary responsibility for physical planning is the municipality. Since 1965 comprehensive municipal plans (*Kommuneplaner*) have been obligatory, binding on authorities, and have provided the framework for local plans. They outline the distribution of land uses throughout the municipality and have to be submitted to the county and Ministry for comment before adoption by the municipality. The 1986 Act sought to simplify the planning system, emphasise co-ordination, social and economic aspects and participation. The municipal plan is now divided into two parts: a longer-term physical plan for about twelve years, reviewed perhaps with electoral cycles every four years, and a four-year action plan linked to the annual budget cycle (Holt-Jensen, 1994). The local plan or regulation plan (*reguleringsplan*) is a different kind of instrument as it is binding on individuals. It has to include a statement of the extent and use of each plot and can also specify more detailed regulations on matters such as layout and design. The structure and contents of each local plan vary according to its needs. In the 1965 Act a municipality was requested to prepare local plans covering its whole area, but this has proved too burdensome and has not been followed in practice.

Finland

During the development of its legal system in the seventeenth and eighteenth centuries Finland was part of Sweden, and even under Russian control in the nineteenth century Swedish laws remained in force; this historical legacy has a continuing influence. The modern Finnish planning system was established as part of the post-war welfare state with its emphasis on creating a network of universal services. This involved a high degree of centralisation, and local administration was seen as the implementer of national policies (Sotarauta, 1994). In physical planning this was embodied in the legislation of 1958 which set up a hierarchy of plans. Regional plans were to provide the guidance for municipal general plans which provided the framework for detailed plans. National government passed legislation and supervised the whole system, although there was no national plan. However it has been pointed out that the reality of planning was not as hierarchical as it might seem (Virtanen, 1994). Historically, detailed planning started first and only later were the broader contextual plans added. A two-way dialogue developed in which the upper tiers gave directions and recommendations and the lower tiers took initiatives, provided information and expressed their requirements.

Much of the supervision is carried out by the offices of the eleven Provinces. However, there are also nineteen separate Regional Planning Associations for preparing the regional plans. The Associations' councils are elected according to the political distribution of municipal councils (Mansikka and Rautsi, 1992). The regional plans set out the broad structure of different land uses and communications for a 4–5 year period within national guidelines. The regional plans have to be submitted to the state for approval and although they are not legally enforceable the Ministry expects municipalities to conform to them.

The 461 municipalities have a strong position in the planning system and prepare and approve their own plans within the national and regional framework. Structure plans (*Yleiskaava*) and their associated regulations cover the whole municipality and are compulsory. They are legally binding on the authorities and set out the broad zoning of the different land uses. Traditionally the state has had the right and duty to ratify municipal plans, but this control is diminishing over time. However, some municipalities still submit parts of their plans for ratification to give them more power and to make them legally binding on individuals (e.g. for conservation areas, shorelines, recreation areas) (Mansikka and Rautsi, 1992). The detailed town plan (*Asemakaava*) is required when development takes place. This zones land into different uses and is legally binding on individuals. Slightly different instruments are used for development in rural areas and shorelines (Modeen, 1986). The municipality has the monopoly for producing plans except for shorelines where landowners can prepare their own plans for submission to the municipality. Detailed plans act as the planning permission but a building permit is also required and further detailed drawings are needed for this. A combined planning and building permission is issued.

During the 1970s there was an increase in grass-roots protest and demands for a more participatory system (Haila, 1990). Delegation and local self-government were key issues in the debates of the 1980s, leading to more decentralisation of decisions to provincial and municipal levels and in 1989 Finland followed Norway and Sweden with experiments in 'free municipalities' (Sotarauta, 1994). Then in 1990 a new Planning and Building Act was passed which included the aims of increasing participation and contributing to sustainable development. A new requirement was that the municipalities should prepare an annual 'planning report' setting out planning projects in preparation and those which had been finalised. The Act also formalised the decentralisation process by reducing the situations in which municipalities need to seek central ratification. It opened up a new procedure for municipalities whereby they can submit their general plan for approval and then formulate detailed plans based on it without submission (Virtanen, 1994). The reform of the planning Act in 1990 not only involved the decentralisation of central state power to the municipality but also involved the speeding up of the system and making it more liberal through more flexible arrangements for granting exceptions (Haila, 1990). In 1993 a further review of the planning system took place which recommended widening the goal of planning, emphasising the environment, participation and the needs of vulnerable groups in society. It suggested further moves in the direction of decentralisation with regional plans being given to central government for examination rather than for official ratification (Virtanen, 1994).

Planning systems in eastern Europe

New planning systems in these countries do not yet exist. The problems in creating a market in land and property, and in many countries the lack of political stability needed to pass the necessary new legislation, have delayed the establishment of a market-oriented planning system. The uncertainties about administrative arrangements add to the difficulties in many countries. Thus at the present time there are various arrangements to deal with the transition period. These often involve using the previous planning approach and adapting it to new conditions. Some illustrations will be given, acknowledging that the situation is one of continual change.

In Poland the responsibility for planning was given to the communes after 1989 but many of these are not able to carry out strategic planning because of their size. The uncertainty over administrative arrangements needs to be sorted out. In the communes planners are still operating under the old laws of 1984. Plans were of two types, a general city plan and a detailed plan, both of which were prepared by city architects who had the power to decide on design details and implementation. The plans created under this legislation were of a rather rigid and master plan type and not necessarily suited to guiding market processes. Many of the new plans still have a strong emphasis on the physical shaping of the city, showing the legacy of the past when planners had considerable powers of implementation. However, in the political turbulence of Poland planning is given low priority and the legislative job is considerable as about four hundred Acts need to be changed. In the absence of any general guidance each city is adopting its own approach in an attempt to relate to the new political realities and market orientation. Perhaps the biggest changes will have to take place in the detailed plans where previous certainties are replaced by a new complex pattern of land and property ownership and uncertainty over who the investors are going to be.

In Łódź, for example, after a competition a group of consultants prepared the new general plan for the city with the objective of showing the local authority commitments in areas where they have a role, dealing with the social and environmental imperfections of the market, indicating the duties of the private developers and creating incentives for desirable activities. The emphasis is on economic development and attempting to establish a 'vision' based on the strengths of the city (Markowski and Kot, 1993) – this will allow it to compete for investment. Similarly in Poznan a new document has been prepared which is aimed at attracting hotel investment and which sets out all the possible locations and their costs and benefits (Poznan Municipal Town Planning Office, n.d.). 'Business promotion zones' are also proposed in the guidelines for the new city plan (Poznan Municipal Town Planning Office, 1992).

However, it has been said by developers that under present circumstances the planning system is irrelevant (Judge, 1994). They say that it is ownership that is currently crucial. If you can manage to establish ownership then you can do what you want. The implication of this is that in complex central city locations not much can be done while ownership is sorted out but that where individual ownership is clear uncontrolled development results. Thus there is much change of use to small businesses and building on the periphery of cities.

This trend for other mechanisms to take the place of planning is evident in Bulgaria. The backlog of demand for services has led to a mushrooming of small shops and cafés. The suburban areas with their numerous blocks of high-rise apartments are particularly deficient in such provision. In response to this demand the spaces on the ground floors of these blocks, for example around stairwells and entrance halls, are being converted for such purposes. However, this activity is not being controlled by the planning system. Instead the tenants' management committees interview potential users and decide who they would like (Thornley, 1993b).

CONCLUSIONS

The similarity and difference in the planning systems across Europe and the influence of national institutional factors will now be summarised. Perhaps the clearest conclusion that can be drawn is the distinctiveness of the British family compared to the rest of Europe. The legal and administrative framework is different and produces a particular kind of planning system. The legal system of evolving case law has its parallel in a planning system in which each planning permission is considered 'on its merits'. Considerable discretion is given to the local planner who has to balance all the 'material considerations'. The basis of the legal system is to explore precedent and this can also apply in considering a planning decision. The decisions on similar cases have to be reviewed as these can be used by the applicant in any appeal argument against the decision. The use of the development plan as only one material consideration contrasts with the approach which seeks to use the plan as a preconceived embodiment of planning rules and regulations. In administrative terms the British family is also significantly different from all others, having much sharper divisions between central and local government and a high degree of centralised monitoring and control. Whereas other countries have exhibited trends towards decentralisation and regionalism Britain has remained strongly centralised with limits on local government autonomy. In planning this is expressed through the lack of a regional level and the strong central controls on local planning through Planning Policy Guidance Notes. The regularly used appeal system also enables central government to intervene in individual decisions and ensures that local planners conform to national guidelines. Thus the discretion is highly constrained. It has been noted (Davies *et al.*, 1989) that municipalities have greater autonomy of policy-making in the rest of Europe with central government objections only being raised when national or regional interests are affected – a form of negative control, rather than one of directing or commanding matters of substance.

The British approach to planning embodies a conflictual style of administration in which the two sides, the local authority and the applicant, are competing to win. This is particularly evident when a decision goes to appeal and lawyers are brought in to argue the case. However, negotiation does take place in many cases and in larger schemes this often involves a discussion over 'planning gain'. In the rest of Europe there is more of a partnership approach, oriented around the detailed plan in which both sides are working together. The strong autonomy which many municipalities have over plan formulation gives them a stronger position in the negotiations and a more equal partnership. A distinctive feature of the planning system in the British family is the separation of different planning functions. Healey and Williams (1993) have described urban planning systems as having three elements: the plan-making function, the developmental function involving such issues as land assembly and servicing, and the regulatory or control function. In many countries the local legally binding plan can be seen to include elements of all three functions whereas in Britain the three functions are usually treated separately, often carried out in different departments of the planning office.

There are a number of common features displayed by the planning systems of the Napoleonic family. There is the tendency to prepare a national code of planning regulations and to create a hierarchy of plans based upon a zoning approach. This conforms to the general Napoleonic legal style. The combination of centralised control plus responsiveness to local pressures creates a complexity of interactive arrangements indicative of the 'fused' administrative approach. The move to decentralisation and a greater regional presence are current developments in this general framework. Planning therefore often takes place within an array of arrangements for vertical and horizontal co-operation. However the Napoleonic family is large and also displays variation in its planning systems within this general picture. To some extent this variation can be explained by the fact that it ranges over the North–South divide. The Netherlands and France present a more systemised version in which the array of organisations, plans, and possibilities for flexibility are clearly set out, while in other countries such as Italy and Greece variety leads to fragmentation and extreme complexity. Then in recent years there have been the pressures for regionalism, particularly in the cases of Spain and Belgium. Here the process has moved close to a federal structure with the corresponding changes in planning powers and responsibilities.

The Germanic family is also imbued with a legal approach which comprehensively codifies the law and this is expressed in the planning system by the rigorously formulated planning regulations. The strong constitution and the federal system results in a strong regional level of planning with its own laws and plans and a set of arrangements for creating consensus between and within levels in the hierarchy. This results in considerable variation in the planning processes between regions but within a strong national framework. The Scandinavian family has probably gone the furthest in decentralisation with planning at the national level reduced to a minimum and regional planning only weakly represented. Here the emphasis is very much on the municipalities and at this level there are similarities with the Germanic family. They both have the concept of a broad general plan, which co-ordinates public sector activity and shows how national interests are taken into account, and a detailed plan, which is the legally binding instrument that is brought into play when development is to take place. The latter involves negotiation between the municipality and the developer and can be viewed as an elaborate kind of planning permission. Countries within the Scandinavian family show a particular similarity which has been fostered by organisations for Nordic co-operation. There has been much cross fertilisation of planning ideas and legislation (Hall, 1991b).

There has been some discussion about whether European countries are moving towards a harmonised planning system (Healey and Williams, 1993) and there are many indications of similar trends operating throughout the countries of Europe, faced with similar broad economic imperatives and common membership of the EU. However, this chapter has shown that there are major differences which are not likely to disappear overnight. These are the distinctions based upon legal and administrative approaches and history. In addition the social and economic processes in each country vary. As a result some countries have coherent planning systems of varying kinds which can be used to constrain the market mechanisms of urban development, others have to respond in a more fragmented and *ad hoc* way. The balance of power between the levels of government involved in urban planning decisions also shows considerable variation. Thus, as Davies (1994) has said, the future is unlikely to produce a harmonised system throughout Europe but, rather, greater mutual learning resulting perhaps in a convergence of planning policies within different legal and institutional settings.

The European Commission Report *Europe 2000+* (EC, 1994b) sets out three themes for exploring the variety across European planning systems: centralised or decentralised, reactive or proactive, regulatory or discretionary. We have already noted the considerable variation in the degree of centralisation in planning which corresponds closely with the legal and administrative characteristics of the different families. Thus the most centralised is the British family and the most decentralised the Scandinavian. In between we have the fused nature of the Napoleonic approach and the federal structures in the Germanic. Spain and Belgium combine elements from both these families. The second theme concerns the degree to which the planning system simply responds to private sector activity. The *Europe 2000+* report contrasts the proactive period of British planning after the war with the reactive approach of recent years. It then claims to detect a continuation of the proactive approach in Europe in the strategic spatial planning undertaken in many cities seeking to create favourable conditions for future economic development. However it could be said that this is still only reactive – a response to the imperatives of international economic competition. This is a theme that will be explored in the next chapter but meanwhile it is interesting to note that certain countries are responding to this increased competition by preparing national spatial plans as marketing tools. This is particularly clear in the Danish and Swiss examples. However, rather than viewing this trend as the resurrection of proactive planning it might be regarded as a shift from a disjointed project-led approach of the 1980s to a more strategic, yet market-oriented, approach. In the future this planning framework is likely to become more important again as broader environmental objectives gain in political importance and as property interests see the merits of a strategic outlook (Healey and Williams, 1993). However, the aims and priorities of these strategies and the interests they represent have to be analysed before one can be clear about the role of planning. Again these issues will be explored further later.

The third theme raised by the *Europe 2000+* report concerned the contrast between regulatory and discretionary systems. The British discretionary approach is compared to one which uses a legally binding plan and the issue usually raised is the degree of flexibility. However, this may not be the most important point. We have already indicated that the British system is distinctive but the most important difference may be its degree of centralisation and the weaker participatory input. This results in more secrecy and less local control. There is no participation in the discretionary development control process or in the planning gain discussions. In contrast the local plans which take the place of development control in most of Europe contain the right for third party appeal and involve a participatory input. There may also be less difference in the degree of flexibility than at first appears. In many systems there are moves to make the plans less specific and to introduce mechanisms for greater flexibility. In many countries the detailed plan is only produced at the time when change is taking place and involves negotiation and participation. It is interesting to compare the power of the local authority in the respective negotiating situations. What actual sanctions and power do local authorities have? It would be necessary to compare control over implementation resources such as finance and land, the degree of independence from central government, whether appeals can be made, who has the right to make the plan/proposal, and whether reference can be made to the public's response. The local authorities' power position is firmly related to the planning procedures and it is clear that the British system is weak in this respect. Other systems of course may be weak for other reasons. The lack of enforcement powers and political clientism in some countries has already been noted. Another means by which flexibility has been introduced into a system is through a general process of deregulation, in which plans are either weakened in content or status. This is often the result of political ideology as in Britain and Denmark over recent years. Mechanisms which bypass the formal processes can also bring flexibility. Sometimes, such as in Britain, this is to reinforce the deregulatory ideology and sometimes, such as in Italy and Spain in connection with sporting facilities, this is simply to speed up the normal cumbersome procedures. Urban planning in many countries, in response to new economic demands in a competitive climate, also shifted from the formulation of plans to the promotion and realisation of projects.

These issues of power in negotiations, the degree of influence over the planning system and the attempts to create alternative decision-making procedures are all illustrations of the way in which the national formal systems cannot be regarded as the only influence. The legal planning framework and the system of plans remain a national characteristic, creating different constraints and opportunities, but we need to explore the role they play in the broader political processes of urban planning. This chapter has set out the national framework within which urban planning operates while the following chapters will look at the way in which this interacts with power structures in particular cities and in particular development projects.

4

URBAN GOVERNANCE AND PLANNING

INTRODUCTION

This chapter extends the discussion from the national level down to the urban. After a review of current debates about urban governance the chapter goes on to analyse the relationships between governance and planning in Frankfurt, Milan, Barcelona, Prague, and Berlin.

In response to the growing complexity of city government the vocabulary of urban political analysis has changed in recent years and the term 'urban governance' has become widely used. Authors claim that it is no longer possible to focus only on the formal institutions of local government and that changing government structures and the growing importance of informal institutions in the government of cities have to be acknowledged (see Stoker, 1992). It is argued that 'urban governance' captures this broader perspective. Cities may be managed through combinations of local, regional, national and supranational agencies and through the direct involvement of private sector actors. The breadth of the concept of urban governance, bringing together many of the themes examined in Chapters 2 and 3, provides a useful starting point from which to understand the changing nature of planning at the urban level.

Changes in formal structures of government include both decentralisation and the fragmentation of responsibilities. We noted in Chapter 3 the major decentralisation reforms in many European countries. The structure of decentralised regional government in Germany was transferred to the new *Länder*, including the new city-state of Berlin. In Britain local councils have been by-passed and government fragmented among quangos, with many urban services provided by private companies. In other countries fiscal crises have brought about both retrenchment in city finances and experiments with privatisation (see Batley and Stoker, 1991; Mouritzen 1992). In many countries there is a continuing search for better central–local arrangements, for example in Ireland (Bannon, 1993), Britain and France, for better co-ordination between lower level tiers, and for more effective urban management. Reforms in France and in some German cities (e.g. Stuttgart, see Carter *et al.*, 1994), indicate a new concern with the co-ordination of policy across conurbations. The complex relationships between increasingly fragmented institutions provide an unstable context for urban planning.

The concept of urban governance includes informal structures and the increasing involvement of private sector interests. In the new context of inter-city competition local politics has become increasingly concerned with economic development issues (Mayer, 1994). It is argued that as nation-states have become weaker, economic actors have taken increasingly important roles in local decision-making. The greater involvement of private sector interests in the government of cities appears to be widespread. Economic change, the transformation to post-industrial urban economies, has effects on the social base of urban politics. The breakdown of old class allegiances (see Harloe and Fainstein, 1992) and the emergence of new classes (Novarina, 1994) have had impacts on the local political agenda. Some trends seem to be international. For example, in the world cities of Frankfurt and London the extreme right has had recent electoral success. In London this is argued to result from the alienation of the local community from the decision-making process of the London Docklands Development Corporation (Colenutt, 1994), and in Frankfurt from the preoccupation of the main parties with middle-class votes and the exclusion of many people from the benefits of the city's economic success (Ronneburger and Keil, 1992). Economic change has also brought forth new social movements some of which have, for example, inserted environmental values into local politics.

The context of change in intergovernmental and public–private relationships is no less significant in eastern Europe. As we noted in Chapter 3, in eastern Europe urban government is characterised by institutional instability. Local governments have tended to fragment, and the lack of working legal frameworks and operational government structures creates significant problems for urban planning (Marcou and Verebelyi, 1993). There is also scepticism about the role of local government. In Brandenburg, for example, disillusionment with politics and the supposed benefits of unification led to resignations from political office and a shortage of candidates for the new local institutions (*The European*, 1993).

Governing cities has become ever more complex. This chapter aims to develop our appreciation of the relationships between contemporary urban governance and planning. It is organised in four parts. Firstly there is a review of literature concerned with the growth of public–private partnership, 'entrepreneurial' cities and institutional innovation in response to inter-city competition. This literature, whilst usefully identifying significant trends, is largely descriptive and the second part of the chapter therefore reviews work which attempts to provide analysis and explanation of contemporary urban governance. It seeks to draw out of this work the range of factors which may account for variation in the form of governance in European cities. The chapter then looks at the interrelationship between urban governance and planning in a number of cities. Finally conclusions are drawn about the key factors shaping relationships between urban governance and planning.

PARTNERSHIP, ENTREPRENEURIAL AND 'SUCCESSFUL' CITIES

One of the ways in which trends in intergovernmental and public–private relationships have been brought together in analysing contemporary cities is through the notion of 'partnership'. It is argued that the combination of economic change, changes in government structures and ideology and the fiscal crisis of local governments have combined to bring forward new agencies to undertake urban development (see Heinz, 1994 for a review of these arguments). The rapid growth and variety of new partnership arrangements have drawn researchers in many European countries into cataloguing the various forms of intergovernmental and cross sector co-operation. The variety of types of co-operation given the term 'partnership' raises problems for the utility of the concept but this has not prevented its international application. The normative component of the concept adds to the difficulty. The literature on partnership is largely descriptive but also prescriptive in its desire to set down rules for better models of co-operation (Mackintosh, 1992; Heinz, 1994).

Many European cities have actively sought to develop, or to borrow from others, models of public–private co-operation which might produce some advantage in inter-city competition. Many British cities drew upon the experience of US cities (see Glasgow Action, 1987) in developing promotional public–private alliances. The UK organisation, Business in the Community, looked at successful European models of public–private co-operation and selected lessons for British cities (Coopers and Lybrand, 1992). The Rotterdam public–private alliance was recommended by the Dutch government as a model for other cities (Netherlands Scientific Council, 1990). Other work has sought to identify best practice across Europe in the institutional support for urban regeneration (Carbonaro and D'Arcy, 1994).

Successful models of public–private leadership and 'entrepreneurial' cities also became a focus for academic study (see Judd and Parkinson, 1990; Parkinson *et al.*, 1992). Parkinson (1992) identified a new breed of 'entrepreneurial' European cities whose local 'institutional capacity', i.e. their ability to create new pro-development institutions, contributed to success in economic competition. Da Rosa Pires (1994) applies this idea of local institutional capacity to Portuguese cities which, he argues, are at a double disadvantage of being both geographically peripheral and lacking the experience and capacity to set up effective institutions. Leontidou (1995) argues that cities in southern Europe have to change their traditional clientelist politics in order to adopt the new urban marketing and management practices which dominate in successful cities.

The recent interest in the contribution of institutional factors to economic success has come from both academics and city governments. One of the 'success' stories of the 1980s was Hamburg (see Coopers and Lybrand, 1992; Parkinson *et al.*, 1992). The experience of Hamburg is interesting because it clearly reveals the need to understand both the significance of political structures, in this case the nature of German city-states, and the role of private sector interests in creating new forms of governance. The federal German structure gives city-states considerably greater autonomy compared to other European cities. Boyle (1991) describes changes in the approach to economic development in Hamburg in the 1980s. Policy changed from supporting traditional port-related industry to seeking to encourage new development in the city. The policy change began in the public sector with the mayor supported by the strong media industry. The city's economic ministry was used to set up Hamburgische Gesellschaft für Wirtschaftföderung (HWF) a joint public–private business development corporation in 1986. It was staffed by local business and the shares in the company were held by the city (31 per cent), the chambers of commerce and industry and, most importantly, local banks. In Germany membership of the chambers of commerce is compulsory and the Hamburg association with 80,000 members (compared to about 5,000 in Birmingham) is a powerful local body (Coopers and Lybrand, 1992). The HWF assesses applications for building approval as well as marketing the city and attracting new investment. It is a strong pro-development agency, sits alongside local public power, and has been a tangible symbol of the city's recent success. The office boom, attracting international and German companies, and the increase in port activity at the expense of its former East German rival Rostock, has made Hamburg one of the more successful German cities (Läpple and Krüger, 1993). Elsewhere in Germany reunification has introduced constraints on local policy initiative. For example, the environmental and social values built into the redevelopment of Emscher Park in the Ruhr depend for implementation on large-scale national development funding. There is concern that since reunification such state investment is being directed to projects in the east (Hennings and Kunzman, 1993).

There were thus interrelated political and economic reasons for the reorientation of the city's development strategy. The HWF exemplified the capacity for institutional innovation, and the 'success' of Hamburg was studied by other cities. Many echoes of such forms of public–private co-operation, developed in the core of the European economy, can also be heard in the periphery. For example, in Bilbao in 1989 the city developed a strategic plan aimed at the reuse of derelict industrial land and attracting service sector development. This replanning has been undertaken by a public–private planning and promotion agency 'Bilbao-Metropoli-30', a grouping of over a hundred members from the universities, local business, the local and regional authorities and some local agencies concerned with social welfare (Martinez

Cearra, 1994). The major role of the agency is promotional and throughout Europe city marketing of this type has necessitated some degree of public–private co-operation, at least to present an acceptable image to the outside world. It was noted in Chapter 2 that marketing and image building have become important features of inter-city competition. However, there are marked differences between temporary promotional alliances and development agencies of the HWF type in terms of the depth of co-operation and their relative permanence.

In western Europe since the mid-1980s cities have been searching for the ideal institutional arrangements which would allow them to succeed in inter-city competition. Western models and techniques have also been transferred to eastern European cities. In some cases western practice has been directly grafted on to emerging structures of governance. For example, the French state-controlled bank, Caisse des Dépôts et Consignations (CDC), has established a development agency based on French practice to redevelop a residential area in the Ferencvàros district of Budapest. An original scheme for the area in 1987 failed because of high interest rates and the weak demand for housing for sale (Aczel, 1993). The new development agency set up by CDC proposed a much bigger scheme which involved developing public spaces to improve the image of the area. The project introduced both market objectives and a new public–private co-operation between the municipality and the banks. However, as noted earlier in Chapter 3, in central and eastern European countries government structures are being reorganised and relationships between city government and the private sector are not always clear. For example, Hudson (1993) describes the city government in Szczecin as playing 'fast and loose' with potential developers involved in a public–private alliance to redevelop a central residential district. Learning the rules of the public–private partnership game is not always straightforward.

Partnership, entrepreneurialism and institutional innovation became dominant themes in urban studies in the early 1990s. Much of the literature described 'success' stories and attempted to draw prescriptive conclusions. There is ample evidence of profound institutional changes both within the structures of government and in public–private boundaries. Both dimensions of change have impacts on urban planning but a clearer understanding is needed of those factors which shape specific forms of governance in European cities and the urban planning that goes with them. The next part of this chapter consequently reviews theoretical literature which attempts to explain the changing roles of public and private interests in the production of new forms of urban governance.

APPROACHES TO URBAN GOVERNANCE

In the late 1980s many researchers (see Lloyd and Newlands, 1990, for example) were attracted to a body of theoretical work, originating in the US, on 'growth coalitions'. This work seemed to apply to the pro-growth attitudes adopted in many European cities. In this section we examine how successful this and other US literature has been when tested in the very different circumstances of European cities. There is a growing body of work seeking to develop comparative perspectives on urban governance and the aim in this section is to draw out of this literature the key factors shaping urban governance and the relationships between governance and planning.

The US pro-growth machine identified by Logan and Molotch (1987) included economic interests, landed capital, developers and local banks. It was argued that city governments supported pro-growth policies for the electoral advantages that economic development would bring and that they used planning and other powers to support private objectives. The media collaborated because of its interest in boosting local sales and advertising. Pro-growth strategies thus had wide support.

The idea of pro-growth coalitions led by private interests had obvious attractions during the 1980s, particularly for British urban research. More generally, during successive property booms and with a liberal ideology in the ascendant across Europe, it was easy to think that public institutions were in retreat. However, despite its initial attractiveness, the bias in this body of work towards private property interests poses substantial problems when applied to European cities where, traditionally, the public sector rather than private interests tends to dominate urban policy-making. Hennings and Kunzman (1993) argue that the equivalent of growth coalitions in Germany are public sector organisations but that they have adopted slogans from the more privatised US versions. Moreover, the US literature focuses on the roles of local economic and political elites, whereas in Europe national interests are likely to play a greater role (see Harding, 1991b on UK experiences, for example). Thus whilst Glasgow Action sought to draw lessons from the pro-growth politics of US cities it relied heavily on the support of the central government funded Scottish Development Agency. In Europe national governments will generally be more important in urban politics. There is also variety in the forms of central–local relationships and in the relative autonomy of city governments. For example, the constitutional autonomy of the German city-states clearly allowed political leaders in Hamburg to take the initiative in redirecting policy in the 1980s. Adapting the growth coalition literature to European circumstances therefore necessitates a strong role for differing national political institutions (see Keating, 1991, 1993a).

A second difficulty in transferring the approach to Europe lies in the emphasis in the US literature on competition for private development finance. European cities may be more interested in competing for central government and European Union finance than for private resources (Levine, 1994). The difference in systems of local government finance also poses problems. Levine (1994) argues that policies will clearly be different in European cities which are not concerned, as are US cities, with their rating in bond markets. Thus

Benfer (1994) argues in relation to German urban policy that cities can concentrate on projects which create jobs over those which raise real estate values. Across Europe sources of public finance play a much more important role in urban development. For example, Kantor and Savitch (1993) estimate that whereas federal aid pays for less than 6 per cent of Detroit's budget about 90 per cent of Amsterdam's expenditure comes from the national treasury. A further factor which marks fundamental differences between US and European city politics is the nature of political parties. In some US cities there are close links between business and local political parties which may be reflected in tightly organised alliances. However, in Europe national party politics and party loyalty are more likely to shape local decisions (see Vicari and Molotch, 1990; Kantor and Savitch 1993; Levine 1994).

In addition to these comparative problems there are also broader limitations in the growth coalition approach. Several authors have argued that the idea of the growth coalition over-emphasises the role of local property interests in urban politics and that the roles of national and international capital need to be taken more fully into account (see Imrie and Thomas, 1993, for example). Another more general criticism is that growth coalitions give a one-sided view of urban governance, being concerned only with political organisation around development-led growth. Other forms of coalition, anti-growth alliances for example, are ignored. A broader view of how coalitions of interests come together in cities is offered by another body of US literature – that on 'urban regimes'. In his discussion of the comparative potential of the 'urban regime' literature, Harding (1994) argues that there is some affinity between the US concept of 'urban regimes' and the concept of 'urban governance' developed in the UK. It is argued that both provide a broad perspective on changing structures and the wide range of interests which influence urban policy. Several authors have been attracted to the potential of cross-national comparison of 'regimes' (Barlow, 1995; DiGaetano and Klemanski, 1993a, 1993b; Stoker and Mossberger, 1994; Levine, 1994; Newman and Thornley, 1995).

According to its proponents a 'regime' is a relatively stable form of governance combining public and private interests. Some types of urban regime resemble growth coalitions. For example, Stone *et al.* (1991) describe a relatively exclusive 'Business-Centred Activist' regime made up of public and private actors pursuing economic development. However, this closed, property-based coalition is not the only type of urban regime. For example, a 'Caretaker and Exclusionary' regime brings together political elites and voters against intrusive development in, typically, suburban locations, and 'Progressive' regimes may seek to integrate social objectives with a pro-development stance (Stone *et al.*, 1991). Indeed, regime theory allows for the possibility that stable governance may be achieved not through any visible form of open combination of interests but by the mobilisation of bias and assumption of common interests between political and business leaders

(Stoker 1993). The regime perspective suggests a variety of forms of urban governance based on the objectives of public and private interests. For example, DiGaetano and Klemanski (1993a) develop Stone's categories by distinguishing between, among others, 'pro-growth regimes' led by either private interests or by government, and regimes which attempt to exact social benefits from urban growth.

Developing the comparative potential of the regime approach, DiGaetano and Klemanski define regimes as 'modes of governance that entail formal and informal arrangements for policymaking and implementation, both across public and private domains and within the public domain, the balance of which differs among nations and over time' (1993a, p.58). This broad-based definition allows considerable flexibility in approaching questions of urban governance. German intergovernmental consensus building would be embraced by this definition as would new forms of privatist decision-making (Barnekov *et al.*, 1989). The inclusion of the public domain in the definition is vital in Europe where, as we have argued, government takes a strong role in urban policy. It is necessary to understand relationships within the public sector, between levels of government and forms of government agency, and those between public and private interests. This broadly based approach to regimes shares much in common with the literature on policy networks (see Marsh and Rhodes, 1992). There is a joint concern to identify the complex networks of relationships between actors. Cole and John (1994) review the theoretical value of local network theory to understanding change at the sub-national level. They argue that analysis of networks should be a starting point for understanding contemporary local governance and that bargaining and conflict between those involved is likely to characterise local decision-making. However, identifying networks of relationships within government and across sectors only provides part of the picture. Fainstein (1994) argues that the regime literature rightly focuses not just on who is involved but on how some actors manage to enforce their objectives. Conflict and bargaining between actors are indeed analytical starting points. In addition to the identification of the constituent elements of stable governance what is also needed is detailed analysis to see which interests are enforced and how that is achieved.

The concept of the pro-growth coalition highlighted the formation of coalitions around economic development. Those authors who have sought to adapt the idea to European circumstances stress the significant role of national political institutions in shaping development coalitions. Supporters of the regime perspective add a further dimension to understanding urban governance. They identify the contribution to different styles of urban governance of the specific institutional forms that urban coalitions take. Stoker and Mossberger (1994) develop the regime literature and attempt a classification of urban regimes based on the ways in which partners come together. The urban regimes in Rennes and Croydon are labelled 'organic', reflecting long-standing close relationships between parties. Glasgow and Sheffield are seen as 'symbolic' regimes reflecting a wider grouping of interests around less concrete objectives, and Atlanta and Birmingham 'instrumental' regimes in which parties pursue limited development interests.

Stoker and Mossberger argue that whereas the US literature tends to stress the role of tangible incentives to regime formation, in Europe 'symbolic' factors, such as values, images and promises, may be more significant in cementing alliances and securing the acquiescence of potential opponents. In Britain in particular the vocabulary of 'partnership' has been extensively deployed in this way to offer a sense of participation to local community groups, for example. Another important means of binding interests together is through adherence to a strong leader. DiGaetano and Klemanski (1993b) argue that cohesive local leadership is important in understanding the relative success of regimes in Bristol and Birmingham. We would expect the variety of local democratic traditions in Europe to offer a range of contexts in which the strong leadership associated with some successful urban regimes can emerge (see Judd and Parkinson, 1990). For example, where local governments have executive mayors the potential for personalised leadership may be greater than in other contexts where formal politics is the responsibility of committees.

The extent to which private sector interests seek to be involved in urban governance varies from city to city. Local economies and market conditions will differ. Kantor and Savitch (1993) suggest that particular forms of political institutions and mechanisms for intervention will influence the nature of public–private co-operation. Different political systems will create different 'bargaining advantages' (Kantor and Savitch, 1993) in relation to private interests. In a strong market developers may be more willing to provide additional social benefits. Rich cities may be able to exact social benefits from development whereas poorer cities may be forced to provide subsidies in order to attract development. The condition of the market introduces a further element accounting for local variation between forms of governance.

Leadership, the ways in which interests come together, and informal politics all contribute to the form of urban governance. The regime literature adds considerably to our understanding of the range of factors involved. Differences between cities will reflect national political structures, the networks of relationships within government and between public and private domains. The literature examined above offers a dynamic perspective on the formation or dissolution of stable governance and the competing interests involved. In many cases coalitions of multiple interests may not emerge (see Barlow, 1995). Coherent city-wide 'regimes' may be found less often than a range of 'development coalitions', 'progressive' or 'anti-growth' alliances, with national and local agencies and interests all competing to determine urban policy and development outcomes. It was argued at the start of this chapter that urban governance had become more complex. The theoretical work examined here certainly attempts to grasp that complexity. The recent revival of interest in the urban level within political science has been accompanied by a theoretical plurality. There are lively theoretical debates between the adherents to various positions (see the discussion between Stone, 1991 and Cox, 1991 for example), but the attempts to develop a comparative perspective on urban governance are more concerned with developing useful frameworks than with theoretical exclusivity.

Some conclusions can be drawn from the foregoing discussion. We must take account of the role of private sector interests in urban decision-making. National political contexts make a difference to the scope of local politics and the types of government agency operating in the urban arena. Indeed, in the UK central government has consciously sought to impose new institutional forms at the urban level. Networks of relationships within government need to be understood as do public–private relationships. Market conditions will have an effect on the emergence and role of private interests. We also need to pay attention to the detail of institutional forms and processes which determine the degree of inclusiveness of decision-making and contribute to the stability and success of particular alliances of interests.

We would expect to find cities continually searching for new institutional arrangements and alliances, striving to draw together the necessary resources to achieve policy objectives. The concept of resources includes not only development finance but also legal frameworks and organisational expertise. Planning systems play an important rule-setting function and the availability of local government planners and other officials will be a valuable resource in establishing successful governance. DiGaetano and Klemanski (1993b), Keating (1991), Levine (1994), and Harding (1994) all acknowledge the influence on urban policy of professional and technical experts in the public bureaucracies of European countries. In order to develop a clearer view of the roles of urban planning in contemporary city politics the next section of this chapter examines the relationships between urban governance and planning in a number of cities. The cities have been selected from countries not studied in greater depth in Part II of this book. Some of the city profiles identify a few key aspects of urban governance, and others are developed more fully to demonstrate the importance of national, institutional and cultural backgrounds. The case of Berlin is examined in some detail as it expresses so many of the themes of this chapter. The new economic pressures on the city, at the gateway between western and eastern Europe, combine with the complete reorganisation of political structures to create a particularly clear example of the forces contributing to contemporary urban governance.

URBAN GOVERNANCE AND PLANNING

Frankfurt: governing the world city

Urban policy in Frankfurt underwent substantial changes in the 1980s. That experience reveals the specific circumstances of city politics in Germany, the importance of understanding the role of economic interests and the political consequences of radical urban change.

Since the 1970s Frankfurt has grown in stature as one of the contestants for world city status. The development impacts of this strategy have been the creation of a downtown skyline epitomising a world financial centre, the suburban development of related economic activities and growth of the airport. The world city oriented strategy was led by the Christian Democrat party (CDU) which was in power from the late 1970s. Party and electoral politics subsequently played significant roles in the history of pro-growth attitudes in the city (Keil and Leiser, 1992).

In pursuit of its world city strategy Frankfurt set up various semi-public corporations alongside its public departments. The Economic Development Corporation, set up in 1987, and bodies managing the airport and exhibition and conference space were given significant autonomy. The development corporation joined with the *Land* government and the chamber of commerce to lobby successfully for the location of the European Monetary Institute in Frankfurt. The growth strategy was managed by the CDU but they were defeated at the elections in 1989. The reasons for this have to do with the way in which political parties adjusted to the new urban geography of the aspiring world city.

The growth strategy had social costs. There was a substantial housing need and the financial centre displaced residents from the central area. The vote for the 'green' parties increased during the 1980s especially in areas close to the financial centre. In 1989 the Greens and the Socialist Party (SPD) gained more seats than the CDU and the ruling party was displaced. The CDU lost nearly 33,000 votes (Ronneburger and Keil, 1992). The new ruling alliance altered the world city strategy by selectively incorporating more interests into decision-making and adding some social and environmental concerns – for example, in positive green belt policies. Political party ideology and electoral systems clearly make a difference to regime formation and planning strategies.

According to Kantor and Savitch's (1993) classification of urban governments Frankfurt enjoys favourable market conditions and local government has relatively strong 'bargaining advantages' in relation to the private sector. However, Ronneburger and Keil (1992) argue that social and economic change, and the fact that their impacts cross administrative boundaries, have created instability in urban government. The evidence of this instability is seen in the political volatility of areas adjacent to the financial 'citadel' and in the absence of a coherent, wider regional approach to growth management. Intercommunal planning in the wider urban area through the mechanism of the *Flächennutzungsplan* agreed in the late 1980s was weak. In some areas developments which would have supported the world city aspiration have been resisted. 'Caretaker' politics exists alongside the pro-growth institutions in the city.

In addition to the impact of political ideology and party politics, governmental structures therefore also play a part in shaping the planning approach in the city. Frankfurt has sought entry to the first rank of European cities. However, the social costs of economic growth – rising homelessness and demand for social housing – have not been met. The city regime found economic development in the centre easier to achieve than broader social planning.

Milan: political influence

The example of Milan illustrates the strong interplay between politics and planning in the Italian situation and also the flexible approach in the use of planning instruments. Milan plays a leading role in the Italian hierarchy of cities lying at the centre of a dynamic economic hinterland. In the various studies of European city league tables it is usually placed in the second rank after London and Paris.

During the 1960s when the city was ruled by the Christian Democrats the major planning problem was the chaotic and uncontrolled growth of the city. The local construction companies were very powerful and had strong links with politicians. However, from 1975 until 1993 the city council was run by a coalition which included the Socialist Party. For all but two years (1985–7), when their partner was the Christian Democrats, the Socialists were in coalition with the Communist Party. During this time the head of the national Socialist Party, Craxi, was a powerful figure on the national stage and yet also kept a close involvement in Milanese affairs with an office in the city and many relatives in local positions of power. During the late 1970s there was a long debate over the New Master Plan for the city, a draft was prepared in 1976 by the new administration and it was approved in 1979. It was, however, already outdated. It responded to the problems of the 1960s and did not take into account the economic shift from industry to tertiary activity. Thus it proposed new industrial zones while the existing ones were falling into disuse. The inadequacy of the numerous implementation plans that followed on from the Master Plan led to a growing dissatisfaction with the traditional planning procedures (Cappellin, 1989).

In 1975 the Communist Party was the dominant party in the coalition and its planning goals were to control urban growth, to reduce the congestion in the centre, and to improve cultural facilities, public transport and environmental quality in the peripheral neighbourhoods (Vicari and Molotch, 1990). This contrasted with the ideas of the Christian Democrats who had been promoting the concept of new radial underground lines reinforcing the role of the city centre. Meanwhile another idea had been proposed by subregional planners back in the 1960s and this returned to the agenda. They proposed a new rail link in a tunnel through the centre of the city which would link several major stations and create new development opportunities on the edge of the central area – this was called the '*Progetto Passante*'. Although Milan's left wing parties were originally against the idea because of the scope for speculation, once in power they adopted it claiming they could control the speculation and exploit the new development potential for social purposes (Vicari and Molotch, 1990).

The year 1980 has been said to mark a turning point in the way the administration viewed the development of the city (Fareri, 1991). This involved a shift from seeking equilibrium throughout the metropolitan area to a concentration on the city centre. From this time a more pragmatic approach to planning developed based on the strong relationships between politicians and developers. Hundreds of amendments were made to the Master Plan with the result that a new planning approach developed without the context of an appropriate plan. The *Progetto Passante* adopted in 1982 in some ways provided a strategic concept. In the following year a group of experts and public officials produced the *Documento Direttore*, exploring the implications of the rail line and proposing the concept of four particular development areas (*Progetti d'Area*), often utilising the derelict industrial areas which had a strategic importance to the city. According to Cappellin (1989) these documents were a new innovation in planning procedures as they placed the detailed administrative planning regulations in some context.

The second half of the decade has been described as featuring a breakdown of the overall strategic framework (Fareri, 1991). The implementation of the ambitious development projects required a considerable amount of managerial and co-ordination skills on the part of the city administration. The tradition of close co-operation between the city authorities and local powerful business interests continued to produce smaller-scale speculative developments but was not up to the bigger task of developing these new large schemes. It was clear by 1990 that the schemes would not be implemented because of the fiscal crisis of the state and the declining interest of investors. Vicari and Molotch (1990) have analysed this period of planning in Milan to test whether the growth coalition thesis could be applied to these events. They conclude that the regime took on a rather different form. Business interests did not involve themselves in the strategic planning discussions and left this to the political parties and bureaucracy. They concentrated their efforts on getting a good deal out of individual development projects. The ease with which amendments could be made to the Master Plan would suggest that this was a sensible economy of effort. Vicari and Molotch also highlight other variations from the growth coalition model. First there was the influence of the national level over planning regulations, particularly for conservation areas. A second factor was the high presence of publicly owned organisations involved in the development process; for example, one of the key development schemes was round the Garibaldi station where 87 per cent of the land is publicly owned. This emphasised the need for negotiation by the private sector. A third influence was the role of political parties with their supporters in the bureaucracy. Local parties had strong links with central figures as illustrated by the role of Craxi. Parties also have their own business interests and organisations. This, with their control of the media, gave politicians considerable power which was independent of the business sector. Vicari and Molotch conclude that rather than the business-led coalitions of the US, in Milan there was a more complex interaction of political and business, national and local interests.

In the early 1990s the corrupt nature of these close links started to be exposed. This particularly, though not exclusively, concentrated on the Socialist Party who were still part of the government of Milan. The corruption scandals included the chief planner of the city and extended to Craxi himself. The investigations revealed a pattern of deals and corruption in development projects. It has been suggested that the property deals in Milan were sometimes used as a way of laundering money. 'This entailed a series of bribes to officials, first to those responsible for giving planning permission and then to those in various state pension funds to buy the buildings for their investment portfolios leaving the developer with a handsome profit' (Muccini, 1993, p.27). The expectation in the property world is that the exposure of corruption will lead to a more market driven real estate sector. Alongside this was the failure to instigate the larger schemes involved in the *Progetto Passante*. Many pointed out that the planning approach based upon big projects, which in any case were not realised, ignored the social needs of communities, particularly in the deprived periphery. In this difficult political climate the council tried to re-emphasise some of the priorities of the mid-1970s with strong environmental and social policies, including the idea of creating green areas from about half of the old industrial areas (Bianchini, 1994).

However, the dramatic new political climate brought the Lega Nord to power in Milan in the 1993 elections. These were held under the new Constitution which aimed to create a majority government and the Lega held 60 per cent of the council seats. In their election manifesto the Lega accepted that the previous strategic approach was not working and proposed the formulation of a new plan. There was also a demand for this, for different reasons, from both big business who wanted a strategic growth plan and environmentalists demanding a more protective approach. However, in office the Lega, who were a populist party with support from small shopkeepers, abandoned the idea. Instead the new chief planner appointed by the Lega wanted to adopt a pragmatic approach utilising previous projects (Bertamini *et al.*, 1994). However, he was frustrated by the actions of the Regional Administrative Tribunal who blocked approval of such projects because of past corruption. Another forty-four existing projects were declared unacceptable by the city council themselves. By 1995 the Lega had lost its popular support and the government of Milan went into crisis as members of the party defected and the overall majority was lost. While government is in a position of stalemate, planning activity has ceased and only limited permissions for development are issued (Estates Europe, 1993/4; Europroperty, 1993). Meanwhile new national Acts have been passed, including building laws, to prevent corruption, but there is a lack of administrative and political structures to adapt to the new situation.

However, there is much activity by the business sector to try to promote Milan more and to encourage central government to start taking an interest in the contribution that Milan could make in the Italian economy. Until the changes in European legislation on open access to business opportunities, embodied in the Single European Act, the Italian and hence Milan economy was a regional economy without outside investment and involvement. In many ways this allowed the closed and corrupt practices to be perpetuated. In the late 1980s there was considerable activity by Italian companies to make the most of their opportunities before the open market competition was forced upon them. The limitations of this closed approach might be argued to have contributed to the failure to realise the strategic projects in Milan.

Since the 1970s big business in Milan had been represented by the Socialist Party, and its collapse left them without representation. Some hopes were placed on the Lega Nord which had a European orientation, seeing opportunities in the EU for their regional approach. However this was more of a populist and small business party and in any case quickly collapsed, whereas Berlusconi's Forza Italia gained 30 per cent of the vote in the Milan constituency in the 1994 general election. Meanwhile numerous collaborative agencies have been set up to promote ideas over the future of Milan (Bianchini, 1994) – these vary from promoting the city as a whole to pursuing particular policies such as underground car parks. One of the important organisations is the Association of Metropolitan Interests (AIM) set up in 1987 and comprising local business interests and banks. The aims of this organisation are to make the city council more aware of the need to promote Milan's position in the climate of European competition and to devise appropriate strategies to achieve this. They have commissioned a series of studies, including a new strategic approach for Milan.

Thus the Milan case shows the strong influence in urban governance of political parties with their connection to national level government and the business sector. This has had a strong influence over planning and has contributed to the weakening of the planning system as constant amendments were made to the official plan and new instruments created to respond to the economic context. However, city governance was not able to carry the strategies through into implementation. The political uncertainty which has characterised the years after the corruption scandals is holding the city back from an international competitive approach with associated planning mechanisms, although there is a developing body of opinion behind this approach suggesting that a new form of governance may develop. Meanwhile the social and environmental needs of the city have not been adequately addressed and are in danger of remaining a second priority to economic development as statistics show Milan slipping down the European league tables.

Barcelona: building on consensus

Barcelona has an impressive record of moving from deep economic crisis in 1980 to a city with a strong image. How has it achieved this? During this period the urban planning of Barcelona has been dominated by the Olympic Games. We will examine the way in which this event affected the mechanisms for planning, the direction taken since the Games, and the kind of urban governance that evolved.

During the 1970s and early 1980s the traditional industrial sectors of the city declined, unemployment and spatial disparity across the region increased (Garcia, 1991). The public and private communication systems were also very deficient and there were severe problems in the 1960s housing estates (Busquets, 1988). At the time that this economic recession was hitting the city the forty-year dictatorship ended and local elections were held in 1979. After a year or so of three-party government, the city has been continuously governed by a coalition of socialists and communists. When the new city government took over in 1980 it was much influenced by the democratic movements of the 1970s and the need to respond to local citizens' demands with redistributive policies. Urban planning became more important and the plans, which incorporated local neighbourhood participation, aimed to upgrade the poorer parts of the city, improve collective services and housing provision and renew the city *plazas* (Garcia, 1991). However, according to Garcia a shift in direction can be detected from the mid-1980s. The previous policies of trying to counteract the effect of economic recession through employment schemes and small-scale projects continued, but these were supplemented by an active policy of economic promotion which attempted to attract investment from the rest of Europe, the USA and Japan. A network of institutions and programmes was set up for this purpose (see Garcia, 1991 for details).

The Olympic Games brought the attention of the world to Barcelona – for example, the city spread out in the sunshine as a backcloth to the high diving competition remains fixed in one's memory. As we shall see the event provided the foundation for a broader city-marketing strategy. However, the decision to bid for the Games in 1980 was taken in the belief that this would would support social aims and contribute to the modernisation of infrastructure. There was also a political motive as there was great competition between the city and the region governed by the centre-right Catalan nationalists (Marshall, forthcoming). Many of the aims contained in the urban regeneration aspects of the bid resurrected older ideas. This particularly applied to the transport system which had fallen into neglect and had lacked resources. The idea within the bid to redirect the development of the city towards the sea also dated from the 1960s. At that time there was strong opposition to the urban redevelopments that were later brought about by the Olympic Games. Garcia (1991) notes how the collective enthusiasm and civic pride generated by the winning of the Olympic competition contributed to their acceptance in the 1980s.

The Olympic bid occurred at a time when the economic strategy for the city was shifting away from consolidating its industrial past towards establishing a financial and service centre on a European scale. In order to implement this strategy suitable sites with improved communications were needed for commercial development away from the congested centre. A strategy in 1986 proposed twelve 'Areas of New Centrality' as part of this aim. The idea was 'to identify intermediate locations, away from the established core areas but not outside the city to which private sector commercial and residential development could be directed' (Riera and Keogh, 1995). The hope was that this would lead to the revitalisation of the surrounding neighbourhoods. Four of the twelve areas established were the locations for Olympic developments and these were linked by a new inner ring road first proposed in the 1960s (Riera and Keogh, 1995). The modernisation of infrastructure and the removal of the railway barrier to the coastal site of the Poblenou seafront opened up the development opportunities in this particular 'area of new centrality'. This area had also been the subject of a Special Plan in the 1960s that was never implemented. It was to become the site of the Olympic village – the principal development in the Olympic plans.

The surge of Olympic-related development was implemented through the special new companies that were set up, supported by the consensus attached to the Olympic effort and the need for commitment to meet the deadline for the Games. A senior planner involved in formulating an indicative Strategic Plan in 1988 makes clear the important role of the Olympic bid in generating consensus for the plan and giving it a sense of urgency not evident in the past.

> The Olympic Games acted as a driving force to arrange and obtain public investment in a short time. At the same time, the necessity of finishing the projects by a pre-set date generated efficiency and control in the execution of projects which would not have occurred under normal conditions.
>
> (de Forn i Foxà, 1993, p.7)

He refers to a 'patriotic consciousness of the city' which allowed for public–private co-operation and strategic thinking among the different agencies. He refers to the Olympic programme as a crisis and believes that 'if a consciousness of crisis doesn't exist, it will be extremely difficult to reach public–private operational agreement, as disagreements on immediate matters will override agreements on general issues' (de Forn i Foxà, 1993, p.17).

A number of new co-ordination and implementation agencies were established as municipal enterprises with managers appointed by the mayor. In 1989 the city came to an agreement with central government over the financing of the works and set up an overall holding company incorporating these agencies, Olympic Holding (HOLSA), with 51 per cent of its money coming from central government and the rest from the city (Vegara, 1992). The manager of this organisation was appointed by central government with the approval of the city. In the case of the Poblenou site, which as well as housing the Olympic village had considerable commercial potential, a municipal company, Vila Olimpica, SA (VOSA), was established in December 1986 to organise the planning, financing and implementation of the project (VOSA, 1992). The Poblenou area, which used to be called 'Little Manchester', had historically developed as a port and industrial area supplemented, as these activities diminished, by industrial dumping. It was an area ripe for development, and the Games provided the opportunity to undertake the expensive site preparations and removal of obstacles to its accessibility. Advantage could then be taken of this after the Games. As VOSA has remarked, 'the Games would generate the impetus needed to overcome a set of technical administrative and community problems that had not been solved in the past' (1992, p.49). By the end of 1988 the land had been expropriated and nearly all the existing buildings demolished. One of the major tasks of VOSA was to promote the real estate opportunities of the development and to attract investment. In 1988 it sold part of the land to private developers for office buildings, a large hotel and shopping centre.

The Mayor set up the Barcelona Development Agency (BDA) in 1987 to

> promote the city overseas and attract foreign investment . . . and deliver to the city those real estate related infrastructures that can help Barcelona to become one of the most appealing cities for business in the new Europe that is being shaped.
>
> (BDA, 1992, p.11)

The comments of this agency illustrate well the links between the Games and the real estate strategy; they describe the Games as an 'excuse' for recovering Barcelona's seafront and gaining a new business and residential district. The twelve new beaches created will obviously add prestige to these developments. However, the slump in the property market since the end of the Games has created difficulties in realising the projects and it is said that there is an oversupply of space especially in offices (Riera and Keogh, 1995). By 1993 only a third of the forty-four-floor office building was occupied – most of that by its developer, the Spanish insurance company Mapfre (Power, 1993). The original idea was that after the Games the Olympic village would contribute to Barcelona's severe housing problems (Kitchen, 1993) and that it would house about 10,000 people. However, the financial realities of holding the Games, and the desire to use the real estate opportunities to balance the books, meant that the housing had to be sold on the market.

Building on the Olympic process, a strategic planning exercise was initiated in 1988 called the Barcelona Economic and Social Plan 2000 (Marshall, 1990). Whereas the Olympic bid was motivated by local political goals and the desire to extract central government funding, the Barcelona 2000 plan was explicitly designed to promote the city. The Socialist project of 1980 was seen as needing some revitalisation and new direction and, building on the consensus achieved through the Olympics, a more moderate approach focusing on the goal of economic growth was adopted (Marshall, forthcoming). This encouraged the involvement of the business sector and the plan was directed by an executive involving the City Municipality, two trade union federations, the university, Chambers of Commerce, the Circle of Economy, Employers' Federation and a number of commercial agencies for particular areas. This combination matches that of a traditional growth coalition. Although attempts were made to maintain openness no efforts were made to encourage wide participation (Marshall, forthcoming). The aim of the plan was to co-ordinate the actions of public and private bodies and promote the city for international investment. An evaluation of the plan in 1992 commented on the slow progress made in achieving any of the social objectives that had been included. By then it was felt necessary to revise the plan, and the body representing the municipalities in the surrounding region also joined the executive, expanding its geographical coverage. This time there was lower public exposure of the plan and the responsibility for its approval was removed from the municipality to a special association. However, the plan was even more oriented towards development and infrastructure and although an early version had mentioned the need for social and environmental sustainability these aspects were removed (Marshall, forthcoming). The plan was seen as a marketing tool and sought a competitive niche for the city as the 'New Centre for Southern Europe'. This involved investment in such facilities as communications, convention centres and hotels. Central government has also played a role through its national infrastructure plan which contains, as one of its six projects of national significance, the Llobregat Delta next to Barcelona. This centrally funded project includes enlarging the port and providing new airport facilities.

So how can the form of governance in Barcelona be described as far as planning decisions are concerned ? It seems that the city, with its consistent political leadership, has played a central role. It has been said that 'the overall operation has always been under public, and basically municipal, control. The private agents that participated in some of the projects had to follow the city's plans' (Llusa, 1994). It is pointed out that the new agencies were led by public sector people and that the city designed and passed all the plans. Central government also had an important involvement through its financing role, both in the Olympics and in the Delta project. However, a change in approach can probably be detected from the mid-1980s. Previously the strategy had been geared to social issues and unemployment, with an emphasis on policies such as training and neighbourhood participation. Although this approach continued it was overlaid by the promotional strategy seeking inward international investment steered by a political and business elite. This is seen as necessary for Barcelona's economic future in a competitive world (Llusa, 1994). The Olympics laid the basis for this approach and although it enabled some social projects to occur it was also used to open up areas for commercial development. As a result new organisations for both promotion and implementation grew up which, although involving the public sector, were outside the normal democratic processes. Their importance in implementing the strategy gave more power to the private sector. It might even be said that there were two planning systems operating during the 1980s – one with a social orientation involving people directly; one geared to the big projects, with special agencies run by political and technical elites relying on the general Olympic consensus for legitimacy. There is evidence of increasing private sector involvement in the formulation of the new promotional strategies. As we know from other cities elite growth coalitions and their tendency to underplay social and environmental needs can lead to social segregation and dissatisfaction. The Olympics brought many benefits for residents as a whole through employment, better environment and road infrastructure as well as the development opportunities for the private sector. It will be interesting to see whether the attempts by the municipality to continue with their social programmes alongside the strategy of opening up the city for international investment will hold the competing interests in balance.

Prague: reinventing urban governance

Urban governance in the cities of eastern Europe is changing as they undergo rapid transformation from their communist form. The structures of government are still to be consolidated and new power relations have yet to stabilise. In the previous chapter we reviewed this at the national level, but the same situation exists within most cities. Meanwhile there are pressures to be faced as property markets are opened up and investors seek to take advantage of the opportunities. This is particularly so in Prague because of its location in Europe and its attractive urban quality. The pressure is illustrated by the number of property advisers producing 'city profiles' for potential investors (e.g. Debenham Thorpe Zadelhoff, 1992; Jones Lang Wootton, 1993b; East 8, 1993; Müller, 1993). Thus in Prague decisions are having to be made using temporary arrangements.

Nationally the political ideology of the Czech Republic since 1989 has been to enthusiastically espouse economic liberalism. This approach was formulated when Thatcherism was at its most influential (Sýkora, 1994) and was strongly propagated by an elite group of influential Czech economists. Public acceptance was based upon an antagonism towards public regulation, including planning. This provided the background to the establishment of various mechanisms to create a market environment. Ownership structures have been transformed through restitution – for example, 70 per cent of the housing stock in the historic core has been returned to previous owners – and small state-owned businesses have been sold through public auction (Sýkora, 1995). Rent regulations have also been lifted with respect to foreigners. A real estate market is therefore emerging with development decisions taken by individuals rather than public authorities.

There has been a reduction of local government's ability to influence the allocation of resources directly. Instead a regulatory planning system has to be formulated to control privately initiated development, but this has been slow to emerge. At the time of writing the approach is in transition and the legal and administrative structures only partially in place. The 1990 Act on Municipalities reintroduced local self-government and made local authorities, rather than centralised bureaucracies, responsible for planning. As a result planning was brought back into the local political arena (Hoffman, 1994). However, attitudes to communist planning and the coerced nature of public participation in the past will make democratic involvement in the new planning process difficult. The problem of generating local democracy is exacerbated by the fragmentation of local government in the city. Although there is a city-wide authority this is divided into fifty-seven communes, some with only about 500 people. These have their own powers, although in many fields the level of responsibility is still not clear. The relationship between the central city authority and the smaller authorities is set out in the Charter for the Capital City of Prague, but this has been changed several times and is under constant political debate. Relationships between local and central government have also been unclear. Soon after the revolution the government encouraged western developers to produce a large redevelopment scheme for Wilsonova station in central Prague. However, the proposals were turned down by the local mayor (Brodsky, 1993). Within the city many of the communes compete with each other to attract development such as shopping centres and do not want to use land for low-value activity such as a city cemetery. There is a strong NIMBY ('not in my back yard') reaction by the communes to undesirable development locally which can be essential for the city as a whole.

Meanwhile the city authority is preparing a new city plan to replace the existing one, prepared in 1986, which still has legal status (Prague City Architect's Office, 1992). The 1986 plan is very inflexible and often irrelevant, and contains monofunctional zoning of areas and proposals for large new suburban housing estates. As an interim measure the city has been divided into two different kinds of area. 'Stabilised areas' cover about half the city and include the historic core. In these areas any development has to respect the existing functional and architectural character. The historic centre has also been subject to increased legal protection since 1971 (Sýkora, 1995). In the rest of the city, the 'unstabilised areas', more flexibility in development is possible but is subject to detailed planning documentation. When the new Master Plan is agreed it will aim to provide a city-wide framework within which investment can be attracted while also dealing with the legacy of problems left by the former regime. Matters under consideration include the controversial routing of a new ring road, solutions to the severe pollution problem, improved living conditions in the large suburban estates, coping with new housing demands, and creating a strategy to deal with transport in the new circumstances of reduced public finance and increased car ownership. Thus a major challenge exists to balance the desire to attract inward investment and promote the city while also preserving the important historic core and finding a solution to the social problems, such as gentrification and insufficient investment in social housing, that the market approach will create.

While these debates over the plan continue development decisions are being made. However, the bureaucratic legacy of the past and the current legal uncertainties make this a very long-winded process (Prague City Council, 1993). First a developer must demonstrate clear ownership rights, which involves obtaining certificates from various public authorities. An application for a planning permit must then be made and this has to include statements from all the relevant agencies. Up to forty-four different statements can be required and the process takes several months. The statements can cover issues such as design, public health, public utilities, agricultural land and conservation. The Chief Architect's Office will then check the proposal, and the associated statements, against the Master Plan, the supporting background studies, and the more detailed zoning plans that are prepared within the context of the Master Plan. Once a planning permit has been granted a building permit has to be obtained, usually from the commune level. This has to be accompanied by a further fifteen statements from agencies covering utilities, planning conformity and conservation. After construction another certificate is needed before occupation. Therefore, notwithstanding the move to a market-oriented system, the procedures remain highly bureaucratic.

Urban governance in Prague is characterised by institutional instability. Planning decisions are taking place within an uncertain local political arena and in a shifting policy framework. Laws are not fully developed and sometimes contradictory, often still relying on those formulated under communism. A private sector is emerging, characterised by joint venture between Czech and foreign interests, which is formulating and promoting numerous development projects in the city. There is clearly a shift in power away from public institutions towards private sector actors who increasingly operate according to international influences (Sýkora, 1994). Meanwhile, tensions within the public sector are reducing its credibility. Local government has been a vehicle for the implementation of the national liberalisation strategy while lacking the regulatory mechanisms to manage urban development. The public sector is operating at two levels. There is the longer-term shift to suitable new instruments and procedures while at the same time the need to react to current commercial pressures. In regulating these pressures in an uncertain policy context, public officials can draw upon strong, internationally supported, conservation controls in the historic core and the continuation of comprehensive bureaucratic checking procedures. However the result is an *ad hoc*, reactive approach which cannot deliver at the strategic level or deal with the uneven development created by the market.

Berlin: world city versus unique character

The collapse of the Berlin Wall in 1989 suddenly changed the planning scenario of the city. The re-integration of East and West had to be implemented in a rapid and detailed fashion. This highlighted the contradictions and conflicts, evident in the planning of any city, in a very clear and dramatic fashion. The contesting forces of economic development, national and local politics and community action were all sharply defined in the debate about the future. As basic social and political challenges arise in the aftermath of re-integration, Berlin expresses particularly clearly the themes of this book. Attempts are made to exploit the new post-Wall economic opportunities, new political arrangements are implemented and the tradition of community involvement is applied to the emerging social difficulties.

Berlin's local economy developed in an unusual way during the post-war period and the separate development of East and West will have ramifications for some time (see Frick, 1991; Häussermann, 1992; Ellger, 1992; Bruegel, 1993; Häussermann and Strom, 1994). West Berlin had been heavily subsidised and as a result an artificial economy developed there. By the late 1980s this subsidy had reached 20 billion DM per year. Much of this went into providing incentives to industry which was at a disadvantage because of the distance from western markets. As a result West Berlin contains a high industrial presence, especially in electrical goods and food products. It also has a high proportion of public sector employment but is low on services such as banks because of its small hinterland. In world city terms, therefore, it has not yet experienced the usual pattern of economic change. It is expected that the economic structure will gradually revert to that which is normal for a city of its size and that high unemployment will occur in the manufacturing sector. In the West the loss of the old subsidies, which ceased in 1993, will lead to closures – although some firms will prosper from new Eastern markets and for a while will be able to benefit from any subsidies available to the new *Länder*. At the same time in the East, although this sector acted as the capital of the German Democratic Republic and the headquarters of the country's political and industrial organisations were located there, uncompetitive factories are expected to close on a considerable scale.

In the West housing developments tended to be at high densities as there was no possibility for suburban expansion. In the East the high concentration of employment, and conditions which were better than in the rest of the

GDR, attracted high inward migration. A big house-building programme resulted in large estates of high-rise blocks. In contrast to capitalist cities, many of these were built in the centre. However, much of this housing was of poor quality and in desperate need of renovation and better provision of local services.

Thus the unified Berlin exhibits a relatively high concentration of industry likely to produce problems of unemployment, a high percentage of public employees, an underprovision in the service sector at all levels, and a high housing density with only 10 per cent owner-occupation. Car ownership has been relatively low due to the lack of supply in the East and the limiting boundaries in the West. A doubling or tripling of car ownership is therefore expected.

International investors and developers saw great opportunities in the new situation. British property consultants rushed to establish offices in the city and to promote its potential. 'The unified city of Berlin is being transformed into the glittering capital of Germany' (Ozanne, 1991, p.32), or in more measured tones, 'Berlin is a new and exciting market which is ideal for developers and entrepreneurs' (Richard Ellis, 1990, p.10). They noted the many open sites right in the centre of the city caused by the 'no man's land' of the wall. They also estimated a huge demand for certain activities that had been underprovided in the past. Berlin was promoted as a candidate for world city status, a gateway to eastern Europe, and as returning to its historic past as a leading European city. New development proposals were formulated to meet the huge office and hotel demand expected. Prime locations were marketed for the HQs of national and international companies, and plans formulated to improve road and rail access to these central locations. For example, priority was given to redesigning the rail infrastructure to concentrate lines on the new central station. The view was widely expressed that the maximum commercial potential of these key central sites should be exploited if Berlin was going to compete for world city status. Some promoted the idea of building architectural images and symbols to enhance Berlin's position. Housing and small-scale services which existed in the centre were seen as obstacles to the growth of the more advanced commercial sector.

The underprovision of shopping in the East was also seen as an opportunity – when the Wall came down 78 per cent of shopping floorspace was in the Western sector. The three large new shopping malls due to open in 1995 in the old centre of East Berlin illustrate this lack of supply. A huge influx of population was also predicted and this, together with the lack of owner-occupied single-storey housing, was seen as presenting a great potential for housing development. Pressure on the periphery is therefore likely to be very great and other developments are also being proposed there. By the end of 1991 the municipalities in the surrounding area had received forty applications for golf courses and twenty-seven for Disney-type theme parks (Ellger, 1992). Modern business parks are also seen as an expanding category and by 1992 over forty applications for grants were made by the municipalities bordering on the city for their new industrial zones (Jones Lang Wootton, 1992).

One problem in controlling such pressures is the lack of a regional authority and regional planning. The city of Berlin is a *Land* in its own right but this forms the hole in the doughnut which makes up the surrounding *Land* of Brandenburg. The municipalities in Brandenburg, exercising their new found powers, are keen to obtain development for their areas, and planning permissions are easier to obtain than in Berlin (South, 1992). However, the projects do not always conform to the plans for Berlin itself. For example, the city has prepared its plans for the huge housing estate of Hellersdorf in south-east Berlin. This was built towards the end of the life of the GDR and is a huge mass-housing area of over 100,000 people. However the services provided were completely inadequate and so the new plan proposes to create new shopping centres in central locations within the estate, well integrated into the public transport network. However, the estate borders on the city boundary and just over the other side the municipality in Brandenburg has agreed a big hypermarket development which will jeopardise the other centres. This is locationally inferior but as there is no regional planning mechanism and the green field site is quick to develop there is nothing to stop it going ahead.

The problems of local administration extend further. Officials in the East are often either discredited or lacking in skills. The sudden introduction of democracy to the East, which previously experienced a highly centralised system, obviously creates transitional problems. It is also felt that the division of Berlin into twenty-three local councils, each with planning powers to prepare legal local plans, creates too much scope for conflict and reform has been proposed. As in the rest of the former eastern Europe a further administrative hurdle is created by the policy of restitution. Finding the former owners and investigating whether their claims are genuine takes a long time; for example, the American developers of a new business centre at Checkpoint Charlie had to settle 220 claims which took four years (*Newsweek*, 14 November 1994). However, there are ways in which the restitution process can be speeded up. The Treuhandanstalt organisation was set up to either float or sell off the former assets of the GDR. It managed much of the commercial land in the old centre of East Berlin and sought to maximise returns on property sales. It was agreed that the former owners would only receive compensation, not the property itself, if the Treuhandanstalt believed that there would be community benefit (for example employment, housing or infrastructure) from selling to a developer (Van den Bos, 1992). A second way round the problem is for the developer to form a partnership with claimants hoping to speed up the property acquisition and development (Häussermann and Strom, 1994).

The decision by a narrow majority to move the seat of national government to Berlin was taken in June 1991. This naturally fuelled property speculation even further as more demand would be created for office space and luxury housing. The new government will require new Ministerial buildings which will be located around the old Reichstag and in two nodes within the former Eastern centre. In order to simplify planning the city has applied special designations to these areas and the surrounding neighbourhoods called 'Development Areas' and 'Adjustment Areas'. These legal designations give the planning authority greater powers, and normal planning procedures involving the local district councils within the city, and citizen participation, are removed. Decisions over these areas are taken by the Berlin Senat. These powers have also been applied to new areas designated for research and development activity. Thus the planning system has been speeded up and the opportunities for consultation reduced (Bruegel, 1993). However, in reality it seems that the full opportunities of these legal instruments are not always adopted and the search for consensus still dominates. The city has set up procedures whereby a coalition of citizens' groups provide information and help to residents of neighbourhoods in the designated areas, where special councils have also been elected. However citizens' legal rights of participation and appeal have been removed and the Senat has the final sanction. Thus, 'the possibilities for districts and citizens' groups to have much influence are greatly curtailed' (Häussermann and Strom, 1994, p.344).

In many quarters the decision in 1991 to move the government was not popular. For example, those with a property stake in Bonn were concerned about a fall in the property market there. As Häusserman and Strom point out the government bureaucrats planning the move were not neutral on the issue and raised practical reasons why the move should be delayed. Those reasons relating to costs had considerable appeal in the poor economic climate. The aim was to stall the process in the hope of a reversal after new national elections. Such uncertainty had its effect on those promoting the Berlin property market who had already lost much of their early enthusiasm. By 1993 an article in the property press began with the words 'an unmistakable shadow of gloom has descended on the once booming property market of Berlin' (Yarranton, 1993, p.26). The reasons given included the lack of a firm timetable for the move of the capital, the slowness of the political machinery to make planning decisions, the general economic recession, and restitution problems.

Competing attitudes have emerged concerning the planning of the reunified city (Frick, 1991). There has been some opposition to the strategy of developing the centre of the city to satisfy world city aspirations. A community tradition exists in the city based upon the high proportion of young people attracted to West Berlin with its two very large universities, the possibility of avoiding military service and the development of an alternative culture. This population has been very vocal, as was evident in the student unrest of 1967/8 and the violent clashes surrounding the squatter movement. The unusual social composition also had a geographical dimension as the parts of the city cut off by the Wall were of little commercial interest. This allowed other uses to grow, such as co-operative housing, workshops, publishing, art centres and counter-culture leisure activities. The difficulty in attracting population led to a big influx of guest workers (e.g. 120,000 Turks) who tended to concentrate in similar parts of the city. As land prices did not dictate uses in the Eastern sector there is also much housing and cultural activity in the former Eastern centre. These alternative communities, concentrations of ethnic groups and central housing areas are all now under threat. Commercial development or gentrification beckons as the former backwaters become strategic locations.

However, an alternative approach to Berlin's development has been proposed which builds upon its past. Here the emphasis is on developing the special character of Berlin retaining housing, vitality and cultural diversity in the centre with an emphasis on green areas. The aim is to make the most of the unique chance to have a vital city centre and not to be pushed into quick decisions by the market, developing all the key central sites for offices. For example, a campaign has been launched against the proposal to put an underground road across the central park (Tiergarten) to improve access to the central development. There was also a very vocal opposition to Berlin's bid for the Olympic Games as it was thought that this would distort priorities. It could be said that there is a healthy climate of open political debate. The twenty-three local municipalities also often disagree with the Berlin Senat.

The conflicting views about the desired future direction are illustrated to some extent by the conflicts between two of the principal Ministers in the Senat concerned with planning. The Senat is run by a coalition of Christian Democrats and Social Democrats; the former control the Department for Urban Development and Environmental Protection while the latter control the Building Department. These administrations have considerable autonomy and have developed rather different approaches. The Department for Urban Development, which is responsible for developing the city plan, emphasises economic growth and the need for more office development. However, it is the Building Department which deals with the detailed aspects of development and has considerable influence through its large budget (Matheou, 1992). It seeks to re-create the old architectural character of Berlin and is keen to include housing in new projects. The new comprehensive plan for the city, the *Flächennutzungsplan,* tries to find a common expression for all these views. It has been through public consultation and is due for adoption in 1994. It attempts to provide scope for growth while also preserving the special character of the city. It allows some exceptional developments in the centre, but the basis of the plan is for commercial nodes on the inner ring road – the *Ring Stadt.*

Berlin gives the impression of a lively and open debate over planning issues. However, there are examples of tendencies to bypass this openness.

One is the use of competitions for the major development sites which are decided by expert committee. Although the winning proposals eventually have to go through the normal planning procedures with their participation opportunities a certain head of steam will have been generated. The manager of Daimler-Benz, who own a key central site, continually makes the point that Berlin would be foolish to waste time on lengthy planning discussions (Krätke, 1992). As we have seen, in certain areas affected by the capital city developments the normal planning procedures have been supplanted by a special instrument to remove legal rights for local involvement. To speed up the planning process another special instrument has been used – the Project and Development Plan (*Vorhaben-und-Erschliessungsplan*). This allows development to take place without a local plan if it contributes to urgent needs for employment, housing or infrastructure. The developer/investor has to be able to carry out the whole scheme and submits a plan to demonstrate this which becomes the basis for a contract with the local authority. The approach has been widely used in Berlin (Berry and McGreal, 1995).

Then there is the establishment of the Stadtforum by Hassemer, the senator for urban development and environmental protection (Matheou, 1992). He is quoted as saying that this 'allows the public a view into the mechanisms of decisionmaking', and it has been referred to as another example of open democracy. However, the group is appointed rather than elected, and although containing some local representatives also comprises large numbers of experts including members of the chamber of commerce, trade unions, architects and planners, economists, journalists, sociologists and a priest. The forum expresses its opinions on planning matters before the senator submits a resolution to the Senat. It has been suggested that although it allows open discussion on design matters the fundamental social and economic priorities of planning policy are not discussed (Krätke, 1992). Some property experts see the Stadtforum as a way of speeding up the planning system, 'it is adopting a fast-track approach to the creation of a new Berlin masterplan, by setting up a special assembly to air planning questions' (Van Den Bos, 1992, p.22).

Thus the debate is still under way in Berlin over whether the city should commence the journey along the road to world city status. Given its history and present position it cannot compete in the first division with New York and Tokyo. It lacks the financial concentration which is decentralised to other German cities. However, there is still a considerable amount at stake if it competes for major office developments with the resulting implications for land use, transport priorities and social division. A frequently stated aim is that Berlin should seek equalisation of conditions for all citizens, but one consequence of adopting a world city approach would be social differentiation (see Krätke, 1992 for a detailed discussion). The problem for Berlin is that this social differentiation is likely to be particularly strongly defined in geographical terms as the two poles gravitate to the two former sectors – maintaining the East–West divisions and generating social tensions.

The re-unification of Berlin has generated a fundamental debate over its future. A wide range of interests are involved in this debate and there is considerable concern displayed by citizens and the local media. The private sector has distinct interests as the historical backlog promises considerable financial opportunities. Some detect a powerful growth coalition as the political leaders of the two parties running Berlin, the Christian Democrats and the Social Democrats, combine with private interests to promote a strategy of attracting international investment and to downgrade democratic processes (Krätke, 1992). However, it does not appear that such a coalition is in a totally dominant position and the political scene is quite fragmented. There are different views within the Senat and conflicting opinions from the surrounding region and the twenty-three lower-tier local authorities. Federal government is also involved through its interest in the capital functions. The relationship between the political strata and the community is also complex. There are strong movements to stop certain developments and to give planning an ecological priority. There are also mechanisms such as the Stadtforum and neighbourhood committees which try to incorporate a range of views and reach consensus. At the same time procedures are being adopted which bypass participatory opportunities. Thus at this stage in the process of re-unification the overall pattern of influence and decision-making power is complex and fragmented.

CONCLUSIONS

The literature reviewed earlier in this chapter pointed to the complexity of contemporary urban governance and suggested some of its analytical dimensions. The cities we have looked at certainly testify to the institutional complexity in which urban planning is embedded. Factors which are clearly significant include intergovernmental relationships, the role of party politics, and the invention of new planning and development agencies to respond to the objectives of urban elites.

The degree of co-ordination between levels of government influences the type of planning which emerges and its relative success. Some levels of government operate more coherently than others. For example, the Frankfurt case reveals the difference between a coherent city regime and the instability of regional administration. The world city functions were catered for in the city but subregional planning was weak. Within the system of national, regional and city government in Spain it was the Barcelona city level which took the initiative in the large development projects, with financial support from the centre. The city-state of Berlin has considerable decision-making autonomy. In Prague, on the other hand, the lack of a clear governmental framework prevented coherent city planning. The bureaucratic culture of the old regime also has a strong influence on current practice. In other cities national frameworks and cultures also clearly define the scope of urban planning. For example, the highly fragmented and uncoordinated laws on planning provide no clear framework for Athens and the very centralised nature of administration gives little autonomy to that city (Delladetsima and Leontidou, 1995).

The importance of stable political leadership is demonstrated in the Barcelona case. In Milan the radical transformation of political traditions has set very different contexts for planning. There is the possibility of new forms of public–private coalition, including new international partners. Successful government also has to be able to manage opposition. Electoral politics brought down the Frankfurt regime, and instability in party politics in Milan has changed the character of city government. In contrast, in Barcelona political stability was achieved through the formation of consensus around the Olympic projects.

It was noted earlier that planning could be seen as one of the resources contributing to stable urban governance. The experience in each of these cities suggests that this can take a number of forms. In Milan we can see planning responding straightforwardly to the requirements of developers. Numerous *ad hoc* amendments were made to the Master Plan to accommodate developers' schemes. More recently a framework is under discussion to support the city's world city aspiration. Such ambitions in Frankfurt and Berlin led to a concentration on the city centres and provision for the development of central sites for commercial purposes. A similar form of contribution to city governance can be seen in the development of new strategic frameworks to support pro-growth objectives. In Barcelona and Milan this involved identifying sites with development potential in the 'Areas of New Centrality' and the *Progetto Passante*. These aproaches have developed into planning strategies which support the international promotion of the cities. However, such forms of strategic planning are limited to economic and marketing objectives. The planners are not engaged in wider territorial planning. Strategic planning in these cases tends to be managed through special agencies, often business dominated, and not through the local democratic machinery.

Thus the cases indicate a significant trend to by-pass democratic procedures and utilise alternative agencies and means of legitimation. There are several reasons for this. In some cities formal politics has a poor image. Generally inter-city competition gives a sense of urgency to local decision-making and formal processes are often lengthy. Bypassing formal structures is justified in terms of the need for the city to be competitive, often supported by appeals to civic pride as in Barcelona. In Berlin the concern to establish a new political culture including opportunities for citizen participation was effectively ignored when it came to major economic development projects.

Planning is a resource for urban governance. In the context of inter-city competition planning is most often used to support public and private development aspirations. The legitimacy of legal planning frameworks supports the development proposals of pro-growth coalitions. At some periods and in some cities different objectives are pursued. In Barcelona and Milan there were periods of interest in redistributive and social aims. In Frankfurt the backlash against the world city strategy brought about a change in city planning. In Berlin the debate between social and economic objectives continues and different kinds of plans are being promoted simultaneously. Thus, planning can be used to identify sites for development, to produce growth strategies and to co-ordinate infrastructure. These activities can be used to support international competition. Such approaches are likely to be co-ordinated through special agencies outside 'normal politics'. The alternative pursuit of social or environmental objectives may involve wider groups of interests and open decision-making. The role taken by planning depends upon the dominant force in the governance of a city.

Part II

NATIONAL PLANNING APPROACHES AND DEVELOPMENT PROJECTS: BRITAIN, FRANCE AND SWEDEN

INTRODUCTION TO PART II

The discussion in Chapter 2 made it clear that the effects of economic forces on the city are complex and mediated by national political and institutional factors. The international level of regulation has become more important, but, despite talk of a return to a Europe of city-states there seems to be no uniform pattern of political relationships between supranational government and cities. National systems of administration and law separate European countries in significant ways. Moves to closer integration of planning systems will be restricted by these differences. In eastern Europe the development of new legal and administrative structures has only just started. The emerging planning systems are as yet unclear. There are significant differences between east and west and between north and south. Planning regulation is less readily accepted in southern cities. The beneficiaries of recent economic growth are, for the most part, northern cities where pro-growth, competitive attitudes and institutions have provided models of urban success. The growth of entrepreneurial attitudes and greater private sector involvement in city governance are mediated by a range of factors. National institutional differences are clearly important to the development of the capacity to govern cities. However, in addition to national differences local factors such as party politics and availability of resources have clear impacts on the development of planning in the cities examined in Chapter 4. It is clear that urban planning in European cities is shaped from the complex interaction of international, national and local forces. Part II of this book develops the understandings from Part I through a more detailed study of both national context and local planning projects in Britain, France and Sweden.

The choice of countries has been made in order to reflect some of the basic differences between the administrative families identified in Chapter 3. The British model was clearly different from the rest of Europe in terms of its legal and administrative underpinnings. The high degree of centralisation of planning since the early 1980s is also distinctive. France exemplifies the centralised Napoleonic model. However, since the decentralisation reforms of the 1980s the Jacobin state has been fragmenting. Decentralisation is most advanced in Scandinavian systems and the choice of Sweden therefore introduces further variety into the detailed studies. The Swedish social market ideology also contrasts strongly with Thatcherism in Britain. In Part II issues and trends in each national planning system are discussed and this is followed by case studies of urban planning and development projects. Politics, intergovernmental relationships and public–private interactions all impact on planning. Styles of urban planning vary from city to city. For each country case studies have been selected from two cities. These are the national capitals and the second rank cities of Birmingham, Lille and Malmö. The choice of cities gives some consistency to the discussion of cases and allows examination of variation within each country. The planning projects examined span a period from the mid-1980s to the mid-1990s. This time-span is important in charting the development of urban planning, at city and national level, through a period of significant change.

5

GREAT BRITAIN

The legacy of Thatcherism

INTRODUCTION

This chapter explores in more detail the changes that have taken place in urban planning in Britain. The dominant factor that has shaped all aspects of public policy during the 1980s and 1990s has been the radical ideology that sprang on to the political scene with the arrival of the Thatcher government in 1979. The stated aim of this government was to attack the whole concept of the post-war Welfare State and replace it with a more market-oriented approach. As the first comprehensive planning system in Britain was established in 1947 as one element in the formulation of the Welfare State it is not surprising to find that planning underwent considerable change under this 'New Right' ideology. However, there was no new comprehensive Act to replace the old ones and instead the changes were of an incremental nature leading to the erosion of the planning system. This is not to say that the changes were insignificant – the constant removal of controls and shifts in priorities have led to a considerable transformation. The flexibility in the British planning system and the power of central government have allowed this major change in planning purpose and role while maintaining the same basic legislative framework.

The main themes of the radical change to British planning since the beginning of the 1980s will be reviewed. Then the directions the planning system has taken in the 1990s will be outlined, looking at the impact of environmentalism, the move to a 'plan led' system, and at urban policy reforms. Throughout the discussion links will be made between the ideological and administrative changes to the statutory planning system – that is, the bundle of instruments that emanate from the comprehensive planning legislation, such as central government guidance, structure and local plans and development control procedures. However, one of the features of British urban planning from the 1960s onwards has been the utilisation of other legislation and initiatives when dealing with the problems of urban regeneration. In many ways it could be said that there are two parallel systems in operation. There is the planning system which is comprehensive, covering all geographical areas and levels of government; but for more immediate and concerted action in urban redevelopment this system is overridden by other initiatives. There is surprisingly little connection between the two systems, and the urban regeneration approach has itself been characterised by its fragmentation and lack of co-ordination (Audit Commission, 1989; Robson *et al.*, 1994). These urban initiatives have been formulated and implemented by various central government Ministries and *ad hoc* agencies and have been in a state of constant flux – new initiatives appearing almost every year. This chapter will also encompass the main initiatives within this urban regeneration programme.

THE THATCHERITE PROJECT

The full debate over Thatcherism will not be explored here (see, for example, King, 1987; Levitas, 1986; Gamble, 1988; Hall, 1988; Jessop *et al.*, 1988; Thornley, 1993a; Ambrose, 1994). Instead, the main themes will be identified and related to planning. Changes in the planning system resulted from two interrelated strands of Thatcherite ideology: the economic liberal strand (which sanctified the market as the best decision-making process) and the authoritarian strand (which included the strengthening of central government). The increased centralised power was used to reduce the involvement of local government in day-to-day decision-making and to set up frameworks for the freer operation of market forces. This combination of authoritarian and economic liberalism has been termed 'authoritarian decentralisation' (Thornley, 1993a) and allows central government to take a major role in ensuring that decisions are decentralised to the market place. An important implication of this strategy is a major reduction in the power of local government and a denial of the importance of local democracy and political representation.

Local government

An important part of the Thatcher reforms concerned local government. Throughout the 1980s a reduction in the power of local authorities was achieved through financial policies (Duncan and Goodwin, 1988). Central government gained greater and greater ability to control the spending of individual authorities rather than just reduce local government spending as a whole. Central government's responses were pragmatic, complex and *ad hoc* within the overall aim of reducing public expenditure, but during the 1980s a more strategic approach developed. Some authors have suggested that a more coherent approach can be identified from 1987 (Stoker, 1991a; Bulpitt, 1989). This involved a move away from pure finance to the restructuring of local government itself. The metropolitan authorities and the Greater London Council were removed as a tier of government in 1986 because they were controlled by the Labour Party and were pursuing popular policies that were seen as an ideological threat to Thatcherism (see Duncan and Goodwin,

1988). The by-passing and replacement of local government activities by quangos has been a continuing trend. The scale of such non-democratic government is illustrated by the analysis of Skelcher and Stewart (1993). They estimate that there are now 17,000 members of appointed bodies compared to 25,000 councillors and that they account for about 20 per cent of public expenditure compared to 25 per cent by local authorities.

In 1991 the government announced a comprehensive review of local government. The review was inspired by the need to reform the community charge but also tackled the structure of local government and its internal organisation (DoE, 1991). The problem of local government finance and opposition to the community charge was quickly resolved. The other themes of this review continue to have significant impacts on the functioning of local government. Change in local government boundaries and structures has had significant effects on planning. In Scotland regional government with regional planning responsibilities has been abolished. In England and Wales the process of reform has been confused, some parts of the country retaining two tiers of government, others adopting a 'unitary structure'. The continued viability of a two-tier system of plans has been brought into doubt.

Internal restructuring has involved the removal of certain services from local government through the separation of the *responsibility* for provision from the actual *act* of provision. This shift has been referred to as one of 'enabling not providing'. A start to this process occurred before 1987 with the 'right to buy' policy for council houses, shifting the responsibilities from local authorities to owner-occupiers. The Education Acts of the 1980s transferred power from the local authority to parents, head teachers and governors. However, latterly there was evidence of a concerted effort to restructure the whole of local government along these lines. This was clearly set out in the pamphlet by Ridley (1988), who was Secretary of State at the time. The strategy was that local authorities should abandon or reduce functions which they did not need to carry out. These should be undertaken by the private sector. This would allow government to operate on a tighter budget and also free resources to concentrate on those activities which only a local authority can do. There would then be a variety of agencies working alongside the local authority, and the latter would no longer take the role of universal provider but instead one of regulating standards and monitoring. 'Inside every fat and bloated local authority there is a slim one struggling to get out' (Ridley, 1988, p.26). The right-wing No Turning Back Group took the idea a stage further and suggested that ultimately a local authority would be run like a company with an annual meeting in which a small number of councillors would meet to decide the allocation of contracts to carry out services. The remaining administrative functions would operate under tight financial controls and be run by a manager – an elected mayor.

The notion of 'enabling' local government was given a different interpretation by some Labour authorities and academics (see Stewart and Stoker,

1989a; CLD, 1995). As well as a positive attitude towards better service provision, ideas of community oriented, participatory local democracy were developed by some authorities. Others were attracted to the idea of local government taking on a strategic and advocacy role on behalf of its citizens (Clarke and Stewart, 1994). However, for all the positive debate about the purpose and future of local government the overall impact of government policy throughout the 1980s and early 1990s was to restrict local councils to the role of agents. Local councils were forced to cut expenditures and adapt to a limited influence over local affairs. The role of democratic local government has been significantly displaced by quangos.

Reorienting the planning system

The British post-war planning system before the advent of Thatcherism has been frequently described as containing three different objectives, contributing to economic efficiency, protecting the environment, and fulfilling community and social needs (Foley, 1960). The history of the period shows that these objectives were not made very explicit – one reason being that they were often in conflict with each other. If these conflicts were brought out then the consensus of general agreement over the need for planning would be disturbed and there would be a need to accept that different and conflicting interests lay under the umbrella of this consensus. One of the results of Thatcherism has been to expose this situation and throw off some of the myths on which the planning system was built. The Thatcherite ideology clearly gave greater importance to market principles in decision-making. This meant that the first of the three purposes, economic efficiency, was stressed; many actions were taken to deregulate the planning system, and as a result developers were given greater freedom to pursue their proposals. Now of course the development process has always been dominated by the property market and so one might ask, what's new? Before 1979, befitting the birth of the system as part of the Welfare State package, it was considered legitimate to pursue social or community objectives as part of the planning process. Admittedly these always had to be fought for against the background of dominant economic interests but at least it was a fight that was considered appropriate. This provided certain opportunities for pressure and action. However, along with the general Thatcherite attack on the Welfare State came the rejection of the social objectives of the planning system. Social or community needs were no longer acceptable planning matters. These would be dealt with through the so-called 'trickle-down' effect, i.e. the benefits from economic development will filter down to improve all aspects of life through the general increase in prosperity.

There has always been a conflict between the economic and environmental objectives of planning. As Fainstein and Fainstein (1982) have pointed out, the same urban space can be a requirement for capital accumulation while also being a territory in which people live. Thatcher governments were not immune from this conflict. They promoted both the ethic and reality of owner occupation and have been subjected to pressure from wealthy residents who want to protect their pleasant environment. This clashed with the government's promotion of economic freedom and the interests that benefit from this. One strategy they adopted to cope with the conflict was to create a dual planning system. Certain areas, National Parks, Conservation Areas, Areas of Outstanding Natural Beauty, Green Belts, were exempted from all the deregulatory measures that had been incrementally passed over the years. Therefore there was a considerable difference in the degree of control over developments in these protected areas and elsewhere. In the areas without this protection the economic objectives were given greater priority and Thatcherism gave legitimacy to this shift in emphasis.

This was not the end of the matter because over and above this duality a further division developed. Throughout their period in office the Thatcher governments set up 'experiments' in the form of initiatives which completely by-pass the normal planning system and have their own procedures. Some of these experiments did not come to much, such as the greater use of Architectural Competitions and Special Development Orders, but others – especially Urban Development Corporations and the simplified regimes of Enterprise Zones and Simplified Planning Zones – grew to such an extent that they could no longer be called experiments and covered a large number of areas. They also had an influence in creating competition and hence encouraging similar deregulatory behaviour and attitudes throughout the system.

Thatcherite urban policy

The ideology of economic liberalism and the 'authoritarian decentralist' approach were also applied to urban policy. Previous initiatives were scrapped and replaced by new ones. Again there was an increase in central government control in order to set up frameworks within which market forces and developers could exert greater freedom. This often involved the removal of local democratic procedures. The key features of this new approach to urban policy can be summarised as a greatly enhanced role for the private sector and a property-led approach to urban regeneration. Some of the key initiatives which reflect this approach were Enterprise Zones (EZs) where planning and tax regimes were eased, Simplified Planning Zones (SPZs) which allow local authorities to stimulate development in an area by simplifying its planning approach, and most important, Urban Development Corporations (UDCs). Eleven of the latter were set up and although the details varied from Corporation to Corporation the same general aims applied. The effect of designating a UDC was to remove the responsibility for the regeneration of the area from local authorities. The membership of the Board which ran the Corporation was dominated by the business sector. Local authorities and local communities had no means of influencing its decisions. This bypassing initiative was used primarily in areas where there was an interest in economic restructuring. Through bypassing the planning system it was considered easier to open these areas up for development without the hassle of local residents or their local political representatives. To support such an approach the 'authoritarian decentralist' aspect of the Thatcherite ideology was employed. Power was shifted to central government, which had very strong controls over planning in these bypassed areas. For example, in Urban Development Corporations central government appointed the Boards which ran the Corporations, determined their finance, and arbitrated over any conflicts with local authorities. This power was then used to ensure that procedures were adopted which give priority to the interests of developers (Brownill, 1990; DCC, 1990; Church, 1988). In these areas the normal participatory rights of communities and individuals in the planning process were removed and local democracy, in the form of local elected councils, was ignored. It was said that development was in the national interest and could not be slowed down or deviated by the parochial interests of local people and their representatives.

The UDCs had a number of mechanisms to enable them to carry out their role of preparing the area for development and then 'selling' its potential to possible investors and developers. They had considerable powers of acquisition of public land – for example, in London land belonging to the local authority or Port of London Authority could be 'vested' in the Corporation. Thus land could be assembled, serviced and passed on to the private sector. This allowed Corporations considerable financial leeway, buying cheaply from public authorities, recouping the benefits of increased land values or selling on to developers at bargain prices. Many Corporations also had development control planning powers; thus they, not the local authorities, determined whether a development would be allowed. Such a decision was supposed to conform to the statutory Local Plan and the formulation of these plans remained formally with the local authorities in the area. However, experience in London showed that these plans played a minor role compared with the promotional development briefs prepared by the Corporation.

Boosterism in the cities

Cities have been heavily constrained by central government's financial controls. The lack of money for capital projects and the changing nature of urban policy under the Thatcher governments set a new context for economic development. Despite the domination of central government throughout the 1980s local initiative has persisted. However, development projects have been chosen primarily on the basis of funding availability. The search for development and regeneration finance has encouraged the new forms of partnership between local and regional, public and private interests.

In Chapter 4 it was noted that many local initiatives in the 1980s were modelled on the apparent success of urban revitalisation in the US. The leaders of Glasgow Action, for example, drew on US experience in setting up in 1985 an alliance of national, regional and local government with Scottish banks, other leading businesses and the media to promote the city and devise a policy for the central area which effectively by-passed the local plan (Glasgow Action, 1987). This form of alliance was echoed in many other cities. Indeed Hague (1993) argues that the fact that 'stolid, reliable' Edinburgh developed a boosterist – Edinburgh Vision – agency in 1989 is proof of the depth and pervasiveness of change in approach in city government throughout the UK. In the mid-1980s Sheffield Economic Regeneration Committee (SERC) brought together the city council, the development corporation, the chamber of trade, the universities and British Steel (Cochrane, 1994). Its objectives were image creation, getting inward investment and acting as an umbrella organisation for various projects. A similar grouping produced a Central Area Study in 1992. The council were compelled to co-operate with the initiative despite their reservations (see Lawless, 1994). Economic decline in the city has forced public and private interests into co-operation. The business sector is weak, however, and local alliances rely on relatively few business leaders and on the expertise of a professionalised local authority bureaucracy.

Elsewhere in Britain national business associations have taken a growing interest in urban policy since the early 1980s. In 1988 the Confederation of British Industry (CBI) produced a strategy document, 'Initiatives beyond Charity'. The CBI saw a need for leadership from business in major cities and argued that regeneration depended on economic development. In the same year the CBI and Business in the Community came together to launch a common Business in Cities Forum. The CBI's strategy was to get local teams of business leaders to take up the challenge of urban regeneration by pulling together public and private partners. One such initiative was undertaken in Newcastle. The Newcastle Initiative (TNI) developed several 'flagship projects' with public sector partners (Collier, 1994). The 'public sector' in the city is fragmented among city council, development corporations, City Challenge companies and the Training and Enterprise Councils (TECs). The leadership and promotional role of the TNI has concentrated spending on a few projects. Business leadership has however largely been spending public money. Public and private interests are committed to TNI, and in the Newcastle case a strong regional identity has undoubtedly aided co-operation. Local elites have at the same time been campaigning for regional devolution (*The Times*, 1994).

In many large cities there has been a proliferation of public, private and mixed agencies. In Bradford for example the Conservative leadership in 1989 set up Bradford Breakthrough as a promotional company run by local business leaders along with the chief executive of the council. The council also set up a construction partnership with Bovis in the early 1990s but the property recession meant that joint schemes failed to materialise. In addition Bradford won City Challenge funding and so has a City Challenge company with private sector representation. In nearby Leeds there is a similar mix of new agencies. The Leeds City Development Company was established in the early 1980s with the council having a majority shareholding. The company engages the private sector in projects to develop the city's landholdings. The Leeds Initiative, on the other hand, has broader objectives and engages a wide range of partners from the chambers of commerce to government departments, universities and the media. The initiative is concerned with promoting the city in Europe and developing the infrastructure necessary to support the city's image.

Public–private lobbying for funding for infrastructure is not confined to the cities. At a regional level local authorities and private interests have come together to attempt to lever development funds from the EU. For example, in the north-west of England, the North West Regional Alliance (representing the local councils) and the North West Business Leadership Team (regional businesses) combined to lobby Brussels over infrastructure investment projects. The combination of public and private interests in land development projects is not new to British cities. Urban and regional corporatism of this sort was behind the major city centre redevelopments of the 1960s and was encouraged by central government (MHLG, 1962). What many commentators regard as different and in some respects new about contemporary promotional activities in the cities, however, is the relative stability of alliances and the wide groupings of public, private and media interests.

Both locally and through national associations the private sector has increased its role in urban policy-making. Most of these initiatives have had public sector representation and projects have depended on public funding either from local government or nationally controlled agencies such as the TECs and development corporations. Since the early 1980s the government's urban policy initiative has been geared to private participation. For local government and private sector alike the route to central government and European capital funding has increasingly been through some sort of partnership. However, the extent to which business controls the local agenda has been questioned. For example, in relation to the pro-growth politics of Manchester Peck and Tickell (1993) argue that business representatives contributed little to the public–private forums and that, as in Sheffield, the local government bureaucracy and political elites steered the city's strategy.

THE PLANNING SYSTEM AND URBAN POLICY IN THE 1990s

The greening of Thatcherism?

For most of her time in office Mrs Thatcher never showed much interest in the environment other than getting upset about the amount of litter on the streets. There was talk at one time of making people who had property fronting on to the street responsible for the tidiness of their patch. However nothing came of this. This lack of concern about the environment was not surprising given her conviction style of politics and imperviousness to the influence of interest groups. The market orientation of Thatcherism did not lend itself to the financial commitments and longer-term perspectives required. The conservation of the environment implies some relaxation of the pure market approach. However, pressure was building up from two directions. The Common Market had been increasingly turning its attention to environmental matters. It had passed the Directive on Environmental Impact Assessment which the British government had adopted with some reluctance, and was exploring other environmental issues. For example, in 1990 it published its *Green Paper on the Urban Environment* in which it analysed the problems of cities and towns and put forward many ideas for priority action. Most of these required a framework of public intervention and planning as they could not be achieved by market processes. Their suggestions included increased urban densities, restrictions on the private car and better public transport, enhancing parks and open spaces and protecting the visual quality and historic identity of cities.

A sudden change seemed to take place in the Thatcher government's approach to the environment in 1988 when Mrs Thatcher started to deliver speeches on the global environmental problem. However, these were directed at large world-scale issues and she urged international approaches to deal with them. She was less interested in supporting policies of a national or local scale as this presented her with greater ideological conflicts. Meanwhile support was at hand in the work of David Pearce (Pearce *et al.*, 1989) who had been asked to advise the government. He came up with policies to remedy environmental problems through taxation and modified cost–benefit analysis. This helped with the ideological difficulties as it suggested that market processes rather than public intervention could still be at the centre of decision-making. Subsequent Ministers have therefore continued to stress that environmental issues can be solved by the private business sector, for example Michael Howard, Secretary of State for the Environment, said in 1992 that he wanted to put more emphasis on economic instruments and 'make the market work for the environment'. However, the European pressure did have some effect and in 1990 the government produced a policy paper (a White Paper) called *This Common Inheritance* setting out its environmental aims and including many of the ideas circulating in the EC. Although this policy has been monitored each year for progress, it has been much criticised for containing nice words but no finance to support them. The government has continued with a low key approach stressing litter, energy conservation and recycling.

Meanwhile a second source of pressure arose. During the 1980s the house-building companies had formed consortiums and used lobby pressure to try to get greater acceptance for the building of new houses in areas containing environmental restriction. This had often taken the form of promoting the concept of new self-contained villages with facilities. However this generated a backlash of reaction in the areas where proposals were made, the well-known phenomena of NIMBYism. As early as 1983 the government had tried to relax the constraints on developers through a modification of the Green Belt policy but this generated opposition from government MPs and members who lived in the pleasant protected countryside. The same reaction occurred with the new village proposals and the government was presented with a split in its supporters between those who lived in protected areas and wanted the full range of controls to be retained and those who believed in greater freedom for enterprise. The problem was brought home to the government in the 1989 elections for the European Parliament when the Green Party gained an unprecedented vote of about 25 per cent in those areas threatened by new development. These people would normally have voted solidly for the Conservative Party.

Thus the combination of grass-roots reaction from their supporters plus the demands from the EU forced the government into acknowledging the need to encompass the environmental issue and to present a greener face. This they have done through a number of statements and policies, such as requiring Local Plans to include a chapter on sustainability. Local government took up the challenge of the Agenda 21 programme from the Rio Summit with enthusiasm. Local government has also pushed central government along the environmental path. However, government has adopted environmentalism without relaxing its commitment to the market as the prime decision-making arena and has not devoted the necessary financial resources to back up the policies. This is particularly evident in its approach to the urban environment where it has done little to take up the EU's Green Paper priorities.

A new plan-led system?

The Planning and Compensation Act of 1991 gave renewed importance to the development plan and encouraged planners to enthuse about the start of a new era. The Act stated that planning decisions should be taken 'in accordance with the plan unless other material considerations indicate otherwise'. The plan is therefore the prime consideration, although the door is still open to uncertainty through the interpretation of the above phrase. The courts will be the final arbiter in the relationship between the plan and 'other material considerations' and it has been suggested that the plans will only have strength if they avoid being vague or ambiguous (Grant, 1991). The government has also extended the scope of the plan by requiring them to include a section on environmental sustainability; it has also reiterated the need for them to be 'efficient, effective and simple in conception and operation' and confined to land-use aspects (DoE, 1992).

The move to a greater emphasis on local level policy implied in the Act had been developing during Mrs Thatcher's last years. It was in 1989 that Chris Patten, the Secretary of State at the time, introduced the notion of 'local choice'. This arose as a strategy to extract himself from the problems over the new settlement appeals. At this time a number of applications had been submitted to build new villages in the countryside near major towns, particularly London. The applications were usually opposed by local authorities and went to appeal where they then had to be decided by central government in a very exposed and publicised manner. Patten found himself caught between two lobbies, both of which were natural supporters of the Conservative Party: the housebuilders and the residents of the shires. As already noted the latter had been showing their displeasure through the Green Party protest vote in the European Parliamentary elections. He was in a no-win situation. The idea of 'local choice' allowed such difficult decisions to be shifted to the local level.

So development plans again became the arena in which the difficult job of balancing different interest groups could take place. The question remained as to how much autonomy they each would have. Central government having devolved this responsibility still retains the ability to control and monitor the process. It formulates the strategic and regional guidance to which the development plan must conform, it can intervene in the preparation of structure plans if the scope of the issues covered is considered inappropriate, and it has the powers to call in the plan if it is considered controversial. Then of course it exerts much influence over the process through the Planning Policy Guidance Notes. Thus although development plans may have regained importance they can only use this power if they conform to the boundaries set by central government. Another conditioning factor on the plan results from the competition between cities described earlier. The desire of a particular city to promote itself in the game to attract investment can condition the role and content of the plan. It may become part of the city marketing publicity, i.e. demonstrating that the city has a plan that encourages investment and provides locational opportunities that match the needs of companies and developers. Too strict a regulatory planning regime could divert interested parties to another city.

Thus the local autonomy that has been awarded local authorities through the 'local choice' approach and the greater importance of development plans has to be regarded as circumscribed. Freedom to formulate policies in the plan is highly constrained – both by the boundaries set by national government, reflecting its ideology, and by the competitive economic environment which often determines the local political priorities. Can similar trends be seen in the more proactive approach of urban regeneration?

New experiments in urban regeneration

During Mrs Thatcher's time policy towards urban regeneration followed a fairly consistent pattern. Central government would announce an initiative which it would administer and control through financial and regulatory powers. It would then use this power to open up decision-making to market influence and reduce local democracy. The Urban Development Corporations epitomised this approach. A major aim behind the initiatives was to provide the infrastructure, financial inducements and decision-making processes which would attract private sector investment. Through creating the conditions which were attractive for the property industry it was expected that development would ensue and that this would create a spin-off effect. This property-led approach to urban renewal has attracted much critical attention (eg. Turok, 1992; Healey *et al.*, 1992; Imrie and Thomas, 1993), in particular for the way in which it ignores many of the dimensions necessary for city revitalisation and its dependence on the cycle-prone property market.

The government has not put much emphasis on the monitoring and evaluation of its numerous initiatives. However, the two government-sponsored reviews undertaken both had critical comments to make. In 1989 the Audit Commission said that the 'the programmes are seen as a patchwork quilt of complexity and idiosyncrasy' and called for a more coherent approach (Audit Commission, 1989, p.1). The government commissioned research to evaluate the success of the various urban policy initiatives implemented during the 1980s and this reported that the economic and environmental emphasis of the policies ignored problems of social disadvantage and that the local voluntary sector and local government should be more involved (Robson *et al.*, 1994). The research could find no evidence that property-led developments had produced any trickle-down benefits for poorer areas.

A new government initiative was launched in 1991 called City Challenge which involved a number of new elements. One of these was the competitive bidding approach, since extended to other initiatives, in which local authorities were invited to enter a competition to try to win a limited number of awards. The approach has been criticised for involving a lot of time, money and effort in producing impressive bid documents and, as there have to be many losers, for often being unproductive. Government sets out guidelines for the bids and of course selects the winners, although it gives no explanation for its choice. In setting out the guidelines the government made it clear that a high priority was placed upon attracting the commitment of private business to ensure good financial leverage and self-sustaining growth for the area. This requirement clearly influenced the choice of projects in the bids – those which would be economically viable rather than those which met the social needs of the area. The focus of the initiative on small areas also meant that problems which might pervade a wider geographical area could not be addressed. However, the initiative also put considerable emphasis on involving other agencies, e.g. the local voluntary sector, universities, Training and Enterprise Councils and the local community. This greater involvement, plus the local authorities' enhanced role in formulating the bid, can be seen as a move away from the centrally directed approach of the UDCs with their divorce from local influences. However, in setting out the brief for the competitive game, central government is still able to impose its priorities.

Many aspects of this new initiative show a shift away from the approach of the 1980s and have been welcomed by commentators, especially the incorporation of different sectors, the role of local authorities and the greater co-ordination between departments. Many of the faults of the previous approach remain, however, such as the concentration on small areas and the limited finance. There has also been considerable doubt cast upon the degree to which the voluntary sector and local communities have been involved. Evidence suggests that, whatever the new rhetoric, the initiative was still geared to property-led physical regeneration.

Although the City Challenge initiative had not involved any new expenditure, being 'top-sliced' from the urban regeneration budget, it fell foul of government public expenditure cuts in 1992. New rounds of the initiative were suspended. Meanwhile another initiative was announced in 1992 called the Urban Regeneration Agency, to be renamed English Partnerships in 1993. This body has statutory powers to reclaim derelict land and property and is allowed to operate anywhere in England. Initially it was feared as a possible 'roving UDC' and Peter Hall, adviser to the government at the time, described it as a major new initiative which could apply the City Challenge approach on a bigger scale (Hall, 1992). However, its importance does not seem to have materialised. Probably of greater relevance is the government's attempt to try to create better co-ordination between the different programmes as suggested in the Audit Commission in 1989. There are two elements to this, a Single Regeneration Budget (SRB) and integrated regional offices of government departments. Both came into effect in 1994. The regional offices prepare an annual regeneration statement setting out key priorities, administer the SRB and continue to be responsible for regional departmental programmes. The new budget, involving no extra resources, encompasses in one pot the myriad programmes of the five government departments involved – Environment, Trade and Industry, Employment, Education and Home Office. The new budget continues the competitive philosophy of the City

Challenge Initiative: the government sets out the guidelines and invites bids. Recipients are expected to make a significant contribution from their own budgets and to maximise contributions from other sources such as the private sector and Europe.

The regional office initiative came from Michael Heseltine while he was at the DoE and his subsequent move to the Department of Trade and Industry eased the integration of these two departments. The reasons for these new offices are many and include a continuing desire (in response to criticism of the lack of co-ordination, see Audit Commission 1989) to integrate policy programmes. They also represent a move towards a regional structure of administration better suited to the planning and funding regimes of the EU. Some local authorities fear that the regional offices will take a greater role in obtaining and allocating EU funds. The better integrated, regional administration also answers demands for greater regional devolution within England, and through the Integrated Regional Offices (IROs) the government can match any *ad hoc* regional alliances set up by local government. The devolution to regional offices, however, has not stemmed the debate about regional government, and the Labour Party has committed itself to devolved government in parts, at least, of the UK. A crucial part of the argument for regional structures is to match those elsewhere in Europe and gain the benefits of accessing EU funds.

Another initiative was also announced at the same time – called City Pride. Manchester, Birmingham and London were invited to compete for resources through preparing a 'city prospectus' in partnership with the business community, the voluntary sector and public agencies such as the Training and Enterprise Councils. In the prospectus authorities were asked to set out promotional activities, a vision for the city over the next ten years and a list of projects and how they were to be funded. Partnership with the private sector and economic development objectives are essential in both the preparation and programmes. At the regional level the Department of Trade and Industry introduced Regional Challenge in 1994. This further competition for development project funding was to be financed by top-slicing European structural funds.

So, do these newer initiatives indicate a change of direction? Certainly there is more devolution of responsibility to local authorities, who are responsible for formulating and co-ordinating the City Challenge and SRB bids. There is also a greater acceptance of the need to involve local communities and the voluntary sector. However such local autonomy is again much constrained. Central government has a strong hold over the process through setting the guidelines and judging the bids and through tight control over expenditures through 'delivery plans'. It could also be argued that the government's priorities are still oriented towards creating the necessary climate for private sector investment rather than addressing the social needs of the areas. If local authorities wish to win in the competitive game they have to show they are conforming to these priorities and the requirement for local partnerships with the private sector will help ensure this. The government is also maintaining its strict control over public expenditure. The SRB included no new money. It also continues old ways of working. The targeted SRB money does not include the much more significant expenditures on housing policy, nor is it co-ordinated with other mainstream budgets (Ryan, 1995). The short-term, annual bidding round also perpetuates the government's hostility to long-term expenditure planning which may even deter some potential private partners from participating. Many of the old problems of British urban policy continue in the new initiatives.

CONCLUSIONS

The advent of the Thatcher government in 1979 brought a new ideological perspective into urban policy. The previous approach was very much led by the state, which channelled resources to specific areas, and decisions were made through organisations closely tied to central and local government. The strategy at that time was one of interrelating the economic, social and physical policies. Thatcherite ideology demanded a different approach in which the aim was to give greater freedom to the market to solve the problems. It was claimed that by liberating the market and making inner city areas more attractive for investment other objectives would be met through the trickle-down effect caused by the rejuvenation of the area. One of the prices to be paid was a commitment and single-mindedness that required the removal of local democracy. The approach was one of central government control imposing the deregulatory strategy on local areas through financial and legal means and by appointments to the decision-making bodies. More recently this approach has been modified with a greater emphasis on the concept of partnership – local authorities and voluntary bodies are being brought back into the picture. However, their freedom is highly constrained and central government is still able to impose its market-oriented philosophy through the way in which it sets out the agenda and priorities for any new initiative. Resources are only obtained through competitive bidding and central government determines the rules of the competition and controls access to EU sources of finance. Thus central government is still operating a strategy of 'authoritarian decentralism' although employing more subtle and less confrontational mechanisms.

These general trends concerning the restructuring of the state can be seen to have affected planning. A change can be identified from the late 1980s whereby central government has been willing to allow greater freedom to local authorities to adopt planning policies. This has been seen in the 'local choice' idea and the renewed power of development plans. It can also be detected in the shift from excluding local authorities and community groups from urban regeneration in the UDCs to the partnership approach of the

City Challenge and SRB. However, this greater local autonomy and flexibility is set within even greater central government constraints. The centre is spending more time in formulating its priorities and strategies to which the decentralised bodies have to conform. These regulatory instruments include strategic guidance, regional guidance, the Citizen's Charter, and guidelines for City Challenge and SRB bids.

The government thus has a range of regulatory instruments with which to ensure that the market-oriented philosophy is maintained. It can regulate this planning market relationship through legislation, Planning Policy Guidance Notes, financial control and through placing private sector personnel in positions to influence decision-making priorities. The virtues of the private sector have been introduced into all parts of traditional public sector capital spending. The government's Private Finance Initiative (PFI) (DoE, 1993c) announced in 1992 aims to bring private investors into most projects. This limits the commitment of public investment. Major infrastructure planning, the Channel Tunnel High Speed Link, and the programmes of the development corporations and housing associations have been subjected to the same private finance disciplines. The long-standing belief in the efficacy of market logic is therefore sustained through the PFI.

Thus, in the post-Thatcher years, the strong market orientation has been retained. This has been part of the central strategic agenda and has provided the framework within which the more autonomous lower level bodies have had to operate. However, we saw earlier that planning has pursued a number of different objectives which often conflict. In the late 1980s the market supportive objective was being challenged by both the amenity protection objective, NIMBYism, and the demands for a greater consideration of broader environmental issues. Such tensions are always likely to exist, but the balance seems to be shifting towards giving greater weight to the environment. This is evident in the policies formulated by the current Secretary of State, Selwyn Gummer. He has given considerable importance to environmental sustainability in his approach to transport and out of town developments (DoE, 1993a, 1994). It is too early to say whether this emphasis will continue and affect other areas of policy which are not consistent with this approach. It may be that it is a response to European pressure and the environment lobby but will be reversed when the property market picks up again. There is also some evidence, with the readmission of local councils into policy-making, that a form of social planning is creeping back on to the agenda.

During the mid-1990s the Labour Party's alternatives have gradually emerged. The party's New Economic Approach looks for ways of making the market more efficient and fair, with partnership being the 'new' slogan which will be applied to inner city and regional policy. However, as we have seen there is nothing new about partnership in British urban policy. The balance in such partnerships as the City Challenge is still weighted towards the private sector, but will the Labour Party version give more power to local authorities and the community? Will this produce a shift in the planning market relationship and reinforce non-market criteria in decision-making? During 1994 a Labour Party inquiry called 'City 2020' gathered views on dealing with urban problems. One of the key messages emerging from this was the need to remove central government interference and to give local authorities and local communities greater influence. The inquiry visited the US and was impressed by the initiatives of the Clinton administration. The central theme in Clinton's policies for the cities was 'creating communities of opportunity', and two new initiatives were started – 'Empowerment Zones' and 'Enterprise Communities' (Hambleton, 1994). The aims behind these initiatives were to ensure that inner city communities benefit from investment and to increase citizen participation. The theme of 'community' seems to have been expanding in recent years at a number of levels, in philosophical literature (e.g. Bell, 1994; Etzioni, 1993), in urban regeneration through Community Trusts (Bailey *et al.*, 1995) and in developers' schemes through the concept of 'urban villages'. Obviously there are many interpretations of the community theme but it does indicate that further shifts in the planning market balance may ensue. The extent of the shift depends partly upon whether it encompasses the idea of 'community empowerment' which many are now advocating from standpoints within particular communities, within political parties and within the European Union (DCC, 1994; Nevin and Shiner, 1993; Labour Party, 1994; Liberal Democratic Party, 1994).

6

ENGLISH CASE STUDIES

The last chapter set out the main trends shaping British planning in recent years; we will now explore some case studies in detail to see how these trends interact with forces operating at the urban level and the resulting approaches to urban development. The two largest cities in England have been chosen for study. The first part of the chapter looks at examples of different kinds of development project and at city-wide strategy in Birmingham. The London case also begins by looking at a number of large planning projects. The chapter then goes on to examine broader issues of urban planning across the London conurbation. London presents in a heightened form many of the general trends in British urban planning. Centralisation of decision-making has been a dominant factor since the abolition of the Greater London Council, and the private sector has taken an increasingly important role in strategic planning and promoting the capital as a 'world' city.

BIRMINGHAM: PLANNING FOR A EUROPEAN CITY

Birmingham is in the second rank of European cities judged on its international position in terms of international exhibitions, foreign banks and international air traffic (see Bonneville, 1994). These factors are different from those which secured Birmingham's place as one of the major cities of the British Empire, although the civic pride which formed an important part of Victorian city government continues to play an important role. The expansion of the airport and construction of the National Exhibition Centre mark a transition in the wider regional economy in the 1970s. More recently Birmingham has sought to generate change within the city. The two most important redevelopment areas, the Broad Street area in the city centre and Birmingham 'Heartlands' to the east of the city centre, provide contrasting models of the management of urban change.

The city centre

City centre redevelopment involving projects of international importance began in the early 1980s and claimed a broad consensus of support. However, this approach now generates controversy. The international reorientation of the city centre was a public sector initiative and local political leadership played a decisive role in forming a new pro-growth strategy. With a short interval between 1982 and 1984 the Labour Party has controlled the city council since 1979. When the party regained power in 1984 feasibility work on the International Convention Centre (ICC) had been completed and the council pushed ahead quickly. The main elements of the city centre development scheme included the ICC (£180 million), the Hyatt Hotel (£31 million) and the National Indoor Arena (£57 million). The city's objective was to change its image from a 'provincial manufacturing centre' to a major European city (Birmingham City Council, quoted in Loftman and Nevin, 1994). A series of other boosterist actions, bidding for the 1992 Olympic Games and establishing new ballet and opera venues, complemented the redirection of the centre towards attracting international investment and business visitors.

The city was committed to its prestigious projects. The city council is one of the largest UK urban authorities with a large staff engaged on planning and economic development matters. The Unitary Development Plan adopted in July 1993 supported the city centre strategy. The government interfered little with the city centre proposals of the Unitary Development Plan. Nor

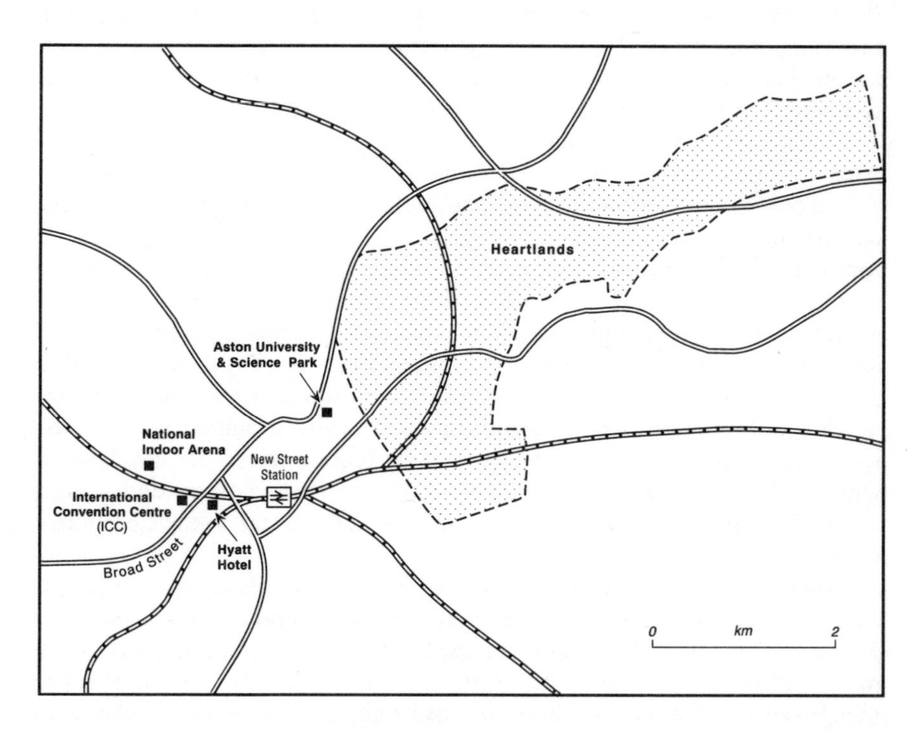

Figure 6.1 Birmingham

was the city constrained by neighbouring planning authorities. Since the abolition of the West Midlands County Council in 1986 neither government policy guidance nor inter-authority co-operation on planning have had much influence over Birmingham's decision-making. The abolition of the county level had the effect of strengthening civic pride and allowed the city to pursue an unhindered boosterist approach. However, central government has continued to play a significant role both in controlling local expenditures and in introducing new agencies.

Within the council, strong officer–member links kept the city centre strategy on course and the appearance of strong political leadership encouraged private sector partners to join in making a strategy for the city (DiGaetano and Klemanski, 1993a). Business interests in the city were grouped together by the Birmingham Chamber of Commerce and, from 1989, by a new organisation named Birmingham 2000. This agency was established with the objective of influencing the city centre strategy and enhancing the city's international image. In common with other Labour-controlled cities, therefore, Birmingham established good relationships with private sector interests. Birmingham 2000 had over 200 members – including local branches of banks and financial services companies, together with professional and media interests. Local business was allied to the council through this organisation. There was therefore a joint approach to dealing with incoming, international business. The large scale of development projects meant that the city had to rely on private sector expertise from the property companies to assess development impacts and markets.

The strength of the public sector and the public–private alliance were crucial elements of the pro-growth strategy. A further important factor was the success the city had in bringing European Community resources into the city centre projects. The city set up a Brussels office in the early 1980s. Its European lobbying paid off in 1988 with agreement to an Integrated Operation package of £203 million (Martin and Pearce, 1992). The UK government matching funds and other resources (including capital receipts and borrowing) brought the total investment to £400 million. The programme for 1994–9 totals £300 million. Birmingham's European success developed its reputation as a leading exponent of urban policies and partnerships from which all cities could learn, including those the city collaborated with in the Eurocities network of second rank cities (Martin and Pearce, 1992). The city's European reputation also secured it a seat on the Committee of the Regions.

However, the huge capital investment in the city centre which was not covered by the EU grants had continuing revenue implications for the city's finances. The city had to manage a large debt and the running costs of the new facilities. According to Loftman and Nevin (1992) this burden was transferred to other council budgets. Investment in housing and schools in Birmingham fell in comparison to other large cities (Loftman and Nevin,

1993). This analysis of the cost of competing for international status was initially dismissed by the city council. The then chair of economic development argued that the convention centre's success could not be judged on only two years' operation (Brooks, 1993) and that the continuing social and economic problems of the city should not be blamed on the town centre investments. Furthermore he said, 'The real issue for cities is to re-create urban confidence' (1993, p.31).

At the political level this confidence in the strategy was short lived. Leadership of the Labour Party changed in 1993. Opposition to the city centre strategy within the ruling party had been quashed by the former leaders and opposition from community organisations failed to challenge the ruling consensus. However, after the change of political leadership the financial weaknesses of the strategy were admitted. The new leader said, 'I want to draw a line. To stop throwing good money after bad' (quoted in Cohen, 1993). The director of the ICC admitted that, 'It's nigh on impossible to break even' (quoted in Cohen, 1993). The shift in policy and doubts about continuing council support led to concerns about the future of other, cultural, elements of Birmingham's prestige such as its international orchestra (*Daily Telegraph, 1994*).

Birmingham Heartlands

The city centre strategy favoured international business visitors and companies. In the 950 hectares of Birmingham 'heartlands' to the east and north of the city centre council policy was oriented towards local and national businesses. Proposals for this part of the city, an area of run down housing, derelict land and former industries, were first developed in 1987. The area was not designated as a central government development corporation but the government allowed a local public–private partnership to develop plans within the context of a relaxed planning regime through designation as a Simplified Planning Zone. The development agency established in 1988 reflects the mix of public and local private business interests. The agency – Birmingham Heartlands Ltd – was 65 per cent owned by mainly construction companies and 35 per cent by the council. The Chamber of Commerce was given one share and therefore the right to vote on company decisions. Three councillors had seats on the board, the other members coming from the private sector, three local development companies and two, Tarmac and Wimpey, nationally based. The city's role in the agency was perceived as contributing land assembly powers and offering some local accountability (Wood, 1994). Despite its political commitment to the project the council found the development agency formula complex. The draft development strategy had to be approved by eighteen separate committees of the council (Wood, 1994).

The agency's Development Strategy adopted in 1988 emphasised economic development, housing improvements and upgrading of the poor-quality environment (Wood, 1994). The vast Heartlands area was broken up into a series of project areas. Some focused on housing estates and government Estate Action funding was used to improve conditions and provide a mix of tenures in the area which had been dominated by council housing (Wood, 1994). Elsewhere the part of the Heartlands next to Aston University and the Science Park is being developed for commercial uses. The city sought to avoid any conflicts with city centre uses and thus the commercial development in Heartlands was for a specific market (Wood, 1994). This area – the Waterlinks business village – has been developed by a joint venture of the construction companies in the development agency with over £7 million government City Grant.

When central government introduced City Challenge as a consolidated inner city funding regime Birmingham Heartlands entered the competition but failed to win. The failure to gain central government's financial commitment to the project brought about a radical organisational change. The local public–private alliance ceded power to central government. The area was redesignated a Development Corporation in 1992 for a five-year period. The Birmingham Heartlands Development Corporation bought out the Heartland company interests, took over nominations to the board and the direction of strategy. Harding and Garside (1995) comment that this new development corporation does not represent a renewed attempt by government to impose the development corporation model of regeneration. The Heartlands Development Corporation should rather be seen as government 'fire fighting' particular political problems. In this case the problem arose when the private sector partners in Birmingham Heartlands had not been able to continue. The project could not be allowed to come to a halt. Following establishment of the development corporation government has invested in the area through its housing and inner city development funds. The Department of Trade and Industry has provided grants in line with the Assisted Area status of east Birmingham. The proposed spine road improving access to development sites is mainly funded by the government. Some landowner contributions have been made to this project but in general no social benefits are expected from developers. The government included the Heartland central site, formerly part of the Leyland-Daf works, in its private finance prospectus in 1993 inviting investors into joint ventures (DoE, 1993c). The government currently plans new investment up to 1997 (BHDC, 1993).

The development corporation gives central government greater control over the Heartlands projects. The continuing involvement of local councillors on the board maintains local accountability. Unlike the earlier development corporations the council retains development control powers. However, the city's policies for the area, for example the restriction on shopping development within the Heartlands project proposed in the UDP, were not supported by the government's inspector at the public inquiry. However, the original Development Framework continues. Each area has a further planning framework but the overall planning controls are loose and flexible giving developers considerable freedom.

City Pride

The internationalisation of Birmingham was further encouraged in 1993 by the government's invitation to the city to enter the 'City Pride' competition. In the government's language the city had to produce a 'vision' and the 'milestones' and 'outputs' towards achieving it. A draft City Pride prospectus was produced in October 1994. The city's objectives were, 'to become an advanced industrial city which is simultaneously the capital and heart of the regional economy, a key player in national economic regeneration, and a competitive member of the network of European cities with a wider international perspective' (Birmingham City Council, 1994). To achieve this status the draft prospectus argued that both physical infrastructure and social and community development were needed. The emphasis on community, training and education reflected new political priorities. Among the council's proposals were new consultation forums and the expansion of the role of the 'social economy', for example voluntary agencies and community trusts. The international economic objective remained fixed. The city has identified economic sectors for expansion and projects to capture more international business. The City Pride prospectus also reminded us that for all the talk of co-operation with other large European cities there is strong competition for international investment.

The City Pride initiative therefore provided the city council with the opportunity to restate its economic objectives. The new language of 'community' and 'empowerment' aimed to avoid the criticisms which were made of the city centre strategy before 1993. The political change after 1993 was one of emphasis. Earlier policies contained community objectives, but these were downplayed. The new approach emphasised the wider needs of the city but the council remained allied to the pro-growth agencies established in the 1980s. For example, the Birmingham Marketing Partnership, comprising public and private interests, was committed to the City Pride process. Further partnerships were developed to support Birmingham's bid to the government's Millennium Commission to fund a technology and training 'campus' in Digbeth next to the town centre.

The Birmingham approach to City Pride was to produce a comprehensive prospectus covering the economy, housing and community issues. It also tackled the process of debate and policy-making in the city. The themes of partnership – the partners behind the city centre and Heartlands projects remained – and city vision formed the government's agenda. The city was following a national policy initiative, albeit limited to three experimental cities. The reorientation of the city's objectives and ways of working was stimulated by central government, and the funding of the prospectus proposals depended on central government's view of how well model city government was expressed. 'Vision', 'partnership', and 'empowerment' provided a formula for replanning the city in which formal plans played only a supporting role.

The Birmingham approach

In its main statement of planning policy – the Unitary Development Plan – the council acknowledged the economic problems of the city and the growing economic and social polarisation among its population. The key to tackle these problems was seen as economic growth and the transformation of Birmingham into an international city. Bonneville (1994) points to a general trend in city government of using international ambition to legitimate large development projects. In Birmingham these projects needed private investment 'on a massive scale' (Birmingham City Council, 1993). The city centre strategy and the Heartlands project provided opportunities for that investment.

In the centre the city took the initiative, brought in private sector partners and then pursued a strategy to attract international business to the city in the conference centre, hotel and cultural projects. However, the construction of similar conference centres in other UK cities raises doubts about the potential profitability of Birmingham's projects. The overall economic and social benefits to the city have been questioned (Loftman and Nevin, 1992). In the Heartlands the council again formed alliances with business elites. This time the lead was taken by the construction companies which might undertake the transformation of this substantial urban area. The development agency originally acted as a focus for promoting and legitimising a range of development projects. The local landowner and developer interests were replaced by government appointees when the government took over the area through its development corporation. Local business and the council had not been able to find sufficient resources to carry through the projects. More central government funding meant greater central control.

In Birmingham, therefore, the council established different coalitions and development agencies to achieve the transformation of different parts of the city. The government supported the internationalisation of the centre and stepped in to support the Heartlands projects when local elites could not secure the resources themselves. The legitimacy of the Heartlands project has not been questioned. Indeed, part of the criticism of the city centre strategy was that money might have been better spent in other parts of the city like the Heartlands. The cohesive public–private alliance which drove through the city centre strategy was founded on local political structures and control of the council's resources. When doubts about the success of the strategy began to emerge the political leadership was replaced. Social objectives now figure strongly in the City Pride prospectus. However the city still has its prestige projects and the financial legacy of the strategy of internationalisation. This objective remained in the City Pride agenda. The City Pride process has also confirmed that partnership and corporate planning will continue to dominate formal planning. Central government also continued to play a strong role, both negatively in restricting local government finance and through managing access to European and national regeneration funds. More recently the government's regional office for the West Midlands has begun to assert its influence through setting regional priorities and through the distribution of the SRB. Local initiative in establishing cross sector partnerships in Birmingham needs therefore to be set in the context of national policy and financial controls. At both local and national levels government has had a strong role in the redevelopment of Birmingham.

LONDON: PLANNING PROJECTS, CENTRALISATION AND PARTNERSHIP

The aspirations and practice of urban planning in Birmingham exemplify general trends in Britain in the development of pro-growth policies, involvement of the private sector and the role of central government in seeking to control the urban agenda. Such trends are also evident in London. One difference between the cities is the absence in London of a single planning authority for the centre of the conurbation. Birmingham City Council provided a substantial resource for those public and private actors concerned with redevelopment and prestige projects. The government of London is fragmented and the three case studies are located in different local political contexts. They cover a ten-year period from the mid-1980s to the mid-1990s. The development of Canary Wharf represents a high point of the Thatcher style. The large commercial development was sited in one of the Urban Development Corporations originally set up in 1981 to by-pass the existing local authorities. Redevelopment proposals for King's Cross railway lands were also formulated during the 1980s property boom. However, delays in implementation meant that the original proposals were never built and subsequently a variety of projects and approaches have emerged. The third case – Greenwich Waterfront – represents a very different approach to large-scale planning with the initiative coming from local government seeking to create a broad-based development partnership. The contexts for the cases represent aspects of the fragmented structure of governance in London. Whereas after abolition of the West Midlands County Council Birmingham itself took on the role of promoting the substantial area of the city council at the core of the conurbation, the fragmented structure of London prevented such a role being undertaken by any of the boroughs. However, strategic planning and promotion of London as a world city have become more important. Following the three cases, our analysis of urban planning in London examines the re-emergence of public and private interest in subregional and city-wide planning issues.

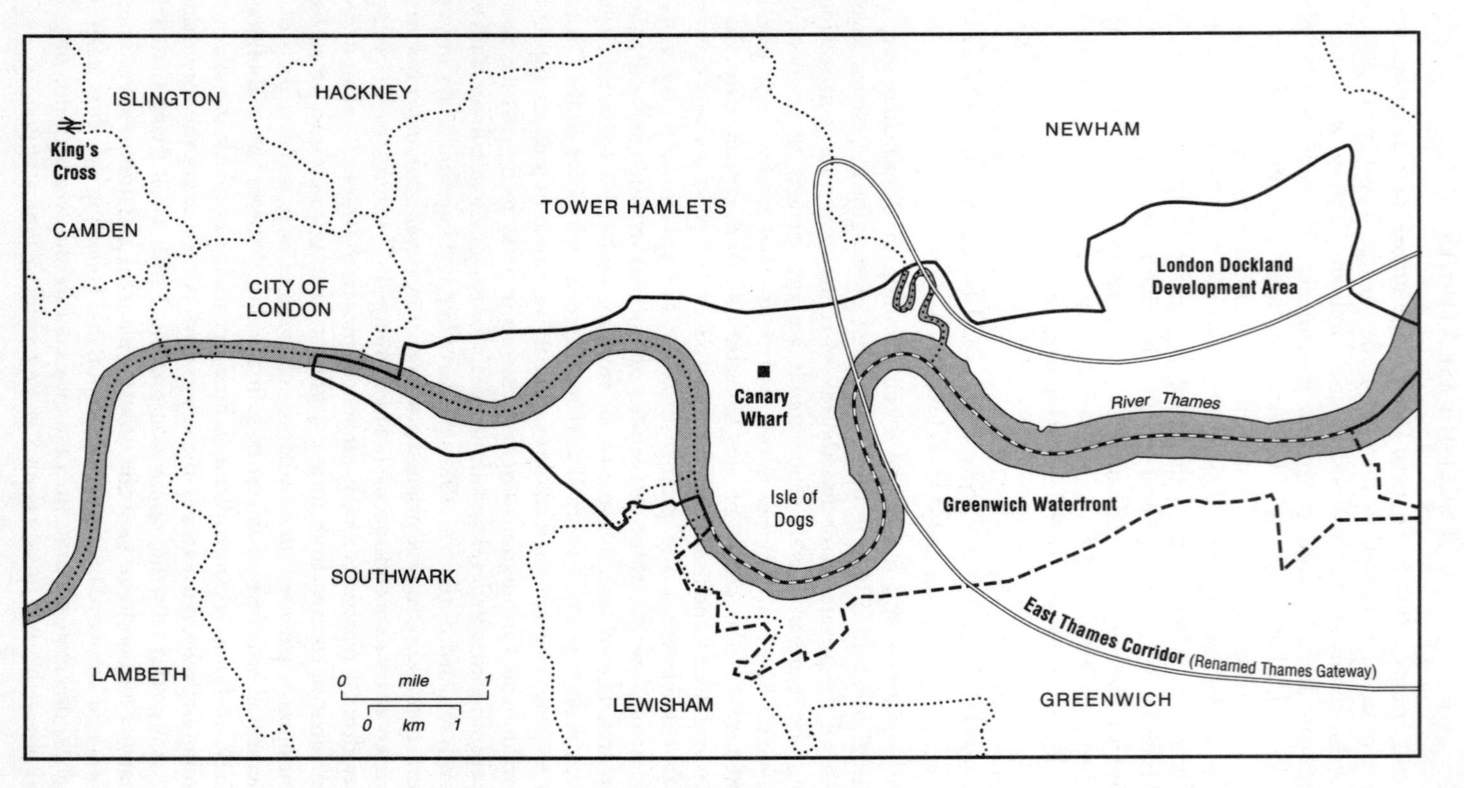

Figure 6.2 London

The story of Canary Wharf

The Canary Wharf development was located in the UDC and partly in the Isle of Dogs Enterprise Zone. The EZ is estimated to have generated £1 billion of tax subsidy for the Canary Wharf project (Ambrose, 1994). There are three main issues in the debates about Canary Wharf (see Coupland, 1992; Thornley, 1992a; Colenutt, 1994; Ambrose, 1994). These are the impacts of the project on the London commercial property market and commercial property planning, the relationship to infrastructure planning, and the impacts on the local communities. Taken together they demonstrate the distinctiveness of a style of urban planning.

The rationale for the development of Canary Wharf was the increased demand for office space expected as a result of the deregulation of the financial markets which was due in 1986. It was estimated that there would be a big demand for a new kind of office space of high quality with large trading floors. At the time annual rents in the City were very high at £600 a square metre. It was also recognised that London had to prepare itself for competition from Frankfurt and Paris in the fight to be the major financial centre of Europe when European Commission controls were lifted in 1992, and indeed the developers of Canary Wharf threatened to take their proposals to one of these alternative centres. Thus the idea of Canary Wharf was a response to these demands; it sought to establish a new commercial area which could complement the City and the West End and act as a third centre for London.

The initial developers could not secure funding and the project was taken over by Olympia and York in 1987. Plans were revised slightly, new architects brought in, and the three towers replaced by one 245 metres high comprising 50 storeys, designed by Cesar Pelli and a symbol for the whole Dockland development. The first phase of the project, involving eight buildings including the tower and a total of 400,000 square metres of lettable floor space, was completed in 1991. During the construction of the scheme a number of changes took place in the property market. The dramatic fall in equities on Black Monday, 19 October 1987, led to many redundancies in City firms. Confidence in future growth was affected and a more cautious approach to expansion resulted. Although the demand for office space continued for a while demand began to fall from 1990 (Cowlard, 1992). Meanwhile there had been a huge increase in supply. The government's relaxed planning regime and the lack of any strategic body to oversee events throughout the capital meant that there were no checks on the total volume of supply. The City Corporation reacted to the threat of Canary Wharf by changing its Local Plan and allowing greater office expansion, and a number of prestige schemes were completed on the fringes of the City (e.g. Broadgate and London Bridge City). By 1991 there was an over-supply of office space; property consultants Jones Lang Wootton estimated that 18 per cent of City office space was vacant. Rental costs in the centre of the City had fallen from the peak of £600 per square metre per year to around £350 per square metre in 1992 – this compared to the £250–£300 per square metre at Canary Wharf. In this climate there were a number of Dockland property collapses, including the Tobacco Dock retail development, Kentish Properties, and the Butlers Wharf Company. Olympia and York were therefore having to argue very hard to persuade companies to take the unreliable Light Railway down to their unknown and rather strange environment rather than stay in the well-serviced, familiar City.

Olympia and York then had to face even tougher problems. The company itself foundered and the project was saved by a consortium of banks keen to avoid a wider banking crisis should the project be abandoned. In the current property market Canary Wharf looks like an expensive building in a location more suited to lower grade activities (see Barras, 1994), but substantial lettings have been made to banking subsidiaries and the managers of the project remain, as ever, optimistic (*The Times*, 1995b).

The Canary Wharf project was conceived and failed in a climate of deregulated planning. At the same time, however, the government tried to exact payments from the company for infrastructure improvements. The original deal struck between Olympia and York and the development corporation and government included a contribution of £68 million towards the total costs of £150 million of upgrading the Light Railway. On the other hand the development corporation also agreed to extensive road works, including the Limehouse Link estimated to cost £560 million in 1989. In the discussions with government about the necessary Parliamentary Bill to extend the Jubilee line the company offered to pay about half the estimated costs. Further studies were then commissioned by government and alternative routes explored. What followed has been described as a high-level poker game between developers and government. Developers with sites on the alternative routes put in bids stating the amounts they were prepared to pay towards the costs, in order to secure the route that favoured them. A protracted period of secret talks then followed but without any discussion about the strategic issues and the best routes for London as a whole. When a decision was taken on the route in the summer of 1990 Olympia and York's contribution was set at £400 million over a number of years. The most generous estimates calculate that this represents just 14 per cent of the development cost and, allowing for inflation over the construction period, the contribution could be a great deal less significant (White, 1995). The government's current total cost estimate is £1.8 million (PFO, 1993) As a result the Jubilee line extension jumped the queue over transport improvements in the capital which were considered by many to have greater priority. In effect the line is being built with public money. The private subsidy is small and deferred until future years.

The final aspect of the Canary Wharf story is the failure of the 'trickle-down' economics behind property-led regeneration (see Massey, 1991 and

Ambrose, 1994 for discussions of this issue). The Canary Wharf project itself has little to offer local residents, being totally geared to its day-time population of wealthier office workers. The shops and services it contains are at the expensive end of the market and there is no incorporation of facilities for the long-established existing community. The scheme has been criticised for it single use approach – 'no-one will live on the wharf, there are no factories or workshops, nothing designed for small businesses, nothing downmarket' (Davies, 1991). There were pressing needs for alternative developments in the area, including affordable housing. One impact of the office scheme has been to encourage the building on nearby sites of private housing for office workers.

The Canary Wharf story shows how in the 1980s central government intervened in local planning in order to create the environment for the market to determine priorities. This is evident in the way so many decisions were made through discussions between Olympia and York (whose office was in Westminster not Canary Wharf) and Ministers, as well as central government control over the London Docklands' Development Corporation (LDDC). Central government provided the strategic planning for London through its allocation of funds and permissions for infrastructure development, and this benefited Docklands. New river crossings, the expansion of the Dockland City Airport to take larger aircraft, and the setting up of the East Thames Corridor as an area for development priority continue these trends. Centralisation has been accompanied by exclusion of local interests. It has been argued that this closed, centralised style of decision-making has had profound effects on local politics (Colenutt, 1994) and on the rise of right-wing parties.

The problems faced by Canary Wharf raise the question of whether the by-passing of planning has actually been in the interest of developers. Olympia and York always claimed that they took a long-term view and were prepared to ride out two recessions. However, it can be suggested that it would have been in their long-term interest to have had more of a planning framework – strategically, planning of office supply could have smoothed out the peaks and troughs of property cycles and, locally, a better external environment and a more mixed pattern of land uses would have given their scheme greater long-term viability. Planning might also have avoided the confrontation with the local community; the image of a divided city did little to help either the developer's image or that of the government's urban policy.

King's Cross railway lands

The ideas for King's Cross railway land which emerged in early 1987 were associated with British Rail's proposal for a £1.5 billion underground terminus for channel tunnel trains. Continuing debate about the project reveals how the approach to large-scale projects in the capital has changed since the 1980s and tells us a great deal about the structure of planning outside the special case of enterprise zones and development corporations.

Edwards (1992) describes the problems created by the nature of the relationship between the landowners – British Rail and the privatised National Freight – and London Regeneration Consortium, the selected developers of the site. British Rail is not allowed to borrow money from capital markets and consequently it is encouraged to exploit its assets in the form of land. It is normal in the British development process for landowners to invite developers to present, in private, competing development proposals and to choose the scheme which offers the best return. This process applied to King's Cross, Edwards argues, created from the start a development project biased towards the most profitable land uses. The cost of developing over railway track would be expensive and this, in part, contributed to the need for a large amount (in excess of 5 million square feet) of office space. The basis of the deal between landowners and developers in this case was that the landowners would get 70 per cent of the surplus profits of the scheme.

In the early stages the initiative rested firmly with the developers. The consortium employed an internationally respected architect to produce a 'master plan' in 1988. The plan outlined an overall strategy of interconnected elements and claimed a unique vision for the future of the railway lands (Newman, 1990). The consortium also undertook extensive consultation and surveys of local public opinion to reinforce their position as the planners of the site. This form of private sector planning was possible because of the weakness of public strategic and local planning. The abolition of the Greater London Council in 1986 deferred any strategic view until the government published planning guidance for unitary development plans. The unitary development plan system was new with little prospect that the local planning authority, the London Borough of Camden, would produce such a plan in the near future. Camden's existing plans and planning studies had little to say about the railway lands or about development potential in the area. In that context therefore private sector planning could be persuasive.

As we saw in the previous chapter local government planning and development powers have been weakened by central government. The council was unwilling to refuse planning permission as this would have led to intervention by the government. A strategy of negotiation with the development consortium was therefore used in order to preserve a role for the council. The council held a series of public meetings and sought to represent local community views to the developers. However, the relative weakness of local government and formal planning encouraged community-based opposition to follow an independent path. Local community associations wanted to negotiate directly with the developer and not allow the borough council to represent their views. One community group, the Railway Lands Community Group, set itself up in 1987 to mobilise local opposition and to sponsor alternative plans. Thus early on in the debate about the railway lands four distinct groups emerged: the landowners, the developers, the council and the community. The community side developed its own consultation processes and forums. Alternative proposals for the railway lands were drawn up. The Railway Lands Group proposed a mixture of 'interim' uses for the sites including gardens, camping sites, and temporary housing (Railway Lands Group, 1993), and submitted their own planning application for development of the sites.

This extensive community planning activity was made possible because the development consortium had run into difficulties. The large redevelopment opportunity of the railway lands was one of a number of such opportunities on the fringe of the city. What enhanced interest in this site in the late 1980s was the proposed location of a terminus for international trains at King's Cross. In addition to their development interest British Rail were seeking parliamentary approval for the new terminus. The process of securing development approval by parliamentary bill is long and complex. Approval for the terminus, and thus an enhanced international role for the railway lands, introduced substantial delay into the development of the consortium's proposals. The local community organisations used the opportunities to object to the proposed train terminus, and presented lengthy evidence to Parliament. Further delay and uncertainty were created. During this time the office market boom which had made redevelopment at King's Cross seem attractive came to an end. Procedural delay, and the market collapse, undermined the consortium's proposals. The development companies who were partners in the consortium were particularly badly hit by the falling office market.

The development consortium's problems were made worse by the failure of government to commit itself to the King's Cross international station. In the process of hearings before the parliamentary committees alternative proposals were put forward. The community groups recommended an alternative to bringing the line into London from the south and tunnelling to King's Cross. It was suggested that a route from the east could follow existing lines and terminate at St Pancras station. In 1993 the government changed its mind about King's Cross and proposed the St Pancras alternative. The development consortium's plans for the railway lands could not proceed if this new railway line was to cut through the sites. The developers' problems, the property crisis and the lack of railway investment planning, became insuperable. They were forced to withdraw their outstanding planning applications in 1994.

The council's position shifted during this period from one of weakness to a relatively more active role in the planning of the railway lands. The council produced a detailed planning brief and revised this document in response to the decision about the St Pancras terminal (London Borough of Camden, 1993). The community planning applications were consequently revised to follow the brief. The council's policy was to declare the railway land a 'special policy area'. Half of all housing to be built was to be 'affordable' – that is, not sold or rented at full market prices; up to 2 million square feet of offices could be built. The London Planning Advisory Committee's advice on strategic guidance for London in 1993 also proposed commercial development on this scale. Relatively late on in the discussions over the railway lands, therefore, the public sector produced a planning framework. The council also rather belatedly took up the theme of partnership which had dominated urban planning elsewhere since the mid-1980s. A cross sector body was set up to bid for funding from the Millennium Commission. The CrossMillennia Art, Technology, and Ecology project brought together Camden and Islington councils, British Telecom, the National Film and Television School, the University of North London and National Freight in a community development trust to develop a media and information technology centre in some of the Victorian buildings on the railway lands.

The context for comprehensive redevelopment was changed by the decision to bring international trains into St Pancras. Re-routing the railway affected the shape of the development site, and brought in new potential developers. The government's approach to a high-speed link from the channel tunnel is to grant a concession to a private company to design and build the railway and run trains. The selected company would pay the construction costs but in return be given subsidy, the fare income, and development land. The railway lands form part of the package. New development proposals can be expected. The government has, however, to go through the parliamentary procedures again to get approval for the high-speed link. Not surprisingly the Railway Lands Group has raised objections. It placed seventy petitions before the House of Commons select committee examining the railway bill. Nearby residents and business lodged a further hundred petitions. The earliest that a new bill can be approved is January 1997. Further delay could arise from the failure of the companies bidding for the concession to build the railway to agree with the government about its exact route.

The recent history of proposals for the railway lands reveals the weakness of public sector planning. Initially there was a lack of a strategic or local planning framework. The government's confusion over railway investment added to the delays of the development consortium. Community intervention also added to delays and brought little comfort to the local planning authority. A second phase of development planning has brought forth proposals for a new railway station, new development interests and renewed community initiative. Neither central nor local government has been able to forge a consensus of all parties and the legitimacy of public and private actors remains contested.

Greenwich Waterfront

Our third London case begins with local government initiative. 'Greenwich Waterfront' extends over seven miles of the river Thames and includes development sites amounting to 500 acres. On the opposite bank is land controlled by the London Docklands' Development Corporation. Most of the south bank is in private ownership and at the end of the 1980s numerous planning applications were made reflecting the development boom in London and in docklands in particular. In the 1980s the Labour-controlled council developed radical proposals for the area, in common with other Labour London boroughs, seeking to create employment for its working class population and build more rented housing. In the late 1980s, however, the approach changed. More ideas for the whole waterfront were developed in 1989 and a planning strategy, The Greenwich Waterfront Strategy, published (London Borough of Greenwich, 1991). The strategy sought to attract development and public investment, in the extension of the docklands' light railway for example. However, of more significance than the strategy itself was the way in which the council sought to approach the replanning of the waterfront. In 1992 the council established a new agency incorporating private and community interests. The Greenwich Waterfront Development Partnership (GWDP) was established to lobby for investment, to attract development and to act as an independent development agency (London Borough of Greenwich, 1992).

The structure of the agency reflected its range of interests. There was a board with overall responsibility, drawing its members from three separate sources – the community forum, a business forum, and the council. Each had equal representation on the GWDP board. The community forum brought together over 250 community groups. The business forum had 120 members, including the Woolwich Building Society and British Gas and organised itself through a set of 'task forces'. The business forum allowed private interests access to public decision-making and whereas the community groups were absorbed by the issue of representation, business interests were concerned with how best to work with the council. The TEC, the University of Greenwich and existing public–private town centre agencies in Greenwich and Woolwich were also brought into the GWDP. The structure of the GWDP was complex and its staff were seconded from both public and private sectors.

For the council the agency offered the potential to influence the large landowners who controlled most of the waterfront. Without control of land the power of the local planning authority was limited. The response to the proliferation of development applications in the late 1980s was not to use development plans and development control powers but to seek to influence landowners by other means. The council, in common with other London boroughs, had limited access to development finance and land acquisition was therefore not an option. As noted in Chapter 5, throughout the 1980s local government had been progressively weakened both by central financial controls and by a general policy orientation to private interests. After the 1987 general election and the return of another Conservative government, many Labour-controlled boroughs abandoned oppositional policies and sought local alliances with business interests. This general shift is reflected in the switch in Greenwich from the confrontational 'People's Plan' of the mid-1980s to partnership with landowners and business.

The council used the GWDP to carry out tasks not already undertaken in mainstream departments, such as promotion and marketing of the area. The GWDP also presented the opportunity for systematically incorporating community interests. The espousal of 'partnership' revealed the weakness of local government but also presented opportunities for action which would otherwise have been economically or politically difficult to achieve.

The overall strategy followed by the GWDP involved a series of projects based around different landholdings and business interests, but the depressed property market contributed to the lack of progress on the ground. A large housing scheme for Deptford Creek planned at the end of the 1980s had failed to get off the ground by 1995. The substantial British Gas scheme for the Greenwich peninsula was also delayed. This 'Port Greenwich' scheme extends over 300 acres. A mixture of housing and commercial uses was proposed. Lobbying by the GWDP and British Gas gained a new underground station for the site as part of the Jubilee line extension, with British Gas making a contribution of £5 million to the cost of the station. However, the weakness of both residential and commercial property markets delayed a start on the project. In Woolwich the Ministry of Defence plans for a museum and commercial and residential uses to replace former defence installations were aided by money from the EU's KONVER programme which supports the reuse of former military sites. The GWDP played a strong role in lobbying the European Commission for this development funding.

The GWDP provided a relatively stable agency through which bids for resources could be channelled (Kerswell, 1994). The GWDP was used as a base for bidding in the City Challenge programme and also lobbied for Objective 2 status as part of the government's East Thames Corridor strategy. Neither of these was successful. However, the GWDP supported a winning bid for Single Regeneration Budget funding for Woolwich. The 'Woolwich Revival' project was allocated £25.3 million from the SRB and this success reflected both the degree of need, with local unemployment rates among the highest in London, and government recognition of the base of cross sector co-operation which the GWDP had fostered.

The council also created a broader based partnership to bid for Millennium funds. The Millennium project included the neighbouring boroughs of Lewisham, Tower Hamlets and Newham, the development corporation, the Port of London Authority, the London Tourist Board and the National Maritime Museum, and British Gas allocated 45 acres of their Port Greenwich site for the Greenwich Millennium festival. The council believed that the festival would have economic development benefits (*The Times*, 1995a). The Millennium project reveals a change in the role of both the council and the

GWDP from that of developer to being more concerned with promotion and marketing.

The planning of Greenwich Waterfront illustrates the weakness of local planning authorities. In the absence of local development finance the council co-operated with private partners and competed for central government money. The GWDP supported a set of private development projects and the GWDP and other local partnership groupings had to compete on the government's terms through City Challenge and SRB for national funds. Success in those competitions depends on a demonstrable local commitment to 'partnership', and the setting up of the GWDP was obviously a strategic necessity. However, little has been achieved on the ground. In response, the discourse of partnership has been amplified to enable the council to engage in international competition for recognition and potential investment.

THE FRAGMENTATION OF LONDON PLANNING

The three London cases are all examples of project-led, localised planning. National policy decisions supported Canary Wharf but failed to get behind the developers at King's Cross. In Greenwich the strategy came from local political initiative, with business partners contributing to a coherent approach to partnership conforming to central government guidance. The cases also fit into a broader picture of the changing nature of urban planning in the capital. The recent emergence of public–private co-operation at local level gives planning in the 1990s a different character to the centralised, deregulated model of the mid-1980s. The increasing rhetoric of partnership clearly echoes national trends. However, the governance of London sets a specific context for urban planning. To understand this context better we need to look at broader trends within the capital. The next part of the chapter discusses the fragmented nature of decision-making following abolition of the Greater London Council and the subsequent emergence of new public and private sector institutions concerned with London-wide issues and subregional planning.

The abolition of the GLC in 1986 resulted in the fragmentation of planning into the thirty-two boroughs and the City which had to set out the strategic framework for their areas as the first part of the new Unitary Development Plans. In addition many other bodies performed not only simple advisory roles but also the provision of London-wide services, creating a very complex picture of governance (Hebbert and Travers, 1988). Most of these bodies were appointed by central government and had little or no local democratic involvement. Skelcher and Stewart (1993) claim that in 1993 there were 272 appointed bodies in Greater London covering such essential services as education, health and transport. The TECs and transport bodies were of crucial importance to urban planning, as were the other unelected bodies, the nine City Challenge Companies and forty-nine SRB partnerships.

Although Skelcher and Stewart's figures may be somewhat overstated the point is well made that these undemocratic bodies cannot be considered exceptions but are central to an understanding of urban governance. As the same authors say,

> the effect of creating appointed bodies to provide local public services is to remove their policies and performance from the local political agenda. It raises the question of the nature of the democratic accountability of these bodies and their relationship to the Londoners whose lives they affect.
>
> (Skelcher and Stewart 1993, p.12)

The picture emerges of a planning approach scattered amongst numerous organisations, many of which are undemocratic. The principal feature which was common to all these bodies, and which potentially provides some degree of co-ordination, was central government control. The local authorities were constrained in their plans by central government guidance – strategic guidance and the planning policy guidance (PPG) – reinforced by the appeal process which gives central government such power in the British planning system. Urban regeneration initiatives were controlled by central government not only through the appointment of members but also through holding the financial power and setting the rules and regulations. It might be said that in the City Challenge Initiative and SRB local authorities gained more power as they initiated the schemes and co-ordinated their implementation, and that they also included community representation. However, local authorities only got the designation in the competitive process if they followed the brief set out by central government which demanded a major involvement by the private sector and conformity with central government policy. The SRB winners had to follow contractual delivery plans. City Challenge Initiatives, and Training and Enterprise Councils, had to be set up as companies limited by guarantee. It is said that this gave them more operational freedom. However, the company status leads to greater secrecy and the interests of company finances came before any accountability to the wider community (Skelcher and Stewart, 1993). Research has shown that the policy of involving the voluntary sector in these initiatives has not been borne out in practice (NCVO, 1993).

The overall picture of London-wide planning was therefore characterised by the seemingly contradictory themes of fragmentation and centralisation. Strategic planning was inevitably weak following abolition of the GLC. Only the London Planning Advisory Committee (LPAC), set up by the government to advise on matters affecting more than one borough, was looking beyond project and local planning. However by the mid-1990s there were several attempts to put together a concerted view of London issues. Interest in the strategic role of London and regional and subregional planning increased both within and outside government.

Public and private strategic planning

LPAC's ability to influence government was constrained. The mid-1990s advice of LPAC on strategic guidance supported a general priority towards investment in east London and other areas in need of, or offering the opportunity for, regeneration (LPAC, 1994). The government's draft planning guidance based on this advice deviated from it mainly in giving greater priority to development and regeneration in and around the central area (GOL, 1995). Until May 1994 the political composition of LPAC was balanced and therefore any statements had to attract cross-party support, thus limiting their scope. Even after the 1994 election in which the Labour Party gained control of many more boroughs, doubts were expressed about the ability of the London boroughs to take on a positive co-ordinating role across the city because of their diversity and rivalries (Biggs and Travers, 1994). The experience of other metropolitan areas suggests that if the boroughs are to do so a greater sense of metropolitan identity and more consensus would be required.

The international economic pressures on London led LPAC to initiate a report on London's competitive position which was co-sponsored by the City Corporation, Westminster City Council, the LDDC, Greater London Arts and London Transport. The chosen consultant's report, 'London: World city moving into the 21st Century', was utilised by LPAC in the formulation of their strategic advice. Gordon (1995) has analysed well the role of this document as London's nearest attempt at a 'boosterist' strategy. One of the significant features of the consultant's study was the limited range of opinions that were sought on London's needs. The survey was dominated by business interests – it included key decision-makers in the global headquarters of major multinational companies, senior executives in thirty London-based commercial organisations and a hundred foreign banks with a base in London. The orientation of the report towards the conditions necessary to retain and attract business investment is clear from this selection of interviews (Newman and Thornley, 1992). Although a range of civic and amenity societies were also consulted there was no means of establishing the views of the various communities within the boroughs that make up LPAC's constituency. It is perhaps not surprising that some boroughs are unenthusiastic about the report, leading LPAC to complain that 'Boroughs have not consistently incorporated the world city needs of the capital into their UDPs' (quoted in Johnston, 1993, p.19).

Co-ordination and overall strategy were weak but gradually the situation changed. The year 1992 seemed to mark the beginning of a new phase in thinking about London government which had implications for planning and development. The problem of the government of London was tackled by both Labour and Conservative parties in the general election campaign. The new agenda for London – transport and strategic planning, economic development

and tourism and promotion – represented a list of issues which had not been adequately dealt with by the boroughs or other successor bodies to the GLC. It was argued that the success of the post-abolition arrangements in many areas of service provision highlighted the failure in strategic areas, and that it was only a matter of time before government would have to respond (Travers *et al.*, 1992).

The debate about Europe, with its increasing references to regionalism, added to the need for a new approach to London. London leaders had always seen the city as a world centre for finance and culture but, with increasing competition between cities, it needed to work hard to maintain this position – especially in the light of the Single European Act. A strong argument made for developing Canary Wharf was the need to ensure that commercial development was not lost to Paris or Frankfurt. More recently the government's pamphlet celebrating London's achievements (DoE, 1993b) reported how other European cities 'are organising themselves to compete more effectively for inward investment' (p.2) – it then referred to Paris, Lyons, Berlin and Hamburg. The fear was that as London had no strategic body which could promote London as a whole it would lose out in this competitive game.

There were public and private sector responses to these concerns about the strategic position of the capital. In the public sector, the Government Office for London (GOL) was set up in April 1994 as part of the national regionalisation of central departments. GOL is supervised by a Cabinet Committee of senior Ministers. A principal function of the regional office was to administer the new Single Regeneration Budget (as discussed in Chapter 5). GOL was also to be responsible for strategic guidance. Its draft issued in early 1995 was clearly centred on the world city role of London. The central area consequently was given special attention as the location of world city functions, and city fringe areas include the 'European gateway' around the terminus for international trains at Waterloo (GOL, 1995). Through the creation of the regional office government signalled its concern with London-wide issues.

In addition to this renewed government interest in regional issues, the private sector took a much greater role in the planning of London. Since the early 1980s this can be seen as compatible with central government's ideology which incorporated deregulation, a greater acceptance of market-led decision-making as in the Canary Wharf case, and encouragement of partnership approaches of the type we saw in Greenwich. It can also be seen in the briefs that were prepared for urban regeneration initiatives, whereby private sector involvement was a requirement for funding, and in the appointment of private sector personnel to quangos. More recently, however, the private sector introduced new agencies into the urban planning arena. In its 1992 manifesto government supported the idea of a private-sector-led organisation to fulfil the strategic need. This was an idea that had been maturing for some time and business interests, such as Stanhope and the

CBI, had been lobbying for some form of strategy led by the private sector (e.g. Robinson, 1990; CBI, 1991).

The LPAC world city report proposed the establishment of a promotional body, called the London Partnership, with the job of 'selling London's enterprise and culture, services and potential to the world at large' (Coopers and Lybrand Deloitte, 1991, p.210). After the election the government set up the London Forum to promote the capital as a tourist and cultural centre and to attract inward investment. Meanwhile the private sector set up its own body, London First, to

> establish a strategic framework for the capital and to pull together current initiatives. London First aims to apply business principles to policy issues . . . in partnership with others, business leaders can provide the spark by developing a credible vision for the future and promoting action towards that vision.
>
> (London First, 1992)

In 1993 London Forum and London First were merged under the banner of London First. The organisation was dominated by private sector interests although it contained representation from local government, churches, the voluntary sector and statutory bodies. One of the roles undertaken by London First was to develop a strategic approach for London and it commissioned various studies for this purpose. Colenutt and Ellis (1993) have pointed to the overlapping nature of appointments to quangos in London and membership of these promotional bodies. They show that many members were involved in the Business in the Community initiative discussed in Chapter 5. These new private-sector-led groups had a privileged position in gaining the ear of government and influencing strategic guidance.

When the City Pride initiative was announced at the end of 1993, unlike Manchester and Birmingham where the city councils were given the job, in London it was London First who were invited to prepare the prospectus. London First then invited the CBI, London Chamber of Commerce, LPAC, London Boroughs Association, the Association of London Authorities, the Cities of Westminster and London, and the London Voluntary Services Council into a 'London Pride Partnership' to prepare a City Pride prospectus. London First had few staff and the City of London paid £300,000 to enable London First to set up a small secretariat to undertake the task. The public and private partners inevitably disagreed about what the document should say. The final report referred to 'interrelated missions' to avoid conflict about the relative priority of wealth creation over social objectives (London Pride Partnership, 1995). Many potentially controversial issues were deferred so as to keep the broad base of partnership. The London Pride Partnership committed its members to a strategy and to continuing to work together. However, the Labour-controlled boroughs were also committed to creating a new-style GLC, and the future of the London Pride Partnership was uncertain. The role of central government's strategic planning guidance became less clear as it had to sit alongside the London Pride Prospectus and the GOL approach to distributing the SRB. The housing targets of the prospectus and the planning guidance differed. The GOL was not a member of the London Pride Partnership, but some form of co-ordination would be required for these initiatives to have a degree of permanence. However, the symbolic politics of the London Pride Partnership may have been more important to central government than the robustness of the organisation. It provided a response to Labour Party calls for an elected London government.

Subregional alliances

In addition to the private sector's interest in London-wide issues business leaders became all the more aware of the competitive position of local economies within Europe (Stevenson, 1994). Many new alliances at subregional level were established. The London First group drew on the Business in the Community experience in other cities and conceived 'wedges' of London in which Business Leadership Teams could co-ordinate activity. In both west London and east London business alliances preceded London First. These subregional groups were incorporated into the London Pride Prospectus and gave a geographical dimension to the proposals. The Thames Gateway Framework published in 1994 can also be seen as evidence of the growing importance of subregional alliances. Various other groupings of boroughs were formed to create joint lobbies in relation to infrastructure investment (Begg and Whyatt, 1994). We saw in Greenwich how four boroughs and private partners joined together for the Millennium project. From April 1995 the GOL's regeneration teams were reorganised on an area basis, ensuring a better response to subregional groupings. The growth of inter-borough and public–private alliances concentrates on those areas where regeneration is needed. Elsewhere strategic guidance refers to areas of 'consolidation' where the role of planning is different. The growing number of conservation areas signals a protectionist local politics in which planners will be responding to residents' wishes rather than the business-oriented objectives of the regeneration and growth areas.

A feature of governmental planning in the 1990s is the increasing involvement of the EU. Central government was compelled to seek Objective 2 designation for parts of London in order to get access to European regeneration budgets. Having succeeded in winning European money for east London substantial SRB allocations were required as matching funds. One-third of the London SRB projects in 1995 were located in the Lee Valley. In supporting these subregional, cross sector initiatives European policy can be seen as an important influence on the emerging subregional direction of planning (Fiddeman, 1994; Stevenson, 1994). This European dimension supplements the moves by the private sector.

Following the fragmentation of London planning in the late 1980s new public and private, city-wide and subregional interests have developed. A complex pattern of agencies has been formed and at the London-wide level the relationships between public and private interests are intricate and in a state of flux.

CONCLUSIONS

The directly controlled Development Corporation model of the early 1980s has been superseded by more flexible forms of central state intervention. In London the London Pride Partnership was given the role of setting an agreed public–private agenda. In Birmingham the Development Corporation solved the political problem of sustaining the Heartlands scheme. City Pride and the SRB regime ensured conformity to the government's overall objectives in the city as a whole.

In both cities central government continued to exert control. This happened through appointments to boards of new agencies, setting overall planning guidance and setting the rules by which local public–private alliances compete for central resources. This strong central intervention favoured economic over social objectives and prioritised development projects rather than territorial planning.

An important difference between the cities was the use of a private agency in London to take on the tasks of strategy and promotion. A further difference in the responses to City Pride was that the London boroughs successfully argued that London Pride should cover the whole of the conurbation and not just, as London First would have preferred, the central area (Newman, 1995). The Birmingham document concentrated on the city and not the West Midlands region. Two very different responses were therefore developed to the City Pride initiative. In London public partners were invited into London Pride, in Birmingham the city council encouraged and used its resources to support public–private initiatives.

There is an important question about the relative influence of public and private partners in the alliances in both cities. Collinge and Hall (1995) stress the roles of local and central government in the replanning of Birmingham. Whilst it is true that the forms of public–private alliance were not US-style growth coalitions, their overall agenda and priorities were those of business interests. The city council provided and organised the resources to implement business objectives. Through its input to the Unitary Development Plan, and through City Pride, central government has also supported the internationalisation of the city centre. The public sector clearly plays a significant role in pro-growth politics, but locally the private sector partners seem to set the agenda. In London public sector actors had significant roles in servicing mixed agencies but in the case of London First it was business leaders who took the decisions (Begg and Whyatt, 1994). Similarly, in local coalitions business leaders saw their role as setting the overall policy direction (Stevenson, 1994). The renewed interest in regional and subregional planning in London was strongly influenced by private sector objectives. The local authorities were left to fill out the details within a framework defined by private partners and central government.

In the 1980s local planning objectives which did not fit government priorities were either bypassed or overruled. Having been substantially weakened local government has been allowed back into decision-making arenas, but on the government's terms and in partnership with the private sector. There was disagreement inside the London Pride Partnership but all sides agreed on the importance of economic development objectives. Social objectives have been recast. The emphasis on 'social cohesion' in the London Pride Prospectus reflected a recognition of the problems of many boroughs but also the philosophy of social responsibility espoused by Business in the Community, and, perhaps most importantly, a necessary acknowledgement of the priorities of the European Commission who may fund more London projects. Whereas in the past there had been conflicts between central government and Labour-controlled cities, Birmingham's approach to international competition meant that it could be entrusted with City Pride. There was a question in Birmingham about how far the change of political leadership in 1993 represented a real change in policy direction. In terms of projects, the SRB regime ensured the compliance of local government with central government's economic and social priorities.

If planning is no longer the sole responsibility of representative government then new claims to legitimacy are required. The inclusion of the London boroughs lent some legitimacy to business-led bodies, both at London-wide and subregional levels. Birmingham councillors retained their seats on the Heartlands development corporation. A second strand of legitimation is the involvement of communities in the new urban politics. Business in the Community has always recognised the symbolic importance of incorporating a wide range of interests. The government's rules for SRB and City Challenge required community involvement in projects. The inclusion of community organisations in Greenwich and their persistent involvement at King's Cross were therefore essential if those areas were to compete successfully for central government funding. However, these community voices have to concur with the overall belief in business-led development planning. For some this incorporation of selective community involvement, within a centrally controlled framework, is not a sufficient expression of local democracy (Dahrendorf *et al.*, 1993). There were indications that the government was becoming sensitive to the democratic deficit in London. The Secretary of State for the Environment, in launching his publicity pamphlet *London – Making the Best Better*, invited all Londoners to write to him and express their views on the future of London (DoE, 1993b, 1994). This could almost be described as a market research approach to democracy. However, it was suggested (ALA,

1993) that the questionnaire was only included at the insistence of lawyers who reckoned that without it the Minister would be open to the accusation that the government was using public funds for political propaganda. One interesting point to emerge from the questionnaire was that, even though there was no specific question on the topic, 31 per cent of the respondents said they wanted some kind of democratic strategic body.

British planning has been profoundly influenced by the Thatcher reforms and their legacy. Political ideology justified abolition of strategic planning and encouraged private sector interests into the decision-making process. The deregulationist approach of the 1980s and the by-passing of local councils has been replaced by a complex set of central controls over urban planning. Central government sets overall policies for development plans. More detailed guidance of local action is set through the rules for bidding for urban development funds. Local government's mainstream expenditure is tightly controlled. In these circumstances business interests have become increasingly involved in setting broad strategy and in local regeneration projects. However, the enthusiastic involvement of local business interests in alliances raises demands for investment in infrastructure. The London Pride Partnership was critical of the government's lack of spending on transport and housing. Local competition for European funding also requires more active involvement by central government in subregional issues. The regionalisation of government departments can be seen as a response by central government to these local pressures. Can central government maintain its current controlling position or will it have to adapt to being a mere partner in intergovernmental, public–private alliances?

7

FRANCE: REORGANISING THE STATE

Planning in France is located in what has historically been a highly centralised and interventionist state. The French state went through a radical restructuring in the decentralisation reforms of the early 1980s in which there was a substantial transfer of powers from national to sub-national government. However, in contrast to the British experience the public sector remains a powerful actor in relation to urban development and planning. The decentralisation reforms and subsequent further reforms of both central administration and central–local relationships represent adjustments within the machinery of government. The boundaries of public and private sectors have not changed in the dramatic ways we saw in the British case. State planning in France has traditionally been concerned with the general economic development of the country encapsulated in the term *'aménagement du territoire'*. National, regional and local levels are all involved in the pursuit of economic development. This chapter starts by reviewing the development of planning competencies in regions and communes following decentralisation. It then goes on to examine significant changes which have occurred at the urban level. Increasing entrepreneurialism in the cities has developed separately from intergovernmental co-operation over urban policy. New institutional relationships have emerged to integrate levels of the state and co-ordinate policy. The decentralisation of planning competencies has raised some concerns about the integrity of the planning system as a whole. Some of the perceived problems concern the detailed practice of '*urbanisme*' – the rules and procedures through which local planning and development decisions are taken. The final section of the chapter looks at the new planning law approved at the end of 1994 which seeks to re-establish the authority of central government. The dominant theme of this chapter is therefore concerned with the reorganisation of governmental institutions following the radical changes of the 1980s. The French case reveals continuing domination of urban planning by the public sector. However, the nature of the state has changed, relationships between levels have become less certain, and there is a continuing search for better intergovernmental relationships and more effective mechanisms for urban development and planning.

DECENTRALISATION OF PLANNING

According to the legal code, planning in France is a shared responsibility between central and local government. The relationship between central and local government must therefore be the starting point for understanding urban planning in France. Prior to decentralisation central government enjoyed direct control of plan-making and development control. National priorities were developed by planners in Paris and they were passed down through local offices of central ministries. In each *département* the Direction Départementale de l'Equipement (DDE), the local office of the ministry, was responsible for planning and contained the technical expertise to draw up detailed land-use plans – Plans d'Occupation des Sols (POS). The government appointed *préfet* of the *département* approved these plans and issued building permits to proposals which conformed with them. The *préfet* and technical experts thus controlled the planning system. In the decentralisation reforms these powers were transferred to the lowest level of elected government, the 36,000 communes.

Local government is highly fragmented. The majority of communes have only a few hundred citizens and thus neither the geographical scope nor the resources to take on planning. Even after decentralisation most communes are therefore still reliant on the DDE. Thus despite the radical nature of the reforms there was some continuity as far as the smaller communes were concerned. In the cities and larger communes the decentralisation reforms can also be seen as the continuation of existing trends. Many planning decisions had effectively been transferred to local institutions before 1982. In four large cities, Lille, Bordeaux, Lyons and Strasbourg, the first *communautés urbaines* (CU) were established in 1966 for the purpose of co-ordinating planning and public works across the numerous communes which made up the conurbations. Elsewhere the larger communes or groups of communes had their own technical staff and in many urban areas Agences d'Urbanisme had been set up jointly with the state to undertake planning work. Throughout the 1960s and 1970s local politicians had progressively been integrated into urban planning (Biarez, 1989). In urban France the decentralisation reforms need to be seen in the context of the progressive growth of local power. However, as we shall see, the transfer of powers from state to local government has had profound impacts on urban planning in France, not only in terms of the ways in which local power has developed but also in the responses of central government to the challenges posed by the decentralisation of power.

The impetus for decentralisation came from the big cities and many of these cities had been in political opposition to the government before 1981. It was the mayor of one of the large cities, Gaston Defferre from Marseilles, who as a minister in the new socialist government pushed through the decentralisation reforms. The initial decentralisation law in March 1982 was discussed through three parliamentary sessions and was subject to nearly 4,500 amendments (Manceau, 1989). The Loi Defferre established the principle of independent local government. The transfer of powers was complicated. The *préfets* remained but with only residual powers to challenge the legality and probity of local decisions, not to determine them. The *préfets*' powers were distributed to different levels of government. Elected councils at regional, departmental and commune level all benefited from the transfer of decision-making power. Their administrations took on many of the tasks formerly carried out by the local services of the state. An attempt was made to associate different functions with different levels of local government. In so far as urban planning is concerned powers were transferred to regions and to communes.

PLANNING IN THE REGIONS

At the regional level the decentralisation reforms brought the most radical change with the creation of a new tier of elected government. Power was transferred from *préfet* to the president of newly constituted regional councils. The former *préfet* was reinstated with the responsibility of co-ordinating the state's local services, that is the branches of central ministries in the regions. The twenty-two regions have powers in relation to the production of regional strategic plans and significant economic development powers, including being able to offer direct subsidy to industry. The position in the Paris region is different. Regional planning power was retained by the *préfet*. The structure of local government in the region is relatively new and central government has historically been reluctant to devolve power in the capital region.

The decentralisation reforms reallocated responsibilities but did not change the structure of local government. The French regions are a long-standing administrative creation. The newly elected regional tier followed established administrative boundaries. The twenty-two planning regions were created in 1964 for economic planning purposes and the intervention of state at this level has steadily increased since the 1960s. In the early 1970s the government set up Economic and Social Committees in the regions, comprising government and private interests, and these were given a continuing advisory role after decentralisation. There is therefore some continuity in regional planning structures.

The first regional elections were held in 1986. The system of proportional representation has allowed minority parties into the regional chambers. For example, the president of Nord-Pas de Calais is a member of the minority Verts party; in Ile-de-France the controlling group is an alliance of government majority and the Ecologistes d'Ile-de-France. Regional elections have been dominated by national politics, and anti-government voting has meant that almost all regions are controlled by the right-wing parties.

Le Galès (1992) points out the various ways in which the new regional councils have taken up the challenge of regional government. There has been substantial growth in regional budgets. This growth reflects both the new tasks given to this level of government in the decentralisation reforms and the way in which some regions have taken advantage of increasing freedom to borrow money in order to intervene in the regional economy. Overall spending rose from 14 *milliards* francs in 1984 to 47 milliard in 1990 (Kukawka, 1993). The regions' mission is geared more towards intervention than managing services and regional finances are therefore directed towards investment rather than staff costs. Expenditure is one measure of the active role of the French regions. Enthusiasm for the regional-level government was also evident in the response of politicians in 1985 to changes to the law about multiple office holding. In France politicians typically hold more than one office; the majority of deputies in the national assembly are also local mayors. The reform in 1985 sought to restrict politicians to two main roles. Where a choice had to be made between *département* and region it was the regional role which tended to be more attractive.

The regions have promoted themselves with vigour and developed pro-European attitudes. Some have forged international alliances. For example Nord-Pas de Calais has formed a Euroregion with Brussels, Flanders, Wallonia and Kent, as have Midi-Pyrénées and Languedoc-Rousillon with Catalonia (Condamines, 1993). In some cases the European ambitions of the new regions have come into conflict with the existing European identities created in the big cities (Ville de Montpellier, 1990).

The large cities tend to dominate regional economies, and the political leaderships in city and region in some cases come from rival parties. The big cities have had to accommodate the new elected tier but in the case of Montpellier, for example, the mayor holds the view that the city is the driving force behind economic growth (*Le Point*, 1994a). Biarez (1993) argues that in reality the regions have proved weak in the face of the alliances between cities, *départements* and central government which have directed urban growth. Ironically the building of new regional administrative centres and council chambers has tended to enhance the economic and political power of the cities in which they are located, not of the regions themselves.

Relationships between city and regional governments vary and are not necessarily harmonious. An equally important issue is the regions' relationship to central government. The roles of state and region are formalised in a regularly negotiated plan of expenditure priorities. The *contrat de plan* co-ordinates most regional investment in co-operation with the state in relation to regular national plans. Each new region began this process of negotiating a contract for the period 1984–8. The *préfet* and president of the regional council are the principal parties in the negotiation, but other local governments in the region and state-controlled enterprises such as SNCF are also involved. The contract then sets out who among the various agencies will pay for what. The government sets overall priorities. For example, the second round of contracts from 1989–93 emphasised job-creating investments

(Chicoye, 1992). EU programmes also tend to be integrated within the contract. In the current contract for Poitou-Charente, for example, 20 per cent of the programmed expenditure is from the EU (*Le Monde*, 1994d).

The distribution of expenditure in the current round of contracts for 1994–8 favours the poorer regions. The biggest contributions by the state per head of population are allocated to Corsica and Nord-Pas de Calais. At the other end of the scale the state has reduced its contribution to Ile-de-France by 10 per cent compared to the 1988–93 contract. In most regions the state investments amount to about half the planned expenditure. The contract for Ile-de-France envisages 68 per cent coming from the region. This reflects the government's current concern to control the position of the dominant region within France. The Ile-de-France case also reveals some of the complexity of decision-making at this level. In Ile-de-France the region's position in the negotiations over the contract was determined by the balance of political power in the regional council. The UDF and RPR parties ensure general support for national policy but their combined strength does not constitute a majority. Support of other minority parties is required and this has come from the ecologists. The region's priorities therefore include both investment in new roads and in public transport. Last-minute deals over the level of public transport investment were struck in order to get the regional council's approval to the programme (*Le Monde*, 1994g). In the period leading up to signature of the contract other levels of local government lobbied for their own specific interests. For example the *département* of Seine et Marne argued through advertisements in the press that the government should recognise the wide intraregional diversity and in particular the unemployment problem of its part of the region in coming to decisions on new investment priorities.

In the decentralisation reforms the regional level was given new autonomous powers, but at the same time new contractual relationships between central and local government were created. The process of decision-making between state and region stresses negotiation rather than hierarchical control but it remains open to question whether the *contrat de plan* represents any more than the state's list of spending priorities. The *contrat de plan* gives the state considerable influence over the regional level of government. The state has always intervened directly to implement its planning priorities and continues to do so. Central planning produced the technopoles (Simmie, 1993), enterprise zones, investment in Disneyland, and regional aid to industry (the Prime d'Aménagement du Territoire) though this is now subject to strong EU direction. In addition the government has its own programme of decentralising government departments and public agencies from Paris to the provinces with consequent impacts on regional economies, and major transport investment, in the TGV network, is determined centrally. There is therefore strong central planning in key areas, coupled with negotiated contracts for infrastructure and investment in the regions. Decentralisation at regional level represents only a partial devolution of power.

Regional planning policy is strongly influenced from the centre. As the new regions were starting up in the mid-1980s so central government policy was changing. The coincidence of decentralisation with preparations for the European Single Market gave a particular emphasis to both decentralised and central policy-making. Both central government and the regions were concerned about their competitiveness in the new Europe (Chicoye, 1992). Central government policy was reviewed in the mid-1980s in the Guichard report (Guichard, 1986). The main idea was that whilst Paris could compete with rival European regions other parts of the country needed support to enable them to compete successfully. It was suggested that twelve urban areas should be identified where the state should concentrate its efforts. The government's central planning agency, DATAR, was given a new role in organising the drive towards European competitiveness. Political control and direction came from the ministerial Comité Interministériel pour l'Aménagement du Territoire (CIAT). A new planning mechanism – the *Charte d'Objectif* – was launched in 1991 and the largest cities were invited into a process of economic planning to be conducted by regional *préfets* with the aim of identifying European-scale projects in each city region. New investment in European projects was to be incorporated into the *contrats de plan*, and the *préfet* would co-ordinate the state's services in support of regional priorities. The significance of the *Chartes d'Objectif* is that they are a further example of the state's strategy of contractualising relationships with local government (Le Galès and Mawson, 1994). Decentralising power to regions has been accompanied by two central–regional contracts through which the state interacts with its new regions.

The policy review of the late 1980s brought forward some new ideas about regional structures. DATAR analysed France in its European context in terms of seven large areas of the country based on their proximity to the core of European economy and the coherence of regional economies. For example, the Arc Atlantique stretching along the western coast and including southern regions and Bretagne was defined by its distance from the Milan–Frankfurt–London economic axis and the danger this posed for its economic future. In addition to the simple geography of the threat of European competition there was concern about the ability of the relatively small regions to compete with larger European rivals. The reforms of the 1980s retained the existing structure of administrative regions and these did not necessarily represent coherent regional economies or have the political autonomy of regions in some other countries (Némery, 1993; Chicoye, 1992). The law on the Territorial Administration of the Republic in 1992 encouraged the idea of forming larger regions and setting up interregional bodies better able to respond to these issues. Since the late 1980s both government and political leaders in the regions around Paris have been thinking about co-operation. In 1992 DATAR produced a document on interregional co-operation – *Le Livre Blanc du Bassin Parisien* – which sought to set the growth of Ile-de-France in a framework of co-operation rather than competition with its neighbours. It was argued that the Bassin Parisien as a whole could match the largest and most successful German or Italian regions. The *Livre Blanc* represented central government's views. The presidents of the regional councils in the Bassin Parisien (Bourgogne, Centre, Champagne-Ardenne, Basse-Normandie, Haute Normandie, Pays de la Loire, Picardie and Ile-de-France) published their own ideas in the following year. The common themes of these two documents relate to the competitive European position of the superregion, environmental issues, and economic and social policies concerned with training the workforce and avoiding social divisions (Unal, 1994). A co-ordinated view of the Bassin Parisien was produced in May 1994 by the regional presidents and the Ministre de l'Intérieur (DATAR/CPPR, 1994). This *Charte du Bassin Parisien* reinforced existing views. The state and the eight regions subsequently signed a joint *contrat de plan*. There were, however, some critical voices to this accord. Environmentalists wanted greater control over regional growth and road traffic and Communist Party members regarded this new super-regional structure as a product of the Treaty of Maastricht, forcing an alien form of administration on France to the detriment of the needs of the component parts of the regions (*Le Monde*, 1994e). However, the significance of this interregional planning lies both in its response to the perception of European competition and in the application again of a contract and a high degree of integration between state and regional level.

The regional councils have taken up their new roles enthusiastically. The state has, however, developed new regional policies and developed new contractual mechanisms for working with the regional councils. We also need to look at regional power within the context of the other tiers of local government and state services at the level of the *département*.

Since the decentralisation reforms government has attempted to clarify the distribution of competencies but without complete success. For example, before the cantonal elections in March 1994 the Conseil Général du Nord encouraged voters to participate in the election of new councillors who would have competence in planning and economic development (Le Nord, 1994). All levels of local government have powers to act in the general interest of their areas and powers in relation to land and property. Despite the transfer of planning powers to the regional and communal levels of government the *départements* still play a role in planning and infrastructure matters. The *département* of Hauts-de-Seine in the Paris region has, for example, used its own development company to intervene actively in the property market and benefit from the property boom of the 1980s. The strong links between local and national politics through multiple office holding have also helped to preserve the roles of the presidents of the Conseils Généraux in the face of new regional-level powers (Mazey, 1993). Relationships between the tiers of sub-national government vary according to the nature of local political alliances.

As well as trying to clarify the roles of local government the state has also made a series of attempts to clarify the roles of its local services and *préfets*. Local state services were destabilised after decentralisation as powers were lost and staff transferred (Le Galès, 1992; *Le Monde*, 1994b). A new group of *sous-préfets* was given responsibility for economic development, the regional *préfet* was given a key role in developing the *Chartes d'Objectifs*. The principle of devolving decision-making to the local level was reaffirmed in the law on the Territorial Administration of the Republic in 1992, and it was hoped that with better co-ordinated and effective state services the *préfets* might be in a better bargaining position in relation to the powerful elected local leaders in the cities.

The process of decentralisation to regions has therefore had continuing impacts on the organisation of the state. New institutional processes have been developed, the state has sought co-operation through contractual arrangements and attempted to reorganise its own services in order to secure stable relationships with the regions.

URBAN PLANNING IN THE COMMUNES

In the decentralisation reforms the communes gained some powers to intervene in the local economy but most importantly they inherited comprehensive planning powers. The mayor of the commune has the power to issue building permits and to organise the production of the POS. Decentralisation created a highly fragmented planning system. This has led to marked differences in the ways in which local government relates to central government in planning matters. The majority of communes are small and rely upon the DDE to continue to provide planning services. Urban communes have been much more able to take advantage of their new powers.

There are of course substantial differences in the degree of financial autonomy enjoyed by small and large communes. The structure of local government finance was not changed by decentralisation. French local government has considerable power to levy local taxes but the quality of the tax base – the amount of commercial property, growing or declining population – obviously varies. Overall almost half of local government revenue comes from taxes. Local governments have four main taxes -the *taxe professionnelle* which is levied on businesses, a domestic property tax and taxes on developed and undeveloped land. Of these the *taxe professionnelle* is the most important, accounting for 46 per cent of tax income in 1990 (INSEE, 1993). Rates of tax are variable and central government intervenes to try to equalise local revenue. Since the new government of centre right parties came to power in 1993 local government finance has been under pressure. As central government reduced its block grants to local government local taxes increased sharply in 1993. Subsequently, local governments were less willing to increase taxes in the run-up to municipal elections in 1995. Each level is also keen to point to high spending elsewhere in local government.

The *taxe professionnelle* produces most income. The base for the tax is local business. Not surprisingly, in many communes the new powers in relation to economic development and planning have been used to attract new business and thus increase local tax revenues. The fragmented structure of local government has meant that competition for development and its revenue benefits have tended to obstruct co-operation between communes on development planning.

Just as the structure of local government finance was unchanged by the decentralisation reforms so democratic procedures remained the same. Subsequently some concessions have been made to improve the availabilty of information, but in general the majority party strictly controls the business of the commune (Borraz, 1994). In Paris, Lyons and Marseilles new political structures were created in the decentralisation reforms. Each *arrondissement* was given a local mayor and thus a new level of political debate. However, in Paris this represented a compromise after the idea of decentralisation of power to the *arrondissements* was rejected by the city council. Strong party control of the city council ensures that this new democratic arena has little impact.

The 'communal public sector'

The political context for local planning needs to be understood beyond the formal structure of local government. Lorrain (1991) and others have argued that the complexity of agencies at local level is a fundamental characteristic of French urban governance. Lorrain refers to the 'communal public sector' as the complex of public and semi-public agencies which deliver urban services. In addition to urban and suburban communes this includes intercommunal agencies, and those services managed by private companies, such as the large water companies, or by mixed companies employing both public and private capital. We need to look at each of these groups in turn.

Firstly, there has been growth in the numbers of intercommunal bodies. We noted earlier the *communautés urbaines* of the large cities. Intercommunal planning is often carried out by a joint Agence d'Urbanisme. In addition to these agencies, 200 *districts* have been set up by groups of communes to carry out joint functions. At the end of the 1980s there were also 14,500 Syndicats à Vocation Unique and 2,500 Syndicats à Vocation Multiple set up to jointly manage single functions or a range of communal functions (INSEE, 1993). Legislation has periodically attempted to both encourage co-operation and make procedures simpler. Fragmented government is seen as a barrier to enhancing the competitiveness of urban areas. The former socialist government introduced legislation in 1992 to encourage a new round of intercommunal co-operation. In each *département* a commission chaired by the *préfet* was set up to propose local structures. The law provided for the establishment in rural areas of *communautés de communes* and in urban areas with a population of over 20,000 *communautés de villes*. These new groupings would have specific responsibility for economic development and spatial policy. The legislation proposed that the whole of the product of *taxe professionnelle* could be allocated to the new grouping. In *communautés urbaines* and *districts* revenue sharing is also possible. The obvious benefit of financial co-operation is the avoidance of communal rivalry over attracting new development. The 1992 law proved popular in rural areas. During 1993, 240 *communautés* were set up. However, few proposed to co-operate in producing their POS (*Le Monde*, 1994a) indicating that new planning powers of the mayors were jealously guarded. In many cases the process of determining inter-communal structures in the departmental commission has proved difficult and the broader approach to planning which the reforms had hoped for has not been forthcoming (de Vos, 1993).

The second component of the 'communal public sector' is not public bodies but private and semi-private companies. Large companies such as *Lyonnaise des Eaux* and *Compagnie Générale des Eaux* have historically provided many urban services (Lorrain, 1991). The recent growth in private management of municipal services may in part be a response to financial restraints on local government. According to Lorrain (1992) it may also result from the financial power and expertise offered by the big groups, including the water companies, the construction company Bouygues, and the state controlled bank Caisse des Dépôts et Consignations. These companies are able to take development risks on behalf of the mayor. However, the comprehensive package offered by the big construction groups – the '*modèle ensemblier*' (Lorrain 1992) – has the effect of removing many decisions from the democratic arena. The objective of decentralisation to enhance local democracy is unlikely to be achieved if more decisions are privatised. The big utilities and construction companies dominate the building sector of the economy and have increasingly internationalised their operations (Drouet, 1994).

The use of mixed economy companies has also increased, especially in implementing urban policy. The Société d'Economie Mixte (SEM) has its origins earlier this century. The law was simplified in 1983 and Devès and Bizet (1991) argue that following decentralisation the use of SEMs increased significantly at both urban and regional level, an increase from 600 such companies in 1983 to 1,264 in 1993 (*Le Monde*, 1993). Normally the public partner has a majority shareholding in the company. Thus political control is retained while the company structure allows for greater operational flexibility, free from the bureaucratic rules of the town hall. The SEM may be set up by the commune or by a private partner. For example, SCET, a subsidiary of the Caisse des Dépôts, has 250 SEMs engaged in development projects (SCET, 1994).

Public and private interests are also represented through the Chambres de Commerce et d'Industrie – 143 have traditionally been involved in economic development. Since decentralisation their roles have changed as different communes or groupings have sought to develop their own links with business and their own local economic development strategies. In some areas local government pursues an independent line, elsewhere there is greater co-operation, for example in the Comités d'Expansion which promote new development. There are numerous local promotional agencies – see, for example, the case of Toulouse (Barlow, 1995).

The context for planning in urban areas is therefore a complex of public and private bodies. However, within this framework of urban governance the power of the mayor of the commune or communal grouping should not be underestimated. The mayor, once elected by the municipal council, has executive powers. The mayor may well represent the commune in mixed companies. Through multiple office-holding many mayors will hold power elsewhere, in parliament, region or *département*, where negotiation on behalf of the commune can take place. The successful mayor can only enhance his or her position and the personal power that goes with it by operating in a number of arenas. These links between levels can also be seen as a constraint on the exercise of local power by committing the mayor to policy decisions taken at other levels. Power arising from political positions also interacts with party politics. On the one hand a strong local base makes the mayor/deputy less easily controlled by party chiefs as in the case of Michel Noir in Lyons who renounced party affiliation yet secured local re-election. On the other hand a strong local base can be a pathway to office in the political parties (see Knapp and Le Galès, 1993).

The internal organisation of the *mairie* is well described by Dion (1986) and Borraz (1994). Since decentralisation mayors have recruited more high-quality staff as more status has been attached to local administrative careers. In SEMs and other bodies national controls on salaries and training are less rigid and high-quality officials have been bought in. Local political advisers may also find themselves moved into other levels of government as their boss moves upwards. Knapp and Le Galès (1993), for example, refer to the transfer of members of Chirac's Paris *cabinet* into national government in the mid-1980s.

The structures of local political leadership therefore provide an important context for planning. Political ideology is also important. The decentralisation reforms coincided with the rise of liberal ideology. Many mayors became quickly associated with economic development and entrepreneurialism (Lorrain, 1994; Borraz, 1994). Borraz describes some ideal types of mayor in the post-decentralisation local government. Firstly, there is the 'entrepreneurial' mayor characterised by mayor Bousquet in Nîmes pursing a pro-growth, image-conscious strategy and employing private sector management methods. We could add the case of Alain Carignon in Grenoble as an example of this type (Novarina, 1994). In contrast, in other cities the mayor may be more interested in integrating conflicting economic and social interests, spending less time on promoting the city and more on arbitrating between the many deputy mayors. Many cases, Borraz (1994) argues, will present a mixture of these types, and the style of city government may change over time. For example, in Rennes and Montpellier policy changed direction in the 1980s as the mayors embarked on expensive development strategies. Such development strategies are high risk and most mayors temper dynamism with the seeking of a wide range of political support.

Styles of urban planning

At the municipal level, then, there is a complex pattern of agencies with, since the 1980s, much greater private sector involvement. Local responses to decentralisation also depend on how the role of mayor is developed. In this context at the end of the 1980s new styles of planning emerged to replace the former hierarchical, centralised system. Let us briefly look at some examples.

In the mid-1980s the Communauté Urbaine de Lyon (Courly) started to discuss a new long-term strategy for the city – 'Lyon 2010'. The documents concentrated on producing a new image of the conurbation in its European context and sought to create a consensus of local views (Biarez, 1990). As an emerging strategic planning document it differed from previous plans which were based on the application of objective standards. The document developed a new language of strategic planning and focused on economic goals and the views of local elites (Biarez, 1990). Padioleau and Demesteere (1992) trace the complex relationships between organisations in the production of this type of strategic planning. The mobilisation of local actors is a central part of the strategy. The Communauté Urbaine de Lyon and the promotional agency, Conférence de la Région Urbaine de Lyon, jointly controlled by the Courly and the CCI, provided the vehicles for new thinking. A grouping of the Courly and sixteen adjacent communes undertook studies, and hundreds of meetings of various political and professional interests were held. The dynamic, pro-growth image of the city became associated with one of the mayor's deputies. The incumbent mayor seemed out of step with the strategy and in the 1987 elections he was replaced by the deputy, Michel Noir, who was better able to take on the image of '*maire-stratège*' (Padioleau and Demesteere, 1992). The new style of planning legitimised the mayor's position at the hub of local economic and political interests. There was a coherence of view between political and economic (the company Rhône Poulenc being among the most influential) leaders, the university and other parties to the growth strategy (El Guedj, 1993a). The 'Lyon 2010' work was completed in 1992 and subsequent revision of the POS sought to secure the land supply to accommodate new development.

The Lyons case presents a particular style of strategy making. It also illustrates a general concern about the roles of the indirectly elected *communautés urbaines* and other non-elected planning agencies. Democratic accountability is obscured in this type of planning. A former deputy mayor of Lyons described the tension between the big budget but low accountability of the *communauté urbaine* as a *'bombe institutionnelle'* in the system of city governance (El Guedj, 1993b). This style of planning has been termed, in *franglais*, *'planning stratégique'* (Querrien, 1992). The resources and investment which cities have been trying to attract can either be mainly private, as in the Lyons case, or public as in Rennes (Le Galès, 1993).

In addition to this strategic type of planning another planning style can be identified at the level of the development project. Verpraet (1992) describes the process of planning associated with a range of large commercial schemes where the property market was strong or where the coming of the TGV had enhanced land values. In these cases the formal planning system became subservient to the demands of the project managers and most decision-making was undertaken by private or mixed companies.

The key role for the public sector in these projects is promotion and the creation of new images of their cities. Investment in opera houses and other cultural activities improves city image and encourages investor confidence in development strategies. In Montpellier, for example, cultural investment has been as significant as economic investment in changing the image of the city (Lenfant-Valerie, 1993). An important part of image building in Montpellier has been the association of an internationally famous architect, Ricardo Bofil, with the mayor's projects. Other mayors, for example Bousquet in Nîmes and Mauroy in Lille, have secured their own prestigious architects (Norman Foster, Rem Koolhass) (see *Le Monde*, 1992).

Image and marketing have been important parts of the strategies of both liberal and socialist mayors. Liberal mayors have used planning powers in particular ways in managing their cities. For example in Grenoble the POS was revised to remove some constraints on development, and this deregulation was accompanied by a strategy of privatising some services as well as delegating to SEMs the running of urban projects. Novarina (1994) argues that the Grenoble strategy oscillated between liberalism and intervention. In some areas such as social housing the joint city–state contract requires intervention and the regime maintains an interest in city-wide planning. In addition Novarina refers to the 'neopopulism' of the mayor seeking legitimacy in the fragmented system of local governance through referendums and other direct appeals to the public. (Some of the mayor's activities in seeking political support resulted in imprisonment on charges of corruption.)

In the context of changing intergovernmental relationships there is evidence therefore of a range of new approaches to planning. These include the strategic planning of 'Lyon 2010', project planning relying on private sector interests, the growing importance of the public sector in creating new images and, finally, the complex contribution of the planning system to liberal city government. One of the emerging problems for central government at the end of the 1980s was how to control the ambitions of the cities and redress the balance in favour of both national planning and social objectives. In part the problem of controlling the cities arose from a separation of economic development planning from wider urban policies.

URBAN POLICY

The pursuit of economic development and property-led growth has preoccupied many French cities since decentralisation. The local search for economic growth is linked to national priorities and the objectives of DATAR in reshaping regional competitiveness. The ambitions of the mayors have, however, made development politics a crucial local issue. The transfer of powers in the decentralisation reforms has obviously aided local policy initiative. However, responsibility for the social, and in particular the housing, problems of cities was not transferred to local government. Indeed, the local governments which had witnessed the state build the vast suburban estates of the 1960s and 1970s with little consultation and often against their wishes did not want responsibility for the consequences of central policy. Since the early 1980s the politics of economic development have become separated from the *'politique de la ville'*. Central government has continued to search for wider solutions to urban problems in a series of interventions since the early 1980s. The substance of urban policy is important but so also are the new mechanisms through which central government has sought to work with local government towards new solutions to urban problems.

The recent history of urban policy in France is well documented (Power, 1993; Provan, 1994). Since the early 1980s successive policies have attempted to target the most deprived areas, and urban riots have periodically brought forth new initiatives. Body-Gendrot (1994) argues that riots aim to achieve political response rather than expressing a politics of disengagement with society. The objectives of urban policy are concerned with social order and the integration of all citizens, including, importantly, minority groups into French society.

By 1987 120 priority projects had been established through the programme Développement Social des Quartiers (DSQ). However problems remained about co-ordinating state activities and tackling the full range of social and economic problems. Towards the end of the 1980s the government therefore made a series of new attempts to solve urban problems, including the creation of a new ministry for urban affairs. One of the last initiatives of the socialist government was to attempt a transfer of resources from rich to poor communes and to prevent the concentration of social problems. Special arrangements were made for transfers from rich to poor in the Paris region. Overall the amounts of money transferred were modest despite the huge political dispute the reforms provoked between the government and the mayor of Paris (Keating and Midwinter, 1994). The Loi d'Orientation sur la Ville in 1991 aimed to create a better balance of land uses, to prevent the formation of ghettos by requiring developers to include a proportion of social housing in new developments or pay a percentage tax to pay for social housing to be built elsewhere. The *préfet* was given powers to force local government to allocate land for social housing. In the event the property crisis dried up the supply of new projects which could have contributed to the policy objectives.

In 1988 government attempted to resolve the continuing problem of co-ordinating its activities by setting up the Délégation Interministérielle à la Ville (DIV) which grouped together various programmes and projects in the ten ministries involved in urban policy. At the same time a new experimental programme for central–local co-operation was begun. The *contrat de ville* programme was intended to be a multi-agency approach to urban problems. In the new contracts the state would commit itself to three years of funding and as the largest contributor to urban programmes the other partners, the regions and communes, would have a more certain environment in which to spend their own money (see Booth and Green; 1993, Le Galès and Mawson, 1994). The *préfet* was responsible for co-ordinating the elements of the contract and the commitment of both local and central government departments. The contracts included housing, training programmes, education and cultural as well as urban redevelopment projects in the framework of an overall plan. The new government continued with the *contrat de ville* programme. Shortly after coming to power in March 1993 the Balladur government held a parliamentary debate on urban policy. The new minister's view was that little would change, and existing programmes would be maintained or increased (*La Lettre de Matignon,* 1994). The sum of 9.5 billion francs was dedicated to *contrats de ville*. In 1995 the new government continued to prioritise policies to relieve social and economic problems of large housing estates and working-class suburbs.

Following the initial experiment with thirteen *contrats de ville* the next round incorporated the DSQ and this previously housing related programme was broadened to include wider urban issues. This had the effect of increasing the number of targeted areas. About 1,500 areas in difficulty were identified in the 210 contracts. The contracts in the *département* of Val de Marne for example identify forty-eight *quartiers* compared with twenty-two in the DSQ programme (*Le Monde*, 1994i). Urban expenditure was thus spread more thinly. One justification for spending beyond the problem housing estates was that investment is required to integrate poor housing areas into the wider urban fabric and that urban policy should treat the whole city.

Since the early 1980s urban policy has continually sought solutions to urban problems. Mayors complain that not enough money is being spent.

The state is still searching for a co-ordinated approach. Le Galès (1992) regards the DSQ programme as a relative failure. The *contrats de ville* have demonstrated better co-ordination between government departments, but the process of negotiation involves a narrow group of local political leaders and government officials. There has been little involvement by community organisations.

The important feature of the way in which urban policy has developed is the negotiation of contracts. Government developed a similar contractual relationship with sub-national governments in the *contrats de plan.* The mutual commitment to negotiation underpins urban policy. In addition to the introduction of this contractual process the government sought to extend its authority by giving new *sous-préfets* responsibility for co-ordinating urban policy, and the local *préfet* became a key figure in co-ordinating the parties to the *contrats de ville*. There was potentially a high degree of integration in urban policy. However, the new roles of the *préfets* were not always clearly understood and the negotiation of *contrats de ville* demonstrated considerable variation between cities. The organisation of urban policy in the post-decentralisation state has yet to be clarified. Coupled with these organisational issues is the problem of the separation of economic from social policy. While the cities negotiated contracts with the state they were also pursuing pro-growth strategies. As well as contributing to the relative lack of coherence of urban policy the economic policy orientation of many cities also presented problems for the legal and technical coherence of the planning system.

PROBLEMS WITH POST-DECENTRALISATION PLANNING

Since the mid-1980s the context for urban development has been dominated by the emergence of new approaches to urban policy and by local entrepreneurialism. Whatever the intentions, relationships between the levels of government were not fixed in the 1980s and new legislation and working practices have continued to emerge in the search for stability in the overall system of urban planning and development. Over the last ten years a number of general concerns have arisen about the ways in which the planning system as a whole has changed. One such concern is about the increased flexibility of planning controls. The POS sets out precise zones and other controls which distinguish between legal and illegal development. The mayor's discretion in issuing building permits is limited. Flexibility in the system is found through the process of producing and amending the POS, and changes to POS procedures in 1983 made it easier for a commune to do this. Once the commune had decided to review the POS the existing plan would cease to have legal force. Development not in conformity with the old plan could then be encouraged and sanctioned in subsequent revision. This flexible plan-making became widely used (Acosta and Renard, 1991) and a cause of general concern. Flexible planning and the potential for extracting revenue from the *taxe professionnelle* encouraged pro-development attitudes in the town halls, and disquiet about the future of the planning system. Such concerns about the post-decentralisation system prompted a review of planning law by the Conseil d'Etat (Conseil d'Etat, 1992). The review expressed fears about declining national standards of administration and about the undermining of the traditional role of the French state in securing equality of treatment across the country. Other critical voices pointed to the loss of objective administration in the new quasi-independent local level and the creation of a *'république des fiefs'* (Mény, 1992).

Renard (1994) argues that decentralisation radically changed the practice of planning. He refers to the Conseil d'Etat's review of public administration and its concern about *'urbanisme clandestin'*, the closed negotiations over development projects and the discrediting of the state because of the variable application of rules. Before decentralisation, the *Code d'Urbanisme* reflected many of the practices and techniques of central planning which the *préfets* and state services put into effect. With decentralisation such a standardised, quantitative approach to planning became unworkable. Renard argues that administration needs to become more flexible allowing local interpretations of rules. Such views bring criticism from the state bureaucracy and provoke fears among those who hold traditional values about the nature of the French state. There was a fear that government would become dominated by the judges. One of the effects of an increasing number of legal challenges was that the courts became more involved in the substance and not just the process of decisions (Lamorlette and Demoureaux, 1994). Increased flexibility also reduced the certainty for public and private sectors alike that projects would go ahead unhindered by legal challenge.

The government tried to adapt the legal system to current circumstances. A new planning law in 1993 sought to clarify some procedures. However, it also had the objective of removing some administrative constraints on development projects which, it was hoped, would help the moribund property market. In addition the legislation retrospectively approved some illegal developments undertaken by individual communes. Because of its multiple intentions the law failed to concentrate on clarifying the uncertainties of post-decentralisation planning practice (*Libération*, 1993).

Related to this general concern with legal underpinning of planning is the wider issue of corruption in the system. Le Galès (1992) points out that corruption did not appear to have increased following decentralisation and may not necessarily be any more a local than a national problem. Multiple office holding makes it difficult to identify local or national culpability. During 1994 corruption did become a major national issue with malpractice exposed at all levels of society. The much publicised lists of the guilty included mayors and local councillors. Mény argued that, in contrast to Italy, it was local political leadership that had been penetrated by corruption and that the confusion of roles arising from multiple office holding and mixing public and private offices may have been responsible for the increase in the problem (Mény, 1994).

There was concern about the legal credibility and probity of the system of planning. There was a parallel concern about the quality of local democracy. Lorrain (1994) identifies serious problems arising from close personal relationships between business and local political figures. Decision-making tends to be closed because, it was argued, if planning projects were opened up to public scrutiny then all sorts of opposition groups would be encouraged, information would get out to speculators and rival projects and political opponents would benefit from the information. It was also suggested that the majority of the population may not be interested anyway. Given this reasoning it is not surprising that secretive, closed relationships developed. The loss to local democracy was offset by a gain in effectiveness, and political leaders chose to be judged on results rather than on the openness of procedures. The excluded citizens could, according to Lorrain, either do nothing or mount legal challenges after decisions have been taken. The weak property market of the early 1990s posed problems for this form of urban planning. If developments did not go ahead or were not successful then the willingness of citizens to acquiesce could be lost and a crisis posed for local democracy.

THE RE-EMERGENCE OF NATIONAL PLANNING

Interrelated concerns have been expressed about the legal basis of planning and about the accountability of French '*urbanisme*'. Many other unresolved tensions within the French planning system were set out for discussion in a 'national debate' on '*aménagement du territoire*' launched in July 1993 soon after the new government came to power. The issues raised (DATAR, 1993) included the new economic geography of France, the balance of economic growth between Paris and provincial France, the relations between large cities and their regions, the competitiveness of France in the Single Market, and the division of responsibilities within the state for economic development and growth management. Current planning issues therefore include both local concerns about the operation of the system and broader national concerns about future national, regional and urban development.

In addition to the substantive issues of regional development there were other reasons for launching a national debate. The presidential election was only two years away. To secure a majority for an RPR candidate votes are needed from across the country. A policy emphasis on the future of smaller towns and rural areas therefore made political sense. In addition, the RPR had a strong Paris base. The Ile-de-France region had one-sixth of the French electorate but one-third of RPR members (Knapp and Le Galès, 1993), and party leaders with hopes of the presidency had strong ties to the region. This perceived bias towards the capital region could be balanced therefore by a policy which aimed to restrict the growth of Ile-de-France. The government's debate was launched in the country, in Mende, at a meeting of the cabinet, eight of whose members were from Ile de France. Subsequently the Prime Minister and Minister for Planning toured regional centres to host debates on the planning of France. The Minister preferred to see the debate in terms of the military concept of *reconquête du territoire*. The regional councils submitted their views to government during 1993 and 1994. One consistent line of criticism was against DATAR's grouping of the regions into seven planning areas. Fears were also expressed about the government's belief that *préfets* should have an enhanced role in planning. Ten years after the decentralisation reforms the major issues of state organisation were therefore being opened up. Generally the mayors, left and right, of the big cities were sceptical. In Nice the president of the *conseil général* was angry that his city was not mentioned in DATAR's document (*Le Point*, 1994b). In Lille the mayor did not like the cover of the report which pictured an idyllic village square, and the contents which failed to recognise the important contribution of the big cities towards regional development (*Le Point*, 1994a).

The debate concluded in May 1994 with the publication of a report of comments received, a discussion of issues, outline plans for the regions (DATAR, 1994) and details of a parliamentary bill to introduce a set of reforms of the system. After six days of debate the 'Projet de Loi Relatif à l'Aménagement et au Développement du Territoire' was adopted on first reading in July 1994. The introduction of '*développement*' into the title responded to demands from the cities who wanted more than just to be planned. The law contained eighty-eight articles and DATAR suggested up to sixty supplementary directives to develop its proposals. The scale of this new legislation was on a par with the decentralisation laws. The parliamentary process was contentious, with a large number of amendments (150 in the Senate) reflecting the local and regional interests represented by the deputies and senators. The law proposed the creation of a Conseil National de l'Aménagement et du Développement du Territoire which would produce a national plan within a year. Each region would hold a regional conference jointly managed by the *préfet* and president of the regional council, and these conferences would produce regional schemes. National planning aimed at strengthening France's position in Europe was less contentious than other parts of the proposed law. The other big issues raised by the debate and included in the law were all unresolved and decisions delayed until after the 1995 presidential elections.

The first of the contentious issues was the clarification of competencies between local government levels. The organisation of state services and role of the *préfet* were also to be clarified. It was proposed that each of the seven large regional areas could be divided up into 350–450 *pays* which would have some economic coherence and form the basis for economic development planning. Each *pays* would be looked after by a *préfet* who would encourage intercommunal co-operation. Understandably the *départements* were concerned about the impact of such a reform on their role and all levels of local government were concerned about growing prefectural powers.

The proposals for financial reform of local government were even more contentious. A separate law on finance was proposed and decisions delayed for a year. The proposals for equalisation of taxes between communes and the distribution of different taxes between the levels of government provoked widespread objections. At present 67 per cent of the *taxe professionnelle* is kept by the communes, 27 per cent goes to *départements* and 7 per cent to regions. The minister favoured allocating the whole of the tax to the *départements*. Not surprisingly the mayors of the big cities objected. Existing arrangements for transferring resources from rich to poor communes have only redistributed small amounts of money. For example, in Ile-de-France in 1990/1 only 547 million francs were transferred from fifty-two richer to ninety-four poorer communes (Marcou, 1994). However, despite the attempt to develop a much more effective system at the end of the debate on planning the minister was forced to conclude that while everyone agreed it should be done nobody knew how to achieve it (*Le Monde*, 1994f). However, there was agreement on the reduction of government grant to Ile-de-France.

The debate on planning promised to tackle many of the outstanding post-decentralisation issues. In the event little was decided. Reform of the system proved highly contentious. The proposed national planning strategy was less of an issue. The government and most of the country were agreed on the need to control the development of Ile-de-France. During the period of the national debate the new *contrat de plan* for Ile-de-France was being negotiated and the regional *préfet* produced a new regional plan. The *Schéma Directeur* scaled down population and employment targets, reduced the amount of land to be developed and reduced the region's share of university students. The plan displayed restraint, though with marginal adjustment to previous targets. The future of Ile-de-France was also set in the new context of the Bassin Parisien and planned interregional growth. There was the appearance at least that the government's national planning priorities were working even if changing the system seemed intractable.

Restraining Ile-de-France was important, both for national party political reasons and to appease the environmentalist lobby in the region – particularly the ecologist members of the regional council. The appearance of restraint on Ile-de-France may not match the reality. Large infrastructure investments were planned – a TGV station at the office centre of La Défense, underground motorways (appealing to both motoring and environmental lobbies in the region), a new private university, and a new national football stadium. There was an unresolved tension therefore between continued investment in projects to enhance the international prestige of the capital region and the desire to diffuse some of the Paris–province rivalry at a politically important moment for the governing party.

ISSUES IN FRENCH URBAN PLANNING

Many of the fundamental issues concerning post-decentralisation national, regional and local planning have yet to be resolved. The national debate during 1993/4 raised questions about overall national co-ordination and roles within the system. Other concerns have been raised about the changing practice of planning on the ground and the loss of some of the perceived virtues of the formerly centralised system such as equality of treatment and the consistent application of rules. In the last few years substantial pieces of legislation on administration and planning have been passed. New forms of contractual relationship have been introduced to integrate the levels of government. The government has attempted to strengthen the local influence of its *préfets*. The context for urban planning therefore is one of continual change.

The entrepreneurialism of communes and intercommunal groupings in the cities has had a strong impact on urban planning. The use of SEMs has increased, with the effect in some cases of fragmenting responsibility among these quasi-private agencies. The big construction groups have also increased their participation in planning and development projects. Formal plans have been adapted to these new circumstances. The mayors play important co-ordinating roles in the new styles of urban planning.

Decentralisation of planning powers has not, however, created an unchecked localism in matters of urban development. Since the Defferre reforms new relationships within the public sector have developed. We have examined the various forms of contractual relationship – *contrat de plan, Charte d'Objectif, contrat de ville* – through which expenditures are programmed. The roles of *préfets* have been developed to co-ordinate economic development and in negotiating the *contrats de ville.* The proposed national planning framework will be used to co-ordinate the activities of different levels of government and may attempt to restrain the independent growth of the large cities. Taken together all of these institutional changes represent a fundamental reworking of relationships within the state. Biarez (1994) argues that the post-decentralisation state has reasserted its power by setting various technical norms and techniques. Contractualisation has been an important part of that process. These institutional changes have taken place within the state, and urban planning in France remains a public sector activity. The construction groups and other private interests have entered more into local economic development but DATAR, the *préfets* and the mayors still exert, sometimes co-ordinated, influence over development decisions and the public expenditures which support development projects.

In the contractual processes between central and local government in relation to urban policy, private sector interests are virtually excluded. There has been very little involvement of private or voluntary sectors in the *contrats de ville* (Parkinson and Le Galès, 1994). Since the early 1980s urban planning in France has adapted to its new political and administrative context. Planning at national, regional and local levels is, however, still being reviewed as the form of central–local relationships within the French state continues to develop.

In the next chapter some urban planning and development projects are examined in the context of these overall trends in the politics of urban planning in France. The choice of cases reflects the main issues discussed. The first case examines the process, style and consequences of large-scale commercial property development undertaken by the city of Paris. The second case, in the inner suburbs of Paris, examines problems of inter-communal co-operation in urban regeneration and relationships between local aspirations, urban policy and national infrastructure investment. The third example comes from provincial France and examines the particular forms of intergovernmental and public–private relationships shaping the development of Lille.

8

FRENCH CASE STUDIES

INTRODUCTION

In this chapter the examination of French urban planning is continued through looking at three case studies. The cases have a common focus on large urban redevelopment projects and their institutional contexts. Thus the replanning of parts of the city is examined to explore how the overall trends identified in Chapter 7 are reflected in the specific cases. The first case study analyses the largest redevelopment project managed by the city of Paris. The case offers an understanding of urban redevelopment in the context of strong local government and a strong development market. In the second case the land and property market is weaker and a more fragmented local government has less ability to control the type and pace of development. The case study describes attempts at local coalition building in Saint-Denis in the northern suburbs of Paris and the impacts of the decision of national government to take over part of the area for the construction of the new national sports stadium. It also examines the degree of integration of national and local urban policy objectives in the regeneration of this part of the capital region. The third case explores the planning and development process behind the construction of the Euralille project in north-eastern France. The case reveals strong, entrepreneurial local government in the particular political circumstances of a provincial conurbation.

There are similarities between the cases in terms of the role of plans and planning and in the institutional forms through which planning and development projects are tackled. Each reveals aspects of the French institutional context, but also shows the considerable variation between cases generated by the specific combinations of local forces.

SEINE RIVE GAUCHE: CITY REGIME AND INTERNATIONAL COMPETITION

Proposals for redevelopment of the area along the banks of the Seine between Austerlitz station and the *boulevard péripherique* were prepared initially in the

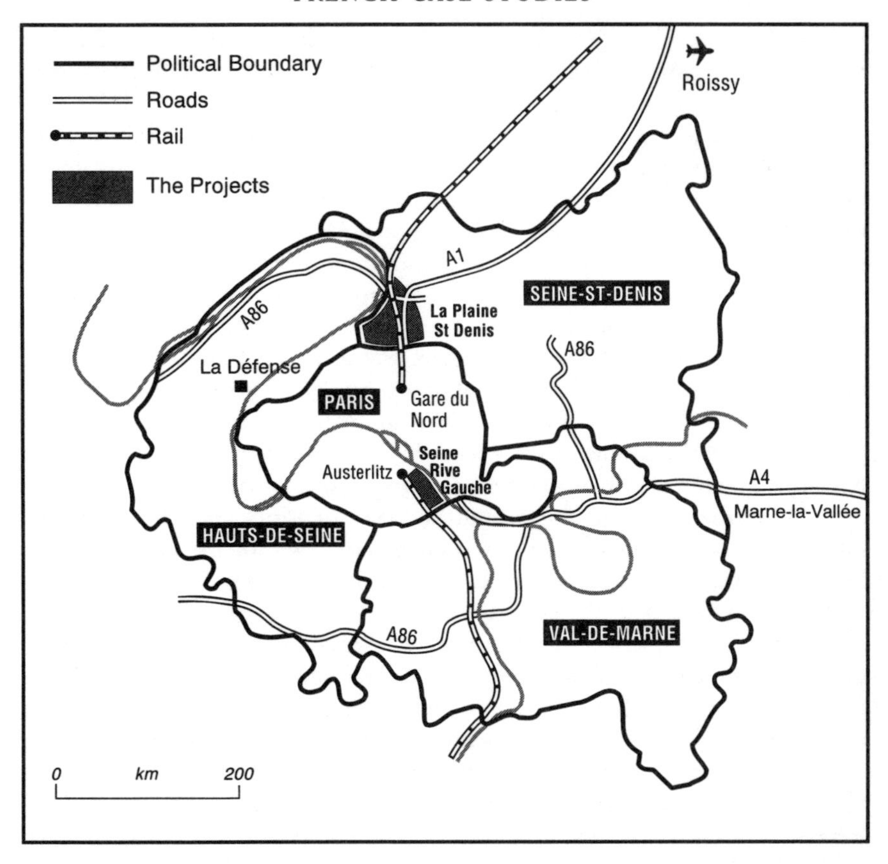

Figure 8.1 Paris

1970s. Savitch (1988) describes the early discussions between the *préfet* and city council over the future of south-eastern Paris and the subsequent debates within the city council in which the principle of allowing office development in the area was accepted. It was argued that because there was likely to be pressure for office development in the future, the state should take the initiative and seek to control the property market.

Local government structures were reorganised in the late 1970s and in 1977 Paris elected its first mayor under the new system. Jacques Chirac supported a comprehensive approach to replanning eastern Paris (*Le Débat*, 1994), where his political support was weak and which housed a greater part of the working class population of the city. The city planning agency, Atelier Parisien d'Urbanisme (APUR) produced studies for redevelopment projects stretching from La Villette in the north-east to Tolbiac in the south-east. In 1983 the city council's document – *Plan Programme de l'Est de Paris* –

proposed large-scale redevelopment including offices on the south bank of the river. Detailed planning was undertaken by APUR in the late 1980s with offices as the dominant land use (Delluc, 1994).

Throughout the late 1970s and the 1980s the city gradually implemented its renewal and regeneration plans for the east of Paris. The key planning mechanism through which redevelopment proceeded was the declaration of *Zones d'Aménagement Concertés* (ZAC). The ZAC procedure combines comprehensive planning of land uses with a programme of infrastructure development (Acosta and Renard, 1991). The planning authority has to produce a *Plan d'Aménagement du Zone* (PAZ) which sets out land uses and has the effect of replacing the POS. The procedures for the PAZ are similar to those for adopting a POS and include holding a public inquiry. The Paris ZACs managed by APUR have imposed a uniformity of architectural style (Dumont, 1994) and one of the effects of the redevelopment of eastern Paris has been a gradual loss of low rent housing and manual jobs (Carpenter *et al.*, 1994). Redevelopment has been at lower densities and cheap housing lost through development has not been replaced within the city. Residents have not always had the right to be rehoused in the city.

Seine Rive Gauche is the last stage in the redevelopment project for eastern Paris. The PAZ for Seine Rive Gauche outlined a large-scale development of over 130 hectares dominated by office development. The scale of office development reflected the original desire of the city council to manage the market but also a more recent objective of extending the commercial centre of Paris in order to keep the city competitive with its European rivals (Technopolis International, 1993). European and international competition became a dominant theme in city policy at the end of the 1980s. French financial markets are dominated by Paris, and Sassen (1994) argues that the city could not avoid the international competition between financial centres.

Seine Rive Gauche figures strongly in the city's international marketing, and, in turn, it is claimed that Seine Rive Gauche will replicate the special qualities of all Paris office locations:

> In keeping with the desire of Paris City Hall and the characteristics of the Parisian landscape, the development of this prestigious quality project will respect the unique diversity of the existing Paris office real estate area in which each Paris business centre is near one of the pleasurable and well-known city landmarks.
>
> (Mairie de Paris, 1993, p.15).

A further objective of the mayor and city council was to create a development project to rival the President's '*grands projets*' in the city (Robert, 1994). The monumental and international role of the city is an integral part of the image which the mayor sought to create in the capital (see Haegel, 1994).

The international role of this part of the city is reaffirmed in the strategic plan for Ile-de-France published in 1994 (DREIF, 1994). At the regional scale Seine Rive Gauche fits into an axis of development stretching from Marne-la-Vallée through the centre of Paris and continuing through La Défense. Whilst, as we saw in Chapter 7, the government have been debating the control of the growth of the Paris region the major development opportunities in Paris have not been restricted. La Défense, the central Paris office core, and Seine Rive Gauche continue to provide the international competitiveness which is at the heart of the regional strategy.

The PAZ for Seine Rive Gauche approved by the city council in 1991 included 900,000 square metres of offices, making a substantial contribution to the office stock and the capacity of the city to attract international business. APUR also wanted to mix this large office element with other uses to create a better balanced new quarter of the city: 300,000 square metres of housing were proposed, two-thirds of which would attract state housing subsidies (SEMAPA, 1993). The new population was estimated to reach 15,000; 50,000 jobs were to be created and a large student population was to be generated by new university buildings. The site is to be serviced by a new public transport interchange linking metro and Réseau Express Régional (RER) services and these projects are programmed in the *contrat de plan* for Ile-de-France. Austerlitz station remains and the greater part of the development site is created by decking over the railway track.

At the centre of the site is the new national library. The library was a late addition to the planning of the south bank and was planned independently by advisers to the President. Other locations, including locations in suburban Paris, had been examined, but the city council eventually persuaded the President to accept a site in Paris within the south-east sector. The benefits for the city's plans were that the new library would bring an early start to development activity and secure the new transport links. This would generate developer confidence in the new quarter emerging among the vast railway lands of this part of the city. The PAZ therefore set out a mixture of land uses, with office development concentrated along a central boulevard – the Avenue de France – running the length of the site.

In much of the redevelopment of eastern Paris, after the land-use plans for the ZACs had been agreed by the city council, implementation was handed over to a development agency, one of the SEMs controlled by the city. This transfer from local government to development agency can include development finance, the commune's powers of expropriating land, and the right to buy any land coming on to the market. The development agency therefore usually inherits substantial powers. Implementation of the PAZ for Seine Rive Gauche was passed to the Société d'Economie Mixte d'Aménagement de Paris (SEMAPA). The city is the majority shareholder in SEMAPA and the city council acts like a holding company in relation to its SEMs which are responsible for development projects around the capital. Implementation of the city's plans is thus fragmented, but political control of the SEMs ensures their ultimate responsibility to the city council. The property recession has forced the city of Paris to merge some of its SEMs in order to keep them under tighter political control.

The city has 57 per cent of the shares in SEMAPA and the Société National des Chemins de Fer (SNCF) is the next largest stakeholder with 20 per cent (SEMAPA, 1993). Almost half of the *ZAC* is railway land. *SNCF* transferred land to SEMAPA and deferred payment until completion of the development. By taking a share in the SEM they have managerial influence over the project without investing any money in the substantial engineering and infrastructure works required and expect a substantial profit in the long term. Other smaller shareholders in SEMAPA include the regional council and the state.

The task for SEMAPA was to provide development sites. The cost of the development was estimated at 25 milliard francs (SEMAPA, 1993). The cost of decking over the railway and other engineering works accounts for 10 milliard. SEMAPA have borrowed 800 million francs to service existing contracts. The city guarantees 80 per cent of the borrowing (SEMAPA, 1993). A group of banks have offered loans up to 1.7 milliard francs to be drawn before 1999 with a five-year repayment period. The huge costs of the development will be recouped by the sale of development rights, especially for offices. However, initially there was a large debt to be managed.

The proposed completion of the library in 1996 provided the opportunity to start the first phase of development and attach it to the library development. A social housing scheme for 350 subsidised flats was started in 1994. A small office development will also be built in this location together with open spaces and other facilities.

The development around the library marked the start of the larger project, but it could be some years before other parts of the project are completed. Work on decking over the railway (30 hectares of podium) was delayed until late 1995. This expensive project may wait until there are signs of recovery in the office market. Potential investors (including Société Générale, the regional council, the university, the Ministry of the Interior) withdrew their original interest in taking office space in Seine Rive Gauche. The bank, Société Générale, preferred to develop two office towers at La Défense rather than locate in south-eastern Paris. The property crisis had a damaging effect on the development. The detailed planning was undertaken during the late 1980s property boom and since then the market slumped. Seine Rive Gauche is far from the prime office location to the west of the city centre and rival office centres in the western suburbs, at La Défense and Roissy, have large supplies of office space. The property agents Jones Lang Wootton suggest it may be the end of the century before the market is ready for Seine Rive Gauche (Jones Lang Wootton, 1993a).

A second source of problems for the development of Seine Rive Gauche has been local opposition to the scheme. Local residents' groups put forward objections in the public inquiry into the PAZ in 1990. Concern was expressed primarily about the proposed mix of land uses. There was a strong desire to see more social housing in order to retain the residential identity of the 13th *arrondissement*. Local groups were not alone in voicing this concern. Associations from across the city supported the call for more low cost housing, with the support of the veteran housing campaigner, Abbé Pierre (ADA 13, 1990). Local residents' groups are linked to the umbrella organisation, Coordination et Liaison des Associations de Quartier (CLAQ), which brings together over sixty campaigns in Paris. Another substantial local concern was that the central avenue would become yet another urban motorway and cut off access to the river. Residents' groups also voiced wider concerns about the gap between the technical expertise of the planners and the limited information available to ordinary citizens which restricts their ability to respond to such large development proposals (ABC 13, 1993b). However, despite this opposition the city approved its plan in 1991.

Opposition groups mounted legal challenges. Two associations, the Verts du 13ème and Tam-Tam, presented their cases before the Tribunal Administratif de Paris. They had two grounds for appeal. The first was that the PAZ did not include the proportion of open space required by the city plan. The second objection was that the technical procedure giving necessary legal rights to the developers had been incorrectly followed. They argued that the Minister of Transport and not the *préfet* should have signed the relevant documents (ABC 13, 1993a). The court upheld the challenges. The effect of the ruling was to prevent any building permits being issued or any land being acquired. Given that SEMAPA's existing debts could only be paid off through the sale of development rights any delay to the project creates financial problems. The city reacted in two ways, firstly by appealing to the higher court and secondly by making some amendments to the PAZ. More green spaces were identified in a new plan approved in September 1993. Although the issue of green space was only part of the broader objections raised by opposition groups it was none the less seen as a victory for community opposition to the city's planning strategy for these sites and a victory over the wider strategy of the city regime. The case also confirms the arguments of Lorrain (1994) and others that legal challenges made after planning decisions have been taken are often the only way in which local residents can hope to influence development outcomes. The legality of the PAZ was subsequently upheld by the court of appeal, the Conseil d'Etat, in December 1993. The court considered that there was enough green space in the original plan and that in addition there were substantial open spaces close by.

Local residents were also concerned that the new public transport links to Seine Rive Gauche would terminate at the proposed office quarter and not go on to link the rest of the 13th *arrondissement* to central Paris. As well as criticising inadequate public transport plans objections have also been raised to the priority accorded to the new road bridge linking the development to the north bank which would encourage traffic and add to pollution in the capital. This growing environmentalist concern contributed to the 'green' members of the regional council refusing to vote through the region's contribution to the 1995 budget for APUR. They argued that APUR had promoted too many large development projects. APUR is funded jointly by central government, the region and the city. The city is by far the largest contributor and proposed to make up the 5 per cent of the budget lost from the region.

Other Paris ZACs have faced similar problems of local opposition. The interventions of residents' associations grew throughout the 1980s and early 1990s (*Le Monde,* 1994c) and CLAQ attempted to turn development controversies into electoral issues. In the 1995 municipal elections the ruling RPR group lost control of the *arrondissements* of northern and eastern Paris. The city regime had gradually increased its political hold on these areas since 1977 as the programme of redevelopment and rehabilitation of the formerly working-class districts progressed. Part of the reason for the electoral shift in 1995 was claimed by residents' groups to be due to opposition to redevelopment. Equally important however may have been the change of mayor. The unique constitutional status of Paris gives the mayor the powers of both leader of a commune and a *département*. The Chirac regime in the town hall strove to enhance both the prestige of the capital and its mayor (Haegel, 1994). When Chirac moved up to the Presidency there was therefore an interregnum and at the same time a series of scandals concerning the management of the city's housing stock. For most of eastern Paris the redevelopment process had also come to an end and the city regime had therefore less to offer voters in terms of new housing or facilities. A combination of factors therefore contributed to the electoral losses of the ruling group.

However, on the south bank, in the 13th *arrondissement* where Seine Rive Gauche is being constructed, the ruling group held on to its seats. The president of SEMAPA, Jacques Toubon, is mayor of the 13th *arrondissement* and a leading member of the ruling majority on the city council, and is thus able to secure the city's commitment to SEMAPA. Following the 1995 elections the deputy mayor of the 13th *arrondissement* was given responsibility for urban planning in the new city council. This role on the city council also gives wider responsibilities. The city has some nomination rights to new social housing, for example, and the local mayor can have many benefits to offer the local population. For local residents, therefore, opposition to projects such as Seine Rive Gauche may not extend to a vote against the mayor. There is a further level of political integration. After the national elections in 1993 Toubon joined the government as Minister of Culture. This ministry is responsible for the new national library. The deputy mayor of the 13th *arrondissement* became a government minister in 1995. Multiple office holding therefore enabled support for the project to be secured at a number of levels. Opposition to the Seine Rive Gauche has had some limited impact. However, apart from adjustment to the PAZ the city and SEMAPA have resisted the challenges to the project. Political control of the project is highly integrated.

Seine Rive Gauche is an ambitious project relying on the proceeds from office development for its success. It is the project of a style of centralised city planning which, through its alliance of technical expertise and political control, has managed the vast project of eastern renewal in Paris. Throughout the replanning of eastern Paris the public sector has managed the property market. Private investors come into the Seine Rive Gauche project at a late stage, to purchase development rights after the new infrastructure is in place. The scale of Seine Rive Gauche and its bias to office development have proved to be weaknesses. There are about 5 million square metres of empty offices in Paris. The viability of the project has been weakened by the property crisis. The city regime is also under attack from critics, local and city wide, of its redevelopment programme and was weakened as a result of the 1995 municipal elections.

LA PLAINE SAINT-DENIS: INTERGOVERNMENTAL RIVALRY

The planning of La Plaine Saint-Denis to the north of the city of Paris takes place in a much more complex institutional environment. This case study examines recent planning issues in three stages. Firstly there is an exploration of the development of policies and proposals for the area by different levels of government. Next the network of implementation agencies is unravelled. Until recently the state has taken a back seat in the planning of the area, but this changed with the decision to locate a new national sports stadium in the northern suburbs. The third part of the case study analyses the relationships between the state and sub-national government in the planning of the stadium and the relationships between public and private interests in the project.

In 1985 three communes, Saint Ouen, Saint-Denis, and Aubervilliers, together with the *département*, Seine-Saint-Denis, formed a syndicate – La Plaine Renaissance – to produce a common perspective on the redevelopment of the whole of the area known as la Plaine Saint-Denis which comprises about 700 hectares. The area has a mix of uses including large areas of derelict land formerly occupied by industry, a large tract of railway land, warehousing, and blocks of flats originally built for industrial workers. The area is crossed by motorways and a canal.

Since the 1960s industry has gradually left the area. As a consequence working class politics has also been in decline. The area used to form part of the '*banlieue rouge*'; a belt of communist controlled suburbs around northern and eastern Paris. The traditional vote for the Communist Party in this part of suburban Paris has declined. Some local communist parties have adapted to these new circumstances, adopting new local policies rather than campaigning for the simple retention or replacement of industrial jobs. The '*rénovateurs*' of the party have sought new voters by attracting new types of economic development and by attempting to diversify the housing stock by encouraging owner-occupation. The mayors of Saint-Denis and Aubervilliers support this new style local communism.

In 1991 an intercommunal plan for the development of Plaine Saint-Denis was produced by La Plaine Renaissance who also organised a series of public meetings to debate the ideas with residents and businesses (*Le Journal de Saint-Denis*, 1990). Further studies were undertaken by architectural consultants and a synthesis of these studies was presented to the public in 1992. The planning principles developed by the joint planning agency have been adopted by the communes. These principles included the creation of employment and training opportunities for the local population, a substantial house-building programme, and improvements to the environmental quality of the area. Plaine Saint-Denis is split into two parts by the A1 motorway which runs in a cutting. The communes' proposals therefore included the covering of the motorway and creation of better east–west links across the area.

In addition to this local planning initiative central government began to take an interest in the area in the early 1990s. In early drafts of the new regional plan for Ile-de-France, Plaine Saint-Denis was identified as an area where growth would be encouraged (Soulignac, 1993). The government set up its own planning office in the area to take forward the detail of strategic planning of the wider area of Saint-Denis and Le Bourget. This local branch of the Direction Régionale de l'Equipment (DRE) reflects some of the uncertainty government had about its planning role following decentralisation. The state had not only retained the regional planning authority in the Paris region but was now actively engaged in local planning. A similar planning office was set up for the redevelopment planning of the former Renault car factory in Boulogne-Billancourt on the south-western edge of Paris, but there the state's intervention failed to clarify responsibilities and the respective roles of the *préfet* and the commune continued to be contested. Following approval of the new regional plan with its designation of Plaine Saint-Denis as a development pole, the communes of Saint-Denis and Aubervilliers have continued to work with the state on the development of subregional strategy. In 1995 a three-year *contrat de développement urbain* was agreed which focused on implementation of regional objectives. The communes organised further public meetings to discuss these proposals.

In Saint-Denis, therefore, a complex map of planning agencies has developed. The communes and *département* joined together in La Plaine Renaissance, the regional planning office had strategic views about the area and its local office produced planning schemes. The local state services, the DDE, also considered that they had a role in any planning and development strategy. In addition the government actively intervened in La Plaine by relocating some government offices there – the Délégation Interministérielle à la Ville and DATAR – which had been included in a national policy of decentralising institutions from Paris.

In addition to this web of planning agencies there were several options about the form of implementation agency to carry forward the planning ideas (Bayle, 1992). The *département* owns important redevelopment sites and had the resources to back two development agencies in the area and could have taken a leading role. Equally the state, having declared Plaine Saint-Denis a development pole, could have introduced a development agency similar to those created to build the new towns or La Défense and steered the redevelopment process itself. The communes, however, were not happy for either the *département* or the state to take over. This was due to party political reasons – in the case of relations with the *département* differences of ideology within the Communist Party, and in relation to central government a mistrust of other parties – and also because, following decentralisation, the communes wanted to make the most of their own powers. In the event the local communes set up a new SEM, Plaine Développement.

Plaine Développement is jointly owned by some of the original partners in the planning syndicate – the communes of Saint-Denis and Aubervilliers (68 per cent), a group of banks (22 per cent), and the state (8 per cent). The key players in the area who are missing from the SEM are the major landowners. The electricity (EDF) and gas (GDF) companies and SNCF own about 50 per cent of the land and have their own planning advisers and strategies. The local mayors complained that public utilities should not operate independently and the mayor of Saint-Denis felt that planning should be 'transparent' and 'democratic' (La Plaine Renaissance, 1992). The city of Paris also owns land in the area and has the resources to play a major role both in the south of the area where Plaine Saint-Denis meets the city boundary and on important sites elsewhere.

Plaine Développement was in a weak position. The company had limited capital and could make only limited purchases of land (Rouzeau, 1993). The state had only a nominal investment in the agency but was unwilling to increase its investment in the company. The banks were equally wary about investing more in property development in this location. The communes are relatively poor. Saint-Denis earns some revenue from its town centre redevelopment projects of the 1980s but the combination of derelict sites and relatively poor population means that it has little money for property investment. Plaine Développement's main strategy is therefore negotiation and co-ordination of the range of public and private actors.

Despite this complexity of local planning and weak public agencies, there have been significant new investments in Plaine Saint-Denis. Electricité de France (EDF) built a large (14,600 square metres) headquarters, and has proposals to develop and manage adjoining land in a mixed use ZAC in partnership with the commune of Saint-Denis. Research activities have located on the more attractive and accessible parts of the area. SNCF located its workshops for the TGV Nord on the railway lands of Plaine Saint-Denis. The number of jobs located in La Plaine has increased since the beginning of the 1990s but, in common with the rest of the region, there are empty office buildings.

The actual redevelopment of La Plaine Saint-Denis has been led by the utilities and the private sector. The communes have tried to co-ordinate an overall strategy and achieve some social objectives largely through their relatively ineffective development agency. The communes have neither the landholdings nor the finance to intervene effectively. However, this process of weak local planning radically changed in 1993 when the state decided to locate the new national football stadium (the Grand Stade) on land owned by the city of Paris in the north of Plaine Saint-Denis. In sharp contrast to the underfunded local planning of the early 1990s this decision brought substantial investment to the area and strong national planning.

The new stadium for the 1998 World Cup in France was originally planned by the Rocard government in 1989 to be sited in Melun Sénart, the new town to the south of Paris. After considering various alternatives, including a location close to Disneyland, the 27-hectare Cornillon site in the north of Plaine Saint-Denis was chosen at the end of 1993 largely for its proximity to the capital.

The decision about location was taken centrally and central government retains the dominant hand in the planning and development of the area around the proposed stadium. However, the agreement of the communes was required. The communes were initially reluctant to encourage the construction of the Grand Stade in the area, fearing a distortion of their own planning objectives and the sterilisation of the area around the stadium except for the limited times when sports events were taking place. However, in September 1993 they changed their view and set out the terms on which the stadium could be accepted, including a role for the commune in whatever company was set up to manage the development. The communes' bargaining position was enhanced by the government's need to have the stadium ready by the end of 1997. The government adopted procedures to by-pass the local plan but this relied upon the agreement of the commune of Saint-Denis not to mount any legal challenges, which would result in delays to the tight programme. The commune therefore argued for the integration of the stadium into the wider regeneration of the area in line with its planning principles.

The government itself wanted an exemplary project, in which the world's press coming to the World Cup finals could see the stadium in the context of a broader programme of urban regeneration. The stadium project could be integrated with the housing and social programmes negotiated in the *contrat de ville* for this part of the Paris suburbs. The priority already accorded to the area reflected concerns about high unemployment, particularly among the young, and the relative isolation of large housing estates from services and centres of employment. Government expenditure was increased to complete projects on nearby estates and to link surrounding areas to the stadium site. For example, a bridge over the canal was proposed to link the large (2,260 flats) Francs Moisins estate to the stadium development. The state is also investing in public transport improvements, better pedestrian connections across the north of the area, and covering a section of the A1 motorway. The transport and estate improvements were all included in a contractual programme agreed between the state, the regional council and the communes. Some transport projects were included in the state–region *contrat de plan*, housing and social programmes in a *Grand Projet Urbain* which includes the communes of Saint-Denis and La Courneuve. The *préfet* of Seine-Saint-Denis manages this network of contracts through which the state invests in the wider regeneration of the area (*Le Moniteur*, 1994).

In addition to housing and transport investment the communes saw the main local benefit from the stadium as job creation. Constructing the stadium was a large project. The stadium itself was needed for the World Cup but also for national football matches and, possibly, as an athletics stadium. There were to be extensive related works – improvement to access, car parking spaces, a heating plant for the stadium, associated facilities for the press, and in the longer term the prospect of facilities to accommodate a future Olympic Games in France. Some new housing was to be built alongside the canal, and offices and shops closer to the stadium. It is estimated that the project will generate over 1,500 jobs. Unemployment in the area is about 15 per cent (much higher on some neighbouring housing estates), and the communes understandably want to see some local benefits (*Le Journal de Saint-Denis*, 1994). The commune of Saint-Denis supported the public–private agency, Saint-Denis Promotion, an alliance of ninety-seven business and other bodies, in its campaign to get local companies involved in constructing and servicing the stadium. In addition to potential jobs in the construction and running of the stadium, improvements to transport connections and the general enhancement of the image of the area could also attract other new employers to Plaine Saint-Denis.

Whatever the ambitions of the commune and local businesses the key decision-makers are the planning and project management agencies set up and controlled by central government. The stadium works and associated transport upgrading represent a substantial increase in the state's investment in the area. The stadium project is controlled by a joint public–private company, Société Nationale d'Economie Mixte (SANEM), set up to manage the whole of the Cornillon site of which the stadium forms a part. The state is the majority shareholder in the company. Other shareholders in the company include the commune of Saint-Denis, the *département*, the Caisse de Dépôts, the electricity and gas companies, and the Paris chamber of commerce. The company has been given until 2000–2002 to oversee construction of the stadium and complete other works.

The selection of a developer to design, build and manage the stadium and related developments was made through competition. The competition jury included representatives of the state, Saint-Denis and the city of Paris as well as technical experts. The jury's criteria included design, the proposed financial arrangements in which the state and the selected developer would share the costs of the project, and the wider question of the integration of the project into the area. It was noted in Chapter 7 that only the large construction companies in France have this comprehensive development expertise. The majority of schemes submitted in the competition were, not surprisingly, backed by different combinations of the same few construction companies. The government had limited time to determine the complex issues relating development finance to the urban impacts of the project. The Prime Minister's final choice allocated the project to a grouping of Bouygues, Dumez (Lyonnaise des Eaux) and SGE (Générale des Eaux). In negotiation with the development consortium the state committed itself to over 1 milliard francs of investment and future financial risk if the stadium were not taken over as the permanent home of a local football club following the World Cup. The *permis de construire* was granted in April 1995.

The communes continued to argue that the stadium project is not generating enough wider benefits. They were worried that only the immediately needed infrastructure will get built. Local residents were also concerned that the benefits from the stadium may be limited (*Le Monde*, 1994h). These concerns may be justified given the likely unprofitability of potential development around the stadium (Acosta, 1994).

The decision to locate the Grand Stade in Plaine Saint-Denis was taken by central government. Local planning principles and national urban policy priorities were incorporated into the discussion of the detailed project. The planning of this part of Plaine Saint-Denis involves both bargaining and contractual relationships between levels of government. The state, formerly unwilling to commit resources to this area, has done so in relation to a prestigious urban project. Detailed planning is the responsibility of SANEM and the construction group which has won the concession to design, build and manage the stadium. In contrast to the weak local planning initiatives of the late 1980s and early 1990s decision-making was transferred to the centre of government: the Prime Minister selected both the winning scheme and large development companies.

The redevelopment of Plaine Saint-Denis is a mixture of central government direction and complex relationships between levels of government, the public utilities and existing and potential businesses. There is a multiplicity of agencies. Some private investors have been attracted to the area and the construction of the Grand Stade should attract more. Local political elites are struggling to retain some initiative and extract local benefits from the process.

THE EURALILLE PROJECT: INSTITUTIONAL RESPONSES TO EUROPEAN COMPETITION

Urban entrepreneurialism was one of the significant trends in post-decentralisation France noted in Chapter 7. One of the provincial cities which typifies this trend is Lille in north-eastern France. Lille has attracted international interest by developing a large-scale commercial quarter – Euralille – around and above the new TGV interchange. As in the case of Plaine Saint-Denis there is local institutional complexity, but, in contrast, there is greater local autonomy in the pursuit of economic growth.

Lille had been designated a *Métropole d'Equilibre* in the 1960s as part of the regional policy of encouraging the development of cities outside Paris. However, it was not as successful as other, southern, cities in attracting new growth. The conurbation, which includes the large towns of Roubaix,

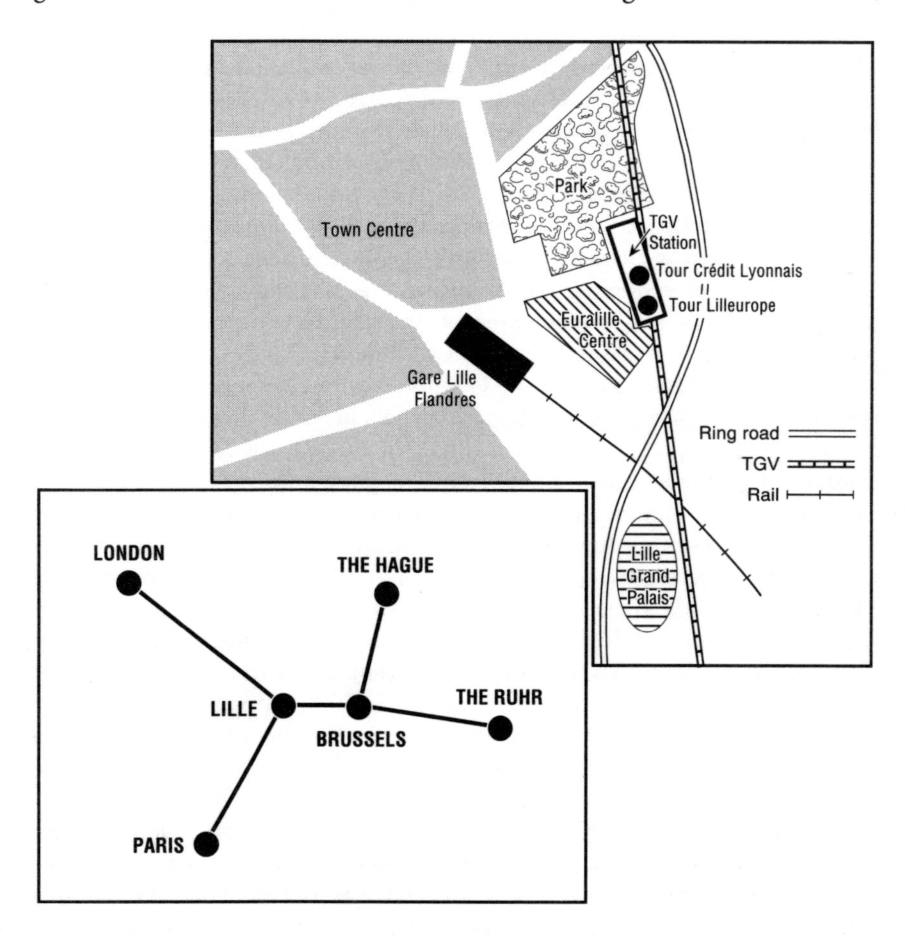

Figure 8.2 Lille

Tourcoing and Villeneuve d'Ascq, suffered from high unemployment as traditional industries declined. However, the national and international economic context changed towards the end of the 1980s. The TGV network improved national communications and the completion of the Single Market and construction of the channel tunnel changed the international position of Lille. The location of a TGV interchange adjacent to Lille city centre created the potential to exploit these trends.

The decision in 1987 to construct a new station in the city, and not as originally planned nearer to Roubaix and Tourcoing, was due to successful lobbying by a public–private group, TGV-Gares de Lille. The mayor of Lille, Pierre Mauroy, a former Prime Minister, used his political influence to obtain this relocation of the interchange (Stevens, 1993). The success of the lobby group created a consensus of both local elites and the state services and *préfet* in favour of the Euralille project (Cuñat *et al.*, 1993).

Lille could expect substantial economic benefits from its position on the TGV network. Investment in high-speed rail links has concentrated in northern Europe. Those cities with the first stations might expect to benefit most from the investment as it will be some years before a wider-reaching network is created (see Nijkamp, 1993). The coming of the TGV therefore represented a significant potential change in the economic fortunes of the conurbation. At this time Lille established a lobby office in Brussels and actively promoted itself as a 'European' city (Balme and Le Galès, 1992).

The change in economic circumstances was accompanied by significant local political change. The outcome of the local elections in 1989 secured political support for the Euralille project within the conurbation. The strategic authority for the conurbation, the Communauté Urbaine de Lille (CUDL), established in the 1960s, is an alliance both of the bigger towns in the conurbation (for example Lille with a population of 175,000) and of smaller communes (for example Warneton with a population of 171) which are controlled by a range of political parties. The CUDL now groups together eighty-six communes. After the 1989 municipal elections Lille took over the presidency in the person of Pierre Mauroy. Multiple office holding allowed Mauroy to be at once mayor of Lille, president of the CUDL and a senator. An inter-party agreement was forged in 1989 which gave the socialist Mauroy the presidency of the CUDL but also significant power to mayors from other parties. As a result of this agreement the CUDL supported the Euralille project as part of a programme of economic development projects across the CU area (Cuñat *et al.*, 1993; Van den Berg *et al.*, 1993; Stevens 1993; Le Galès and Mawson, 1994). The other development projects were largely industrial and distribution centres, and the programme was linked to an idea of a distribution of functions between different towns. Lille, for example, was seen as a business centre and Roubaix with its teleport as a centre of communications (Stevens, 1993). The new programme of the CUDL included social expenditure as well as economic development projects.

Revisions were made to the strategic plan for the conurbation to accommodate new development proposals. A new Agence de Dévéloppement et d'Urbanisme was established in 1990 to co-ordinate the various aspects of the economic development and planning strategy of the CUDL.

Within the city of Lille the socialists were supported by the 'green' parties. Early proposals for the new development project around the TGV interchange were scaled down in response to criticism on environmental grounds. However, the project, extending over 70 hectares, remains substantial. Euralille is made up of a number of elements developed by different agencies. The project includes an office tower and a business centre funded by French banks. A large shopping/services/hotel/leisure centre was located between the two railway stations. The Euralille project includes the adjacent development of the Lille Grand Palais, an exhibition and conference centre which opened in May 1994 and was financed by local and central government and European structural funds. The proposed Parc Matisse adjoining the development is intended to contribute to the quality of life enjoyed by new workers and residents in the centre of Lille (Mauroy, 1992), and its inclusion in the overall plan for Euralille reflects the importance of local 'green' politics.

The CUDL approved a development plan for the project in 1990. At the same time responsibility for implementation of some elements and overall co-ordination was passed to the company Euralille. Euralille was constituted as a *Société d'Economie Mixte.* The public sector partners in the SEM Euralille were the CUDL, the *département* of Nord, the Regional Council, and the communes of Lille, La Madeleine, Roubaix, Tourcoing and Villeneuve d'Ascq. The regional, departmental, local and intercommunal authorities held a little over 53 per cent of the shares. The remainder were allocated to a mixture of public and private interests including local and national banks, SNCF, the local chamber of commerce and industry and Société Centrale pour l'Equipement du Territoire (SCET). A small number of foreign banks took nominal shareholdings in the 50 million francs of capital of Euralille (Euralille, 1994). However, despite the small scale of investment the presence of these foreign banks was important for the prestige of the project and Euralille wanted to attract new international investors, both to augment its capital and increase foreign interest.

There was then a complex arrangement of public and private partners in the project, pursuing commercial and public interest objectives. The management of the project was equally complex. Pierre Mauroy became president of the SEM Euralille, thus ensuring strong political direction if only indirect political accountability through the local government shareholders. The main task of the agency was management of developers and contractors. Most of the work on the project was contracted out to around 1,000 staff in architectural and engineering consultancies and other companies (Brillot, 1993). The SEM contracted work to technical consultants, management consultants and to the CUDL for those aspects of the project affecting the road network.

SCET advised on financial management. Work on the station itself was the responsibility of SNCF, and the CUDL was responsible for planning the extension of the metro through the site. In addition to this management role the SEM had some direct development responsibilities – it was one of the developers of Lille Grand Palais and responsible for car parking and some other works. Project management was thus a complex business involving a variety of relationships with the developers of the main constituent parts of the project and management of a range of other agencies and sub-contractors. The task of Euralille was not surprisingly seen as a technical challenge of co-ordination and management, and its success defined in how well these challenges were undertaken.

The director of the SEM had experience in other large-scale urban projects in France (Baïetto, 1993a). However, management fashions had changed since the building of new towns in the 1970s and the development process in Lille owed as much to private sector management techniques as to public bureaucracy. 'Quality circles' – groups of selected experts to advise on economic development, communications, and cultural developments in the project – were established. Most important was the 'circle' of 'urban and architectural quality' which consisted of experts drawn from architecture, journalism and government departments (at both regional and state levels), the chamber of commerce and industry, and local government. This group was initiated at the start of the project and originally had twenty members chosen by the the project manager. The number was increased to thirty-eight in 1993 (Letruelle, 1993). The 'circle' acted as an advisory committee. Both the project manager and the architect claimed that the 'circle' gave them a source of informed opinion and support in the vast and complex task of running the project (Baïetto, 1993a). Meetings were held every two months, with the participants meeting on site to discuss a variety of projects within the overall scheme. However, another important function of the 'circle' was to promote Euralille by involving a large number of urban experts in the details of the project.

In addition to this incorporation of expert opinion the project was promoted in a number of other ways. One of the first actions of the political leadership was to secure an architect with an international reputation. The choice of Rem Koolhaas and his company, the Office for Modern Architecture, lends prestige to the project. In the choice of architect the president was seeking commitment to the project before any drawing was started. In a similar way Euralille tried to encourage an international hotel group to join the company, to get commitment to the project before starting to negotiate about the details of hotel development. It was thought that getting an international hotel company on board would encourage other international investors to risk money in the project.

The SEM Euralille also invested directly in promoting the project. During the planning work of the first two years of the project one-third of Euralille's budget was spent on information and publicity. Euralille's marketing strategy won an award for urban marketing in Europe in 1992 (Engrand, 1992). The publicly funded Lille Grand Palais was the first part of the development to be completed, and this had a promotional role in bringing business exhibitions and conferences to Lille. Architectural prestige, the support of experts in the 'quality circle' and the nominal shareholding of foreign banks were all important to presenting Euralille internationally as an exemplary development. The local population was targeted through publicity and project open days. It is argued that the primary role for the public sector in this sort of large urban development project is promotion (Verpraet, 1992).

There has been no significant challenge to the image presented by Euralille, and few have acknowledged the lack of local opinion in the debate about the scheme (Letruelle 1993). Broader notes of criticism, about conflicts with environmental values for example, have been raised in the professional press but not presented as significant issues (Baïetto, 1993b).

Euralille is a mixture of smaller projects funded by private investors within a master plan defined by the public sector and co-ordinated by a nominally mixed, though public sector controlled, management agency. Promotion is part of the task of the SEM, whose other role is the smooth management of the overall scheme for the benefit of individual investors. The management task is not simply routine but involves negotiating with investors and ensuring a strategic direction to the project.

These tasks of promotion and management were undertaken at arm's length from democratic bodies. Euralille was created by the CUDL and managed by an SEM which was a further step removed from democratic accountability. The legitimacy of the project rests on arguments of technical and managerial competence in the SEM, on the international reputation of its architect, and on the political reputation of its president.

Pierre Mauroy was at once leader of the commune, the CUDL and the SEM. He argued robustly for the economic development of Lille (Mauroy, 1994). Multiple office-holding enabled Mauroy to draw together the various strands of the coalition and ensure consistency in its direction (see Newman and Thornley, 1995). As a former Prime Minister he had considerable influence on the national stage which he used, for example, to influence the location of the TGV interchange. As mayor of Lille he had control over the local political scene and through his presidency of the strategic authority a strong influence over regional politics. Thus he was in a position to try to integrate the three political levels. As president of the SEM he was also able to carry this influence through into the implementation agency. Multiple office-holding and the local political settlement in 1989 created the opportunity for personalised leadership to legitimise the development.

The economic success of the development was far from assured. Lille Grand Palais and the commercial centre opened in 1994. The main shopping attraction was a supermarket which, on opening, faced competition from other new regional centres, for example, Cité Europe at Coquelles near Calais. International shoppers from England and Belgium were needed to make the Lille centre profitable, but this group could be attracted elsewhere. The bulk of the office space was not built for specified tenants. However, Euralille provides a relatively small addition to the office stock (45,000 square metres) and recession in the commercial property market was less of a problem in the Lille conurbation than in other French cities.

In addition to commercial concerns about the project questions were also raised about its political leadership. The political settlement in Lille which brought forth the Euralille project was threatened by the outcome of municipal elections in 1995. The CUDL had been dominated by the larger communes. However, as a result of the new planning law passed at the end of 1994 each commune had to be represented in the CUDL's decision-making body. Rival parties could thus have overthrown the majority established by Pierre Mauroy. Socialist Party gains in the elections secured the continued leadership of Mauroy.

The project's legitimacy rested upon its commercial success and the political authority of its leader. Decision-making was in the hands of the independent SEM and therefore relatively closed. This was not a transparent, accountable development process but one which claimed its legitimacy through its managerial success, its vision and leadership and through its international prestige. The development-led strategy in Lille resulted from what the project manager called '*la guerre de villes*' (Baïetto, 1993b). The winners need to show evidence of success.

APPROACHES TO URBAN DEVELOPMENT AND PLANNING

The three cases share aspects of a common institutional context in which the integration of levels of government and the use of quasi-autonomous SEMs are significant features. However, the cases also reveal the complexity of local planning circumstances and considerable variation in the detail of intergovernmental relations.

The ability of the mayor to integrate levels of government was crucial for the development of Euralille. The achievement of Mauroy in drawing the TGV line into Lille laid the foundation for the Euralille project. In the Paris case political power was equally important. However, Seine Rive Gauche depended not on the ability of the mayor to co-ordinate differing levels of government but on the special status of the office of mayor and the high profile which Chirac had given the role. Large-scale commercial redevelopment fitted well with the international image of the city which the mayor sought to create. In contrast to the other cases, in Saint-Denis there was conflict between levels of government. The communes failed to get the support of other levels and there was some hard bargaining over the impacts of the Grand Stade. The pattern of intergovernmental relations therefore varies between the cases. Several factors seem to account for this variation.

Intergovernmental co-ordination is influenced by party politics. In Saint-Denis party politics weakened the influence of local leaders at other levels of government. The state was at first unwilling to support local initiatives but then added its own parallel agency and a national SEM to control the development of the area around the Grand Stade. Party political conflict also lies behind the uncertainty about the role of Lille in the development of the regional economy of north-eastern France. It was a socialist government which sanctioned the TGV interchange in the city. The urban development strategy of the socialist mayor of Lille was, however, challenged by the centre-right national government elected in 1993.

The local development market also seems to have been an important factor in determining differences of approach in the three cases. In Paris strong city government managed a traditionally strong development market. The development process was dictated by the city, and developers and builders provided with a steady stream of redevelopment projects. In Plaine Saint-Denis the communes could not compel the market to develop the sorts of projects they favoured or determine their location. The effect of planning in a weak market was the proliferation of agencies and plans. In Lille the prospect of economic growth following from investment in transport infrastructure attracted private finance. Substantial parts of the project were, however, funded publicly by the CUDL, the state and the EU. Central and local government also paid for infrastructure in Paris and Saint-Denis. In Paris there was a prospect of getting back the investment in Seine Rive Gauche through the sale of development rights.

Public money therefore played a crucial role in all three redevelopment projects. Local and central government also co-ordinated investments by the utilities, SNCF in particular, and the state banks. In the weaker market of Saint-Denis the utilities and banks were less willing to join in local government initiatives. The apparent strength of the public sector in France is due in great part to public investment in infrastructure. Developers and private investors follow public initiatives.

The financial base of local government was also clearly a significant factor. In Lille it was the conurbation-wide body the CUDL whose resources were directed at Euralille. Intercommunal co-operation was also necessary in Saint-Denis. However, the communes there lacked development resources. The tax base of these former industrial areas was weak. The *département* of Seine-Saint-Denis accounted for less than 10 per cent of the tax potential of the Ile-de-France but had 13 per cent of the population. In the city of Paris, however, 21 per cent of the regional population accounted for 35 per cent of the tax potential (Marcou, 1994). The dependent population in Seine-Saint-Denis had to pay high taxes. In Paris rates of tax could be low because of the high potential tax yield. The unevenness of the tax base therefore substantially influences the ability of local government to engage in urban development projects. The issue of equalising local government revenue has concerned government for some years. The impact of decentralisation and the competition between communes for commercial development and the tax it generates has exposed a fundamental weakness in the French approach to *'aménagement du territoire'*. The new planning law raised the issue but the search goes on for a politically acceptable formula for redistributing wealth amongst local government.

The three cases differed in the form of intergovernmental relations, the influence of party politics, and in the relation to local market conditions.

The use of semi-autonomous, public–private SEMs distinguishes the process of redevelopment in French cities from other parts of Europe. However, the SEMs work in differing ways and with a range of impacts on the planning and development system. Plaine Développement was weak because of its lack of land and limited development finance. This contrasted sharply with SEMAPA which was able to secure almost 2 milliard francs in bank loans. In the Lille and Paris cases the SEMs inherited the responsibility for large projects. In Plaine Saint-Denis the SEM's primary role was negotiation with limited direct intervention.

SEMs offer considerable operational flexibility outside the direct control of the town hall. In Lille the mayor took a strong leadership role in partnership with the technical director. In Paris it was the local mayor who presided over SEMAPA with the agency's director taking responsibility for running the project. In Paris the regime was less closely associated with the day-to-day management of development. Indeed, the mayor stated that he was not *'interventionniste'*, giving the city's development agencies complete independence (*Le Débat*, 1994, p.56). He did not interpret the attack on APUR as a personal challenge. In Lille the closer association between the SEM and political leadership could prove a political disadvantage should Euralille run into economic difficulties. In Plaine Saint-Denis the SEM Plaine Développement was created to demonstrate the political independence of the communes from other levels of government, but its role was more technical than political and it was not a vehicle for promoting the personal authority of the mayors. The national SEM behind the Grand Stade was an operational agency, directed by a former *préfet*, at some institutional distance from political power.

The degree of political control of SEMs varied and they had differing relationships to formal local government. On the one hand the Paris SEMs fragmented urban development authority, but on the other each integrated development partners and committed public and private interests to the projects. The functions of SEMs are complex and our cases reveal some of that complexity.

In Saint-Denis the partners were ambivalent about the project. A more significant integrating force in the redevelopment of Plaine Saint-Denis was the various contracts between levels of government. Since decentralisation contractual relationships have replaced the centralised direction of local development. The state had less ability to compel lower level governments but exerted its power through contracts. The *préfet* of Seine-Saint-Denis co-ordinated the local, regional and central partners in the *contrat de plan de développement urbain* which pulled together public investment in Plaine Saint-Denis. The state's power was exercised through the combination of *préfet* and contract. The process of decentralisation was much more complete in Paris where the power of the mayor and size of the city's budget guaranteed local autonomy.

The processes of urban redevelopment in our three case studies raise questions about the accountability and legitimacy of planned intervention. In Chapter 7 the criticism was noted that *Communautés Urbaines* concentrated resources and planning powers but had only indirect accountability. The Lille case would confirm such views. In Lille the responsibility for development was passed from the CUDL to the SEM. Democratic accountability was thus further removed from elected local government. The legitimacy of the agency rested on the personal authority of the mayor, claims to technical competence and the reputation of its architect. SEMAPA similarly claimed technical competence and political authority through its president. Objections have been raised and doubts expressed about this relatively closed style of development agency. The experience of the extensive consultation during the development of the intercommunal policies for Plaine Saint-Denis contrasts with the closure of planning in the other two projects. Consultation could be seen as a positive response to circumstances in which little direct intervention was possible. In the other cases where land and development finance were available then the relatively closed alliance of technical and political leaders prevailed. In Saint-Denis there was a contrast of styles between the development of intercommunal policy and the operation of SANEM. Local residents were not convinced that SANEM would deliver local benefits.

Where a challenge to planning authority was made, it highlighted the distinctive characteristics of a system of codified planning law. In Paris objectors challenged procedural irregularities in the plans for Seine Rive Gauche. The disputed issues were the co-ordination between plans and whose signature was on the relevant documents. The resulting delay to the project might have proved expensive if the market had been buoyant and there had been any short-term prospect of sales of development rights. The relatively closed process of planning through intercommunal bodies, or semi-autonomous agency in the case of APUR, and the exclusiveness of the SEM restricted objectors to legal challenge after the event. It was also the possibility of procedural legal challenge and delay which won the commune of Saint-Denis a seat on the government's development agency.

In Saint-Denis the objections to the Grand Stade project concerned the distribution of social and economic benefits. In Lille and more particularly

Paris, environmental concerns have been raised. Environmentalists and local residents challenged the proposals for Seine Rive Gauche. Environmentalist members of the regional council forced the withdrawal of support for APUR. In Lille the local influence of green politics in the ruling majority scaled down the ambitions for Euralille.

In all three cases formal planning fell into line with large-scale redevelopment projects. In Paris the city plans supported the renewal of the eastern part of the city. The new regional plan for Ile-de-France recognised the international importance of redevelopment in this part of the city. The regional plans also supported growth in Plaine Saint-Denis, and the intercommunal work of the late 1980s was incorporated into the POS. The government's stadium necessitated special procedures to change that plan. In Lille the strategic plan reflected the set of development projects agreed in 1989. In general we could conclude that formal plans function to support the development ambitions of mayors, intercommunal bodies and the state. However, plans provide legal authority and legitimacy to the process of development. The formal planning system sits alongside claims to authority made on the basis of the personal and positional power of the mayor, technical expertise, or prestigious architecture.

The prospect of new national planning guidelines and the reassertion of central authority may change the context for urban planning. The growth targets of the regional plan for Ile de France were scaled down in response to the government's desire to contain metropolitan growth. However, as was noted in Chapter 7, large infrastructure investments will still be made in the capital region, and the development projects we have discussed are unaffected. In Lille however there was conflict during the national debate about the future role of concentrated urban growth. Seen in its European context, DATAR regards the future growth of the Lille conurbation as uncertain (DATAR, 1995).

The projects examined in this chapter span the 1980s and the 1990s. During this time circumstances have changed. In Plaine Saint-Denis the weak market has been boosted by the state's new investment. Investment in the stadium may draw in private investment. In contrast to the history of the late 1980s the government is now attempting to integrate local and national initiatives, urban policy and large-scale redevelopment. In Lille the economic prospects of growth generated by the TGV interchange have not yet been realised. In Paris the city's planning and development style has been severely questioned, both by the property crisis and by local political opposition.

The three cases display the distinctiveness of French approaches to urban development – the interconnection of levels of government, the strong public sector role, the use of quasi-autonomous agencies, and claims to legitimacy based on political and technical authority as much as democratic process. The cases reveal not one but several ways in which urban redevelopment is undertaken in France. The interaction of levels of government varies, relations between public and private sectors differ, local residents may or may not be excluded, and some redevelopment projects seek to integrate urban policy and economic development objectives.

In all three cases the projects are seeking to replace the industrial base of the city. Paris is competing for world city status, Plaine Saint-Denis is providing space for regeneration through prestige development, and Lille is building on the new European communications network. Each has high national and international ambitions. The distinctive approach of each development project, however, reflects the interaction of the national institutional context with local politics and local economic conditions.

9

SWEDEN: FROM 'MODEL' TO MARKET

This chapter explores the trends in the Swedish planning system over the last fifteen years or so. The aim is to show how the distinctive features of the Swedish approach to the welfare state have moulded the planning system. In recent years there has been considerable discussion over the future of this approach against the background of economic and political uncertainty.

ECONOMIC CRISIS

After the Second World War the Swedish economy experienced an export-led expansion leading to full employment. An approach involving close co-operation between labour, business and the government spearheaded a period of prosperity throughout the 1950s and 1960s. The 1960s have been described as the golden decade. This prosperity enabled the construction of an elaborate welfare state with an emphasis on social equality and full employment. However, the Swedish economy is highly dependent on international trade. Increased competition, the oil crisis and weaknesses in the international economy created the first signs of strain in this 'Swedish model' as four devaluations of the currency took place in the 1970s and early 1980s. The internal structure of the economy was also undergoing major changes. Many of the traditional industries were based upon raw materials, such as timber and iron ore, and after the Second World War industries such as chemicals and car manufacture were important. However, the older industries were in decline and along with most developed countries a process of deindustrialisation was taking place (Fournier and Axelsson, 1993). In this context the rapidly rising costs of such an extensive welfare state were questioned. Most services were delivered through local government, and therefore their budgets made up a large proportion of this figure. As municipal and county taxes absorbed about 30 per cent of a person's income there was little leeway for further expansion of the tax base. This fiscal crisis became a major feature of central–local government relations and, as we shall see later, influenced the way in which local authorities were to operate.

However, the 1980s featured a temporary economic boom linked to an upturn in the international economy and the devalued currency. The banks, which previously were strictly controlled, were deregulated and an 'orgy of lending and spending' took place in the second part of the decade (Branegan, 1993). This was to have a particular impact on planning because of the huge investment in real estate. Not only did Swedish financial institutions make big forays into European real estate markets, they also invested heavily in commercial projects in Sweden. From the mid-1980s there was therefore a period of major building activity, particularly in urban areas. But the bubble burst in 1990, commercial real estate crashed and banks' loan portfolios were wrecked. This triggered a banking crisis and two of the five largest banks collapsed and had to be saved by the government. It has been estimated that up to three-quarters of the credit loss sustained by the banks was a result of the property crisis (Kalbro and Mattsson, 1995).

In 1991, the Social Democrats, who had dominated post-war politics, lost the election to a centre-right coalition led by the conservative Moderate Party. The economy moved into the deepest recession of the post-war period. The new government, with the agreement of the Social Democrats, put together economic 'crisis packages' involving major cuts in expenditure to try to deal with the problem. However, both the government's budget deficit and unemployment continued to rise rapidly and the currency was allowed to float. The elections in 1994 saw the revival of the Social Democrats, who were the original architects of the post-war welfare state programmes. However, during the election campaign the head of the Wallenberg family made a rare public statement saying the family firm, whose companies account for 40 per cent of the Stockholm stock market, would move some of its companies abroad unless the government made deep cuts in public expenditure. Divided opinion was also reflected in the referendum on joining the EU in November of the same year which resulted in only a narrow majority in favour of joining (52.2 per cent). Again industrialists and other commercial interests made it clear that the only salvation for the Swedish economy was to join the EU.

These major economic changes can be related to the significant shifts that have also been taking place in the political and administrative arenas. We have seen how the economic difficulties have led to a questioning of the role of the state, a shift to the right in politics, a fiscal crisis in local government and fluctuations in the real estate market. All these factors have important influences on the way urban planning itself has changed. So before moving on to planning we will explore these contextual factors a little more.

THE CONSENSUS CULTURE

Dahrendorf (1990) has suggested that the Swedish 'model' exists in the minds of political reformers rather than in the messy reality of life. However, there is general agreement that there is something distinctive about the Swedish approach to government in the form of a consensus-seeking approach to politics. In the past this approach has been used to show that there is a middle way between capitalism and communism, for example by Anthony Crosland in his classic statement of socialist reformism (1956) and more recently by Gorbachev. This Swedish approach has been given a variety of labels, including that of 'corporate pluralism' (Castles, 1978). The corporate part indicates the structured nature of the consultation process, and the pluralism part the opportunities for a wide range of interests to be involved (Heclo and Madsen, 1987).

For example, at the national level there is the extensive system of official investigatory commissions which has long historical roots. The deliberations of government are supplemented by these commissions in which specialised groups, professions, experts, interest groups and so on can work out the proposals for a particular problem before they are submitted to Parliament. Every bill of any importance will have attached to it such a commission. In theory proposals are constantly being fed backwards and forwards between the commission, the constituency of the various interest groups and government agencies, although in practice most consultation takes place after the commission has produced its report. This report gives an exhaustive account of the present state of affairs, a justification for the changes, and proposals with commentaries on each aspect. The reports are published and sent out to relevant authorities and organisations for their comments. The government then prepares a bill which is submitted to Parliament. It is not surprising that a considerable degree of consensus will be expressed for the end result and that, as many of the implementation agencies have been involved, the process of implementation can be fairly smooth. However the process has also been criticised for too much government control. Central government can influence the outcome by its choice of members for the commission and the scope of the brief.

Such consultation requires a lot of organisation. It means that social and economic interests have national bodies and these have developed well staffed research departments on permanent salaries. Such organisations may exist in other countries, but in Sweden they are more highly institutionalised and given an important status in the decision-making process, dominated by the principle of consensus formation. This culture of consensus formation clearly places considerable importance on opportunities for participation. Access to information is a basic resource for meaningful involvement and Sweden has strong laws giving the right of access to public documents to both citizens and foreigners. This applies to all public documents except those relating to national security or containing personal information on individuals, such as health records. The preservation of this right of access was an issue in the referendum on joining the EU.

The consensus culture can be considered either within the political system or within society at large – the former often being a reflection of the latter. Even within the political system one can discuss different levels. In the post-war period the political consensus has focused around the economic approach, in which efficiency is sought from co-operation and rationalisation. This kind of consensus extended into the 1970s even when the alliance of 'bourgeois' parties managed to dislodge the Social Democrats from office in 1976. They were keen to demonstrate to the electorate that they were going to continue economic policies within the consensus culture and were unable to take the strong measures that the declining economic situation required. Although there was consensus at this broad level, differences existed on particular policies even in the 1960s, such as on pensions and Europe. It was the Social Democrats who perhaps epitomised the culture – they regularly obtained over 40 per cent of the popular vote and could claim to be the party that best represented the population as a whole. An important political change took place in the 1991 elections when the opposition parties campaigned together, won the election and formed a government. This time they set out a more radical ideology and tried to break free from the consensus culture of the past. However, although there are many factors involved, the return of the Social Democrats in 1994 suggested that within the population at large this consensus culture was still seen as important.

THE IDEOLOGICAL SHIFT TO THE RIGHT

We have seen how Sweden has been suffering from economic problems, making it difficult to maintain the high public expenditure necessary for its extensive welfare state. Significant tax reforms were implemented and a change in mentality had to take place. It could no longer be assumed that the state could deal with all problems. This created splits within the Social Democrat Party, some wanting to stay close to their roots linked to the trade unions and give priority to full employment and social equality, while others wanted to put more emphasis on efficiency and expenditure cuts. During the 1980s this latter strand gained ground under the direction of the Finance Minister Kjell-Olof Feldt. There were significant changes in policy such as the deregulation of the finance system, a shift in taxes from incomes to goods and services, liberalisation of the protected agricultural and retail systems and moves to privatisation of some of the state-controlled industry (Taylor, 1991). In the campaign for the 1991 election the conservative Moderate Party and the Liberals combined on a common platform of 'a new beginning' which sought to 'promote a strong market economy, increase competition and stimulate much more individual ownership and saving'. A reform of the welfare state was promised with greater freedom of choice and cuts in public expenditure, thus also allowing for tax cuts.

The 1991 election result demonstrated that the population was attracted by this new beginning. One unusual feature of the election was the new populist right-wing party called the New Democrats, set up by two people – one a member of the old aristocracy, the other the owner of a leisure park.

To begin with this party was considered a joke with their extreme policies of freedom (e.g. doing away with all parking controls in the city), and it often gave the appearance of being racist. However, the party got a significant vote in the elections (6.7 per cent), enough to have twenty-five MPs; they were not, though, invited to take part in the new ruling coalition which was made up of the Moderates, Liberals and Centre Party (the old Agrarian Party). The Liberal politicians would have nothing to do with the New Democrats because of their racist leanings. However, the popular appeal of the New Democrats gave confidence to the new coalition, led by the Moderates, in their move to a more right-wing approach. Many of the New Democracy ideas were implemented and some said this contributed to their subsequent decline – they failed to gain any seats in the 1994 election.

DECENTRALISATION OF POWER?

Although the move to greater market freedom has been one of the trends of the recent decade there has been another which has also dominated debate since the 1970s. This has been the questioning of the centralisation of power and the promotion of decentralisation in administrative and political structures. However, as we shall see, these trends are linked in many ways.

In June 1985 the ruling Social Democrats set up an inquiry into power and democracy in Sweden. The purpose of the five-year inquiry was to explore the possibilities for citizens to influence decisions in the light of changing economic and social circumstances, although it did not explore specifically either local power relations or urban planning. The brief for the study was to explore a number of interrelated questions about the prevailing distribution of power in Swedish society and how this was changing, particularly in the business sector, interest organisations, the public sector and the media. What influence did individuals and different groups have over their own lives? The government was concerned that the traditional principle that 'all citizens have the right to participate as equals in the development of their society' had been eroded by contemporary events. The turnout at elections in Sweden in recent times has always been around 90 per cent and research has shown that voters are well informed and make their choice on a reasoned and consistent basis (Holmberg, 1984, quoted in Petersson, 1987). Thus on the level of representative democracy the position looked very healthy. However, fears were expressed that the rationalisation of municipalities into fewer, larger entities had reduced citizen involvement at the local level. Nevertheless research had shown that representative democracy at the local level was also operating well and that 'citizens seem better informed, more interested and – contrary to what is often claimed – more active on municipal issues than before' (Westerståhl, 1987, quoted in Petersson, 1987).

The concerns about democracy also related to the broader arena of participatory democracy which has been part of the Swedish 'democratic ideal'.

According to Petersson, Sweden is an example of 'how grassroots "popular movements", an extensive system of organisations and popularly supported political parties can elevate groups into political consciousness and political involvement – groups that would otherwise have been condemned largely to political silence' (Petersson, 1987, p.18). He goes on to point out that the existence of these channels of democracy can help to overcome social imbalances in a way that individually oriented forms of democracy cannot.

The main conclusion of the study (final report published in 1990) was that Swedish society was moving into a new era. The old social system, culminating in the early 1970s, was characterised by a concentration of production, urban growth, a consensus about the expansion of the welfare state, and a corporatist approach to the involvement of special interest organisations, particularly unions, in the political decision-making process. The political culture was oriented to agreeements and compromises built upon principles of centralism, universalism and homogeneity utilising concepts such as 'social justice', 'solidarity' and 'social equality' (Petersson, 1986, p.12). The 'new era' was brought about by the end of economic growth, the questioning of large-scale solutions and the influence of certain special interests, especially unions, and a more sceptical attitude to authority. The economic trends showed increasing reliance on the service sector and increased international dependence. Social changes included multi-ethnicity, a greater role for women and an increase in the elderly. According to Petersson, 'society is becoming less uniform, more fragmentized and more complex. Particularism, variety, non-predictability and entropy are increasing' (1991, p.9). A major aspect of the 'new era' was the financial burden of the public sector. The study concluded that the government would have to find a 'new balance between society and the individual, between social welfare based on collectivist principles and the desire for individual freedom' (Taylor, 1991). In other words the 'Swedish Model' was dying. Politically the messages in the report presented a threat to the Social Democrats as the party most closely connected to the old era of corporatism and contributed to the general reassessment within the party.

One important dimension of the distribution of power in society is the relationship between central and local government. In Sweden the tradition of local government independence has a long history. Compared with Britain local government has a much stronger position in both legal and financial terms. Municipalities can raise and spend their own local tax. As we saw in Chapter 5, Britain has been centralising government authority; however, Sweden has been following the more widespread European route of decentralisation (Crouch and Marquand, 1989). The main trend is towards transferring further powers from the centre to local authorities (Reade, 1989). As part of this trend an experiment in 'free' local authorities was introduced in 1985 in which a select number of authorities were allowed to apply for the removal of state controls as a test for increasing local self-government (Stewart and Stoker, 1989b; Gustafsson, 1991). The lessons from these experiments were incorporated into the 1991 Local Government Act.

The question of the relationship between central and local government has been under discussion for most of the post-war period. Reforms encompass a number of different, possibly contradictory, objectives. These include greater efficiency, better management methods, cutting expenditure, being more responsive to users of services and creating more democratic involvement (Wise and Amnå, 1992). From the Second World War to the 1970s the issue of efficiency predominated and the number of municipalities was progressively diminished from about 2,500 to 280. The aim of the central government programme was to create more professional local administrations capable of implementing national welfare policies and school provision. This sparked off a reaction regarding the loss of democracy and the movement for greater decentralisation of power, one response being the ability to set up neighbourhood councils since 1979. Some functions were decentralised to the local level, such as civil defence, public transport and environmental protection, and local authorities expanded their own activities (e.g. in the field of local economic policy) (Elander and Montin, 1990). During the 1980s central government passed a number of new laws which removed the detailed regulatory approach to various services and replaced them with 'framework' laws which gave local authorities more discretion.

However, while this decentralisation process had been occurring some authors have pointed out that there was a parallel process of centralisation taking place (Elander and Montin, 1990). This was driven by central government's priority to reduce public expenditure. As local government is such a large part of public expenditure (e.g. in 1980 25 per cent of the national labour force was employed in local government) means had to be found to control their finances. Thus, increasing central interference in local government finance took place, culminating in 1990 with a law preventing local authorities raising their local taxes any further. Elander and Montin have summed up the situation as 'centralising [financial] power – decentralising responsibility' (1990, p.165).

A further development in the 1990s has been that of bringing the market into the state (Montin, 1993). This is based on the view that market principles will bring efficiency and quality. One element of this process has been the introduction of the 'client-performer' model in which elected representatives set out what should be done and other units are made responsible for the production of the actual services (Häggroth, 1993). The idea is that the producing (or performing) units are free from political interference. The ordering committee enters into contracts and agreements with the producer units. About twenty local authorities have adopted this model and many others have started to use the terminology. The producing units are generally formed from within the municipal staff, but it is only a small step to then privatise this function (Montin, 1993). The pragmatic element in the Social

Democrat Party and the coalition government elected in 1991 both supported the idea of privatising some elements of local government. So far this has centred on emergency adult health centres and some schools. Another feature of the 1980s was the increased use of municipally owned companies to remove services from the political arena. The general effect of these initiatives was to blur the distinctions between public and private bodies. The Citizen's Charter is seen in Britain as one of the few original ideas of Mr Major. It is interesting to see that the same approach has been adopted in Sweden (Elander and Montin, 1995). Here also citizens are being addressed as consumers and the role of local politics diminished. However, in Sweden, according to Elander and Montin, there is also evidence of a reaction to this with an increased interest in collective participation in local matters and a growth in the co-operative movement – perhaps further evidence that there is an enduring consensus culture within Swedish society.

In December 1992, during one of the economic crises, the government set up a commission, chaired by a renowned economist, Professor Lindbeck, to draft some guidelines for economic policy. The report, which was published a few months later, is interesting for the radical nature of some of its proposals and for linking changes in the political system to economic solutions. At the level of central government the commission proposed a greater concentration of power with the strengthening of the Cabinet, Prime Minister's Office and Finance Ministry. The number of ministries should be reduced and MPs cut by 50 per cent. It suggested withdrawing the freeze on local taxes, but subjecting any increases to a referendum, an improvement in the auditing and control of local finances and the abolition of the county level of government. It also said that 'interest groups should not be involved in decisionmaking in the public sector'. Thus it was attacking the consensus-forming approach to decision-making and suggesting a more streamlined and centralised approach. The Swedish model incorporating compromise between labour and capital utilising the big associations, political parties and the public sector was seen as too rigid (Norell, 1993).

These trends in local government show a number of different elements which reflect the tensions operating in Swedish society. There is the move to efficiency and cutting of public expenditure to deal with the economic crisis, the ideological shift to accepting a market-oriented solution, and the perpetuation of the consensus values which lead to support for greater democracy and the retention of welfare programmes. Montin (1992) has shown how these different objectives reflect the broader political strands. He identifies four positions: those in the Moderate Party and some Liberals advocating privatisation, based on the rolling back of the state and reducing the involvement of politicians, and three competing positions relating to the Social Democrats. Here he sees traditionalists upholding the current system of representative democracy, decentralists seeking a better relationship between local government and popular movements, co-operatives and other local groups, and a final position wanting to make public services more efficient. This final position accepts both decentralisation and selective privatisation as long as they lead to efficiency. Thus these differences are being played out at a broader national political level both between and within the parties. We now explore how these contextual issues have impinged upon urban planning and development.

THE SWEDISH PLANNING MODEL

The development of the post-war planning system was of course embedded in the broader ideals of the Social Democrats and their approach to establishing the welfare state. However, until the 1970s there was considerable agreement on planning issues between the Social Democrats and the Moderates, who were the 'natural' party of big business. The initial planning Act was passed in 1947 and included the concept of the master plan with more detailed plans for densely built-up areas. The plan was to set out the land uses for the future and would be revised every five years. It could be given legal force through ratification; until 1970 this was officially given by the King and afterwards by the Cabinet. However, most municipalities were satisfied with using it as a guidance document which gave them greater flexibility (Rudberg, 1991). Many of the plans were drawn up by architectural consultants who often produced very similar plans for different towns.

During the boom period of the 1960s, changes in approach took place. The government had adopted a strong social housing policy for the whole population and believed that people had the right to a good home regardless of income. In contrast to most other developed countries the Swedish government sought to control the whole housing stock and not just that for lower income earners. A massive programme of suburban high-rise development took place called the Million Homes Programme, and large numbers of existing town centres were also redeveloped. An organisational structure was set up to deal with the huge programme. This resulted in wide-ranging controls on residential development which extended far beyond the traditional planning system. The production of housing was steered by the public sector with the aim of removing speculation in land and housing. Central government was responsible for regulations and the supply of resources while municipalities ensured that building took place. Central government controlled housing production through its system of housing finance and the municipalities undertook the planning and land provision aspects. State housing loans were subsidised and therefore beneficial to developers, but all recipients, regardless of the tenure of the housing built, had to abide by the contracts agreed with the municipalities. The 'contracts do not only specify when, where and how the housing is built, but also its final price including the land acquisition element' (Dickens *et al.*, 1985, p.85). For the developer, the advantage of the loans was not only financial but also in providing stability as state loan schemes gave ease of access to developable land and a guaranteed market, until this was oversaturated in the early 1970s.

Another aspect of this housing picture was the development of a number of large housing organisations with a close relationship to the municipalities. Most municipalities formed their own companies for the contracting of construction and management of rental housing for a full range of income groups. Many would use their powers to ensure that the housing stock was owned by either municipal companies or, later, co-operatives. The national umbrella group the National Association of Municipal Companies (SABO), together with the two major national housing co-operatives, the National Federation of Tenants' Associations and trade union organisations, formed a powerful grouping (Elander, 1991). This demonstrates the corporatist nature of housing provision in Sweden, but even though most new housing was promoted by the municipal companies and co-operatives the actual construction was nearly always done by private companies (Barlow and King, 1992). Mergers and take-overs have meant that a few firms now dominate the construction scene with the largest Skanska building about a third of total housing output.

A further dimension in the control of development is through the public ownership of land (Duncan, 1985). Central government encouraged the process of municipal land banking and suggested that every municipality should have reserves for ten years' future development, and expropriation laws were extended to allow this (Kalbro and Mattsson, 1995). Municipalities could use their land-ownership to decide who develops and owns the housing. Once the building had started most municipalities would sell the land off to co-ops, housing companies or private owners. It was more complicated to try to control tenure through the planning system. Municipal land-ownership was sometimes used to get certain land uses in a development, such as small shops in housing areas. Rather than selling the land a few municipalities (e.g. Stockholm, Västeras and Örebro) established a leasehold approach which allowed them to obtain a regular income from their land holdings. A 'land condition' was introduced in 1974 which said that housing under the state housing loan should be built on land owned by the municipality at the time. However, exceptions were allowed and these were widely used (Vedung, 1993). Nevertheless the threat of the 'land condition' would often be enough to gain conformity to the policies of the municipality. The removal of this condition in 1992 placed more reliance on the planning monopoly and development agreeements. The public ownership of land varied considerably from city to city, with Stockholm having a particularly large landholding.

Thus, during the 1960s and early 1970s when the Social Democratic Swedish 'model' was at its peak, the state at local and national levels had considerable powers to control development. This was not only through the planning system but also, and perhaps more importantly, through housing finance and, in some areas, land ownership. The period can also be described as one in which much discussion and bargaining went on within the public sector bureaucracies. The planners in the planning department of local authorities would spend a lot of time dealing with central government or municipal and co-operative companies. The plans, which would be of a detailed nature, would be prepared by architect-trained planners who would usually be able to impose their design ideas, although until 1987 they had to be ratified by the state to gain legal status.

THE 1987 BUILDING AND PLANNING ACT

During the 1970s the adequacy of the planning system was under question and the rigid master planning approach considered inflexible. Municipalities had started to experiment with other types of plan, and the new Act which eventually appeared in 1987 took on board this developing practice. It also reflected the decentralisation trend of the time, providing 'framework' legislation which left most responsibility to the municipality. Its aims were to simplify the process, allow greater scope for variation to local circumstances and increase participation. Greater opportunities for participation were given to social groups such as environmental movements and tenants' associations. It symbolised a shift from the confident bureaucratic style outlined above to one which saw planning as more integrated into the political process (Holm and Fredlund, 1991). The consensus culture described above leads to a form of participation which is biased towards corporate institutions. Although individuals and small groups have a right to make observations on commissions and committees the regular consultation with organised interests dominates. At times grass-roots movements, based upon NIMBYism or such causes as environmentalism or public transport, would rise up and challenge the formal system. The 1970s was an example of such a time as exemplified by the demonstrations in Stockholm against the inner city redevelopment. This was the background that led to increased participation opportunities in the 1987 Act.

However, the political changes at the national level through the 1970s and 1980s delayed the formulation of the Act and led to its compromise nature. Ödmann (1987), writing at the time the Act was passed, feared that the resulting vagueness would be exploited by developers. She predicted that the interpretation and implementation of the Act would be prey to particular local political and business interests. The Act was passed by an alliance of the Social Democrat and Centre Parties with the free-market-oriented Moderates and Liberals objecting. They claimed that planning in the public interest was not in accordance with the European convention on civil rights.

The planning system which was set out in the Act has already been outlined in Chapter 3 (for fuller details see Kalbro and Mattsson, 1995). Central government confined its involvement to matters of national interest, giving responsibility for all other planning matters to the municipality. This has caused anxiety for some on the left who fear that without national norms and standards people without capital or landed interest may be left undefended against a right-wing municipality. Others (e.g. environmentalists) think that the national government needs to give a stronger lead on setting national environmental values (Hall, 1991a). The municipalities are required to produce a comprehensive plan for their whole area – some say that this was to ensure that they took their new powers seriously. However, the importance of these plans was much reduced during the process of passing the legislation (Hall, 1991a) as the legally binding nature of the plans was removed by legislators at a late stage in the process. According to Ödmann (1987) the Swedish Constitution does not allow general legally binding codes on planning matters to be adopted by municipalities. As a result the comprehensive plan does not have much importance in the development process, often being a description of current uses with vague designations of 'areas for further research' covering potential development areas. Fog *et al.* (1993) have criticised these plans for still taking the old technical approach and not utilising the opportunity to present political alternatives and engage public debate; without this connection to reality the plans become irrelevant. There are, however, examples where the plan has been used as a focus for discussion within the municipality and with citizens. The latest comprehensive plan for Göteborg focuses on the issue of a sustainable city contrasting this with an alternative vision of a competitive city. The plan looks at water and nitrogen cycles, toxins and energy issues (Berggrund, 1994). Göteborg featured as the Swedish showpiece at the OECD project 'The Ecological City' in Madrid in March 1995. The municipality of Lidingö has also put a lot of emphasis on defining ecological areas and corridors (Malbert, 1994) and this provides a link with the matters of national importance which central government still controls through the Natural Resources Act.

There has been some criticism that these comprehensive plans are too broad and vague, and in the larger cities more detailed insets to the plans are often prepared for key development areas. The comprehensive plans provide a framework for integrating public programmes and are only guiding documents. They are often rather thin on policy statements and provide a better framework for protection than development. The central instrument in the Swedish planning system is the legally binding detailed plan which is only prepared when development is expected. The planning authority has the monopoly in deciding whether a plan is necessary and in adopting it. Any developer who wishes to carry out development has to discuss their ideas with the municipality and get the agreement to proceed with the plan process. The actual plan is usually drawn up by the municipality but can be prepared by the developer or consultants. It will contain considerable detail on the use, location and layout of development although plans vary according to the nature of the development. As a minimum all plans must set out public land, use of land and boundaries of building sites and an implementation programme. The property owner cannot demand that a municipality should draw up a plan for his/her land and there is no appeal against a refusal to do so (Kalbro and Mattsson, 1995). Linked to the detailed plan in areas with complex ownership patterns is a subdivision plan which co-ordinates the subdivision of property units and easements. A building permit application is then made and if this conforms with the detailed plan, and subdivision plan if there is one, permission will be given.

An alternative to the detailed plan is also available to give legally binding effect in particular areas. This alternative is called 'special area regulations'. It can be adopted for limited areas not covered by a detailed plan where only a few aspects need to be controlled. It can also be used by central government to safeguard a national interest. Whereas a detailed plan can, for particular reasons, conflict with the comprehensive plan the special area regulations must conform to the comprehensive plan. The special area regulations might be used to preserve the line for communication facilities, prevent the conversion of second homes, or impose broad conservation policies.

The initial stage of preparing a detailed plan often involves preparing a programme which sets out the current situation, the purposes of the plan, and a timetable for the planning process. In some situations this is compulsory, though the comprehensive plan can sometimes fulfil the need. Both comprehensive plans and detailed plans require consultation exercises (for full details see Kalbro and Mattsson, 1995) and in both cases central government through its county offices will check, through compulsory consultation, that the plans have conformed to policies of national interest. For comprehensive plans residents can appeal against procedural issues but not content. In the case of detailed plans property owners and other affected parties who have submitted objections in writing have the right of appeal, first to the county and then to central government.

There are provisions in the Act for a simplified approval procedure where a detailed plan or special area regulation is considered to have limited significance and is of no interest to the public at large. Compatibility with the comprehensive plan and national interests through county scrutiny is still required. In the simplified procedure those with an essential interest in the development are still consulted and given an opportunity to make objections. A report on observations is appended to the documents submitted for adoption.

THE 1980S BOOM AND 'NEGOTIATION PLANNING'

Much of the impetus for the 1987 Act lay in the participatory climate of the 1970s. As already noted, by the mid-1980s the country was experiencing a development boom. In the earlier decades housing dominated building activity and the planning and control of this activity was set within a rational plan-led philosophy with the added controls of financial subsidy and land-ownership. As we have seen housing development was promoted and built by a small number of organisations, many of which were public bodies, closely connected to the central and local government bureaucracies. Much of the discussion that took place therefore occurred between these various bureaucracies. The 1980s saw a shift to commercial development backed by a huge influx of investment by banks and financial institutions. As elsewhere these developments were partially an expression of the de-industrialisation processes and advantage was taken of central sites previously used by docklands or railways. The expansion of commercial development led to a different set of organisational relationships set within a different, more flexible, planning regime. Thomas Hall describes this new organisational and planning approach in the following way:

> a consortium, consisting of building contractors and insurance companies, negotiates with the municipality. The consortium provides something which the municipality wants, while the municipality exploits its planning monopoly in granting generous building permits, and the insurance company takes care of the financing.
>
> (Hall, 1991, p.240)

This process has worried many people because of the secrecy involved and because it reduces the planning and political process to empty formality (e.g. Hall, 1991a; Fog, 1991; Tonell, 1991). Others (Snickars and Cars, 1991; Cars *et al.*, 1991; Snickars, 1991) see merits in the new approach to planning which they call 'negotiation planning'. However, they accept that it has been criticised for its lack of democracy and for leading sometimes to inefficiency as solutions become difficult to implement. However, it is claimed that the negotiations themselves are not the problem, but the way in which they are handled by the parties involved, for example the lack of preparation or openness on the part of the municipality (Bejrum *et al.*, 1995).

One question needs to be asked in relation to 'negotiation planning' – what exactly is new? Negotiation has always taken place within the planning system. It is not negotiation *per se* but rather the form it takes and the power of the various parties sitting at the table that has changed. In negotiations over housing developments during the 1960s and 1970s public authorities at national and local level had considerable power. In any negotiation they could use their sanctions over finance, land-ownership and planning. However, even during this period negotiations over commercial development could be very complex and political, as Hall's interesting and thorough account of Stockholm's city centre redevelopment shows (Hall, 1985). The negotiations over commercial development in the 1980s occurred in the context of the weak financial position of municipalities. They were also less likely to own the land in central locations as previous land banking had been geared to potential areas for housing growth. In dealing with commercial development they also lacked the kind of powers they could draw on to control housing developments. The financial pressure could cause a municipality to seek contributions from developers to help with their budgeting problems (Cars *et al.*, 1991). There was also the developing attitude, linked to the move to the right in politics, that planning was too rigid and inflexible. It could be said that as a result of these conditions in the 1980s there was a shift in the balance of power from public authorities to the private sector. The initiative came from private entrepreneurs who often offered packages to the municipality (Engström, 1991). As a consequence of this shift, the planning process, built around the principle of increased participation, was sometimes by-passed and manipulated. Let us look at this in more detail.

According to the legislation which seeks to ensure political control and participation, the planning process follows a sequence from idea through preparation of the programme, the detailed plan, public participation, adoption to implementation. Cars *et al.* (1991) have described the logic behind this process in the following way. In the early stages the political priorities are established and the broad parameters of the development identified, and this provides the basis for drawing up the detailed plan. This is then discussed in public and perhaps modified before final adoption. Then the implementation process begins as the developer applies for a building permit. However, according to Cars *et al.*, certain changes took place during the 1980s. First, as there was an increase in commercial development, the initiative would often come from the investor or developer, perhaps by sounding out politicians about a development proposal. The developer is then likely to be strongly involved in the programme stage leading to agreements between the developer and municipality. In the booming real estate climate there was also an urgency to complete the planning process quickly.

'Development agreements' provide a vehicle for conducting negotiations between the developer and the municipality outside the formal planning process. They are conducted in secret, do not involve participation and can be used to manipulate the planning system (Mattsson *et al.*, 1989). These agreements are contracts between developers and the municipality to allocate the costs and responsibilities for infrastructure and public facilities. However, they can also cover matters such as land uses, design standards, tenure, density, etc. Now the crucial aspect of these agreements is that they are often prepared before the detailed plan is drawn up. Thus there is no legal planning framework which they have to follow. The comprehensive plan is advisory and the detailed legal plan is formulated after the negotiations over the agreement. These negotiations will have taken place behind closed doors and once they have been concluded the local authority will have a considerable interest in ensuring that the subsequent detailed plan conforms to the agreement. It is possible that the participation process involved in the detailed plan could upset this. However, first of all there has to be sufficient public awareness and interest and, secondly, objection has to be strong enough for politicians to feel electorally threatened and thus reject the approach they have adopted in the negotiations. These earlier commitments mean that they are deterred from any change at the later planning stage. Thus while the planning system of the 1987 Act seems to provide a comprehensive, integrated approach which protects long-term interests and provides for public discussion, the economic realities of decision-making during the 1980s seemed to have undermined these aims.

The project which is often referred to as the first to express the new form of negotiation planning is the Vasa Terminal, which included the World Trade Centre, in Stockholm (Cars *et al.*, 1991). The site involved some sophisticated engineering as it was largely on a deck over railtracks next to the Central Station. The city council needed a new bus station on the site but because of these engineering difficulties it had always been too expensive. In the new favourable commercial climate the idea of building offices above the station was mooted as a means of paying for the public facilities. The state railway company and the city council joined forces and launched a competition in the autumn of 1982 which covered not only design aspects but financial viability. The winning consortium then entered into complex negotiations and eight agreements were signed, covering such matters as finance, ownership, rights and responsibilities. The essence of these agreements was that the state railway company and the city did not contribute any of their own money, the public facilities being financed by the yield from the offices. A more notorious example of negotiation planning was the Globe sports complex project also in Stockholm; this will be described in detail in the next chapter. However, such projects were not confined to Stockholm and most large towns had their own version. For example, in Malmö there was the Triangle project. A site became available in the centre of the city and the original intention, which was the subject of a competition, was to have a new concert hall, some offices, pedestrianised open space and an indoor bus terminal. The fate of this scheme illustrates the way in which decisions were made in negotiation with developers outside the normal planning process. The municipality could not afford the concert hall and came to an arrangement with a developer on another site. This site was allocated as a trade centre with a height limitation of eight storeys. However, there was no commercial interest in this use. Under the deal the municipality agreed to let the developer build a fourteen-storey hotel in exchange for building a concert hall as part of the development to be rented out to the municipality. Meanwhile, back on the Triangle site investment consultants, who became familiar actors during the 1980s, were drumming up support for a new package. This included a considerable amount of indoor shopping and the consultants took politicians to the United States to show them what was possible. Eventually a scheme was put together dominated by shopping and offices and including a twenty-two storey hotel. Controversy over the scheme contributed to the failure of the Social Democrats to win the 1985 election after sixty-six years in power. However, the new council only made small modifications to the scheme.

Thus the planning approach that evolved during the 1980s created a number of concerns. These focus around the question of where the power lies in the decision-making process and whether there is sufficient accountability. It is often difficult to identify who is responsible for a decision as political and economic considerations become more and more entwined and chambers of commerce, banks, corporations and interest groups all become involved (Tonell, 1991). The division between the public and private sectors becomes very blurred (Hall, 1991a).

ENHANCING THE MARKET IDEOLOGY

We have seen that planning controls in the past were accompanied by controls through housing subsidies and land-ownership. This meant that residential development was directed to a large degree by the public sector. We have also seen that during the 1980s there was an increase in commercial development where controls are weaker. Then during the 1990s a more market-oriented approach to housing developed (Elander, 1994). The housing subsidy programme was under pressure in a faltering economy, as it was a large item of public expenditure, and even the Social Democrats were accepting that cuts had to be made. However, the arrival of the non-socialist coalition government in 1991 brought a stronger ideological dimension to the process of deregulation and one of their key aims was to create more 'free choice'. As an expression of this one of their first moves in office was to break up the Ministry of Housing and allocate the responsibilities to seven different Ministries. The Minister responsible said, 'partitioning the Ministry of Housing is one step towards the abandonment of the special treatment given to the housing market. The liquidation is one stage in the development of the market economy even as regards housing' (quoted in Elander, 1994, p. 104). The new government also thought that a major shift was necessary in the housing legislation and set up a Commission to examine this. As a result municipal housing companies will have to operate in the marketplace with no special privileges, and the allocation of housing will be increasingly privatised. The Commission also looked at simplifying the development process and recommended streamlining the building permission process. It also suggested that building permits should not be needed when there was a detailed plan. The philosophy was that quality control should be left to developers and not be the concern of municipalities.

Another Commission was therefore set up to review the 1987 Building and Planning Act. However, as the government was made up of a coalition of three parties they all had an influence on the Commission and had different perspectives. The chair of the Commission and the Minister for the

Environment were both members of the Centre Party who wished to strengthen environmental controls. The Minister of Culture was a Liberal and wanted more conservation safeguards. The Minister of Trade and Industry was from the Moderate Party, the leading party in the coalition, which supported the interests of the construction industry and developers. Thus the Commission was given a compromise, and probably contradictory, set of objectives; to give greater consideration to the environment, strengthen public participation and explore means of deregulating and simplifying the regulations. The Commission was also asked to make the distinctions between the roles of central government, municipalities and the private sector clearer. The Commission reported in several stages.

The first report looked at the question of simplifying the building permit and suggested dividing the permit into two parts. The first would be a locational decision and kept as part of the Act. The second part would cover the technical standards and this would be considered in a new Construction Act. Municipalities could then decide in their detailed plans whether a locational building permit was needed or not, thus there is potential for removing this requirement. The second, technical, aspect would be the subject of discussions between the developer and municipality. The developer could set up their own internal arrangements, appoint a consultant or use local building inspection.

In the second stage the Commission looked at planning aspects and discussed a number of ideas. It was asked to look at the effectiveness of participation and the appeal system as central government was having to deal with about 800–900 appeals a year. It was suggested that participation could occur at an earlier stage in the process – at the idea or programme stage. This is currently optional and would be made the rule. Discussions are under way about linking this to a limitation of participation opportunities in later stages of the plan process to those with a property interest. Currently all plans have to go to a full council meeting, and the Commission explored the possibility of delegation of some cases to officers.

It was thought that municipalities could use the comprehensive plan more effectively to allow speedier local plan procedures. The suggestion was made that the plan area could be broken down into different zones with differing degrees of planning control. In some zones the simplified planning process mentioned earlier, that exists in the Act for small schemes with no broad impact, could apply and thus be more widely used. In other zones the municipality could set standards but not require a detailed plan. Other zones could continue to require a detailed plan although in some the locational building permit would not be required. Thus a more differentiated system could apply with opportunities for greater deregulation.

Previously municipalities could prohibit food retailing in a detailed plan in order to preserve existing town centre establishments. This was made illegal in a 1992 Act on businesses which aimed to extend the principle of free competition. The Commission was asked to investigate whether this should be applied retrospectively to existing plans. On the environment front the Commission looked at what was needed to integrate with the EU requirements on environmental impact analysis (EIA). However, further work in this field was restricted as the government was also working an a new 'basic' or framework law on the environment with which all other acts would have to conform. This Act will enable central government to set out environmental quality standards for the physical environment. The Commission was also asked to explore how to deal with issues of national heritage. Although the Natural Resource Act of 1987 empowers central government to set out issues of national importance this is rather vague and in practice a lot of disputes have resulted which only tend to be resolved through detailed plans. County administrations have to check these plans against national policies and are often caught in disputes between municipalities and central government.

There were some general trends evident in the thinking of the Commission. One was towards greater simplification with less control on the part of the planning system, the second was less public participation even though the brief was to expand opportunities. It is well known that people are less inclined to participate at the general level where the new opportunities are being proposed, and they may have fewer opportunities to do so when the details are known. There is also the possibility that although environmental matters would be enhanced through the new suggestions this would lead to greater centralisation. However, the change in government in 1994 means that it is unclear how many of these ideas will be pursued – for example, the Social Democrats have indicated that they intend to start from scratch in thinking about the environmental framework legislation.

So far we have been looking at how the increased emphasis on market approaches has affected thinking at the national level. How has this affected the municipal level? In some cases, such as Malmö, where local Moderates have been in control, the market orientation can also be observed in local political approaches. Local politicians in the Moderate party are often less ideological than their national counterparts however, as they are more aware of implementation problems and have in many cases co-operated in the past with local Social Democrats. This can be illustrated by the example of Tyresö municipality which lies about 15 kilometres south of the centre of Stockholm (Hellsten, 1993). Here the council changed in 1991 from a Social Democrat-led council to a Moderate-led one, and the new council was keen to introduce a new market-oriented approach. They had ideas to free housing from municipal control by abandoning the housing programme, turning the municipal housing company (TYBO) into a private company and encouraging public housing tenants to buy a stake in their flats. However, in the event they perhaps introduced even greater controls as they were subjected to local NIMBYism. Financial problems also caused difficulties in implementing their approach. About half of TYBO's income came from state grants which could not be replaced by increased rents as this would have involved an impossible 50 per cent rise. TYBO therefore faced bankruptcy and the municipality was dragged into the problem as it provided the security on loans and the whole municipal budget was threatened. The idea of selling off part of the property rights in the flats, changing the tenure from municipal to co-operative, failed because the banks did not think that this was a secure enough deal and would not lend people the money. The sensitivity of the banks has been a crucial issue since the collapse of the banking system. Some people have suggested that it is the banks, not the developer or the municipality who now control the planning system (Wretblad and Lindgren, 1993). Before they will provide finance for a scheme they require very detailed financial information and evidence that contracts exist with future tenants.

The organisation of the planning function at the local level can however display evidence of a greater market orientation. We noted earlier the move to increase market criteria within local government administrations, often using the producer/client approach. Sollentuna, a municipality with a Moderate Party-led coalition just to the north of Stockholm, claims to be one of the first to adopt such an approach (Eriksson, K., 1993). It aims to clarify the relationship between politics and the market, reduce staff and open up opportunities for privatisation. The administration has been reorganised with a central board looking after council-wide issues, including plans and municipal real estate. There are then five other boards covering the council's work, one of which is the Urban Development Board serviced by about eight staff. These boards take their orders from the council which sets the brief and allocates funds. The boards then organise the work programme. This is done by buying in a project leader through open tender, though in practice this has been an internal person. The project leader is then responsible for carrying out the brief within the allocated budget. He/she is on a fixed-term contract and is responsible for appointing a team from either within the council or from consultancies. The council has retained a core staff of planners for statutory activities such as carrying out participation. The update to the comprehensive plan is, however, being carried out by consultants.

The last twenty years or so have seen some major changes in the economic fortunes of Sweden and some fundamental questioning of the way decisions should be made. This has involved ideological shifts within the Social Democrat Party and also an increase in support for more market-oriented parties. From a fairly stable system of power in the immediate post-war period there is now considerable flux in the relationship between central government, local government, the private sector and the public. Such uncertainty has provoked considerable debate and has had its impact upon the planning process. During the 1960s there was much negotiation in planning decisions but political leadership was strong and there were clear principles about social equality. The prodigious house-building programme of this decade came to an end in the 1970s and there followed a period of political caution, and demands for participation, with much public protest against development. However the climate changed again in the 1980s with a big upsurge of private sector interest in property investment. This time the form of negotiation differed as municipalities had less money and the private sector took more of the initiative for development. This shifted power towards developers while politicians acted as bargainers not leaders. The goals of development also changed, with less concern for social issues and more emphasis on investment returns and image. A further shift to an emphasis on market stimulation and infrastructure projects took place in the 1990s as the development boom collapsed. The decision to join the EU has boosted the emphasis on marketing cities and regions to attract investment. As we shall see in the next chapter, Malmö has joined Copenhagen in promoting their region as Scandinavia's bridgehead to Europe. Similarly there has been activity to promote the Stockholm–Mälar region. A voluntary council of counties and municipalities has been preparing a strategy to market the region as one with great potential for business investment (Stockholm County Council, 1994). The good infrastructure, quality of environment and concentration of research activity are highlighted. Alongside this emphasis on economic development there has been increasing attention to environmental aspects both nationally (Ministry of Environment and Natural Resources, 1994) and locally. By the end of 1994 about 200 municipalities were actively taking part in Agenda 21 work (Elander *et al.*, 1995). The next chapter will explore some key projects which illustrate in more detail the themes of the 1980s and 1990s.

10

SWEDISH CASE STUDIES

INTRODUCTION

This chapter explores three case studies in detail. As with our examples from England and France these case studies are drawn from the capital city (Stockholm) and a second order city (Malmö). The case studies give greater detail concerning the themes and trends identified at a general level in the last chapter and, as they span the 1980s and 1990s, illustrate the changing emphases and issues of concern over recent years. We start with the Globe project in Stockholm which is an example of the property-led period of the 1980s and a well-known case of 'negotiation planning'. One of the most controversial schemes in the capital during the 1990s has been the proposed road building programme which has caused much concern over democratic procedure and environmental impact leading to demonstrations in the capital reminiscent of the 1970s opposition to city centre redevelopment. The road proposals have been incorporated into a transport programme called the Dennis package, and this is the subject of the second study. One of the arguments used to support the package has been the need to upgrade the infrastructure in the light of European competition. This argument is even more central to the third case study, the project to build a bridge over the Öresund from Malmö to Copenhagen. The joint region is promoted as an important European bridgehead for Scandinavia. The planning implications of this proposal will be examined.

NEGOTIATION PLANNING OF THE 1980s: THE STORY OF THE GLOBE

Some of the themes of the 1980s, outlined in the last chapter, will now be explored through the example of a particular project, the Globe Arena (Cars *et al.*, 1991; Sahlin-Andersson, 1992). This development has included the construction of an 85-metre-high dome which has been described as the world's largest spherical building. It is visible from a large area of Stockholm and has become a major city landmark. The implementation process has been described as an early example of the new negotiation style of planning.

The city had been looking for a new sports arena, and the search acquired some urgency because of the wish to host the World Ice Hockey Championships in 1989. Real estate concerns had been trying to interest the city in their particular projects, but in 1982 the political decision was taken to locate the arena in the Globe area which was owned by the municipality. This area covered 16 hectares and already had an older ice hockey arena and football stadium. The current comprehensive plan dated back to 1954 and provided no particular guidance, simply zoning the area for its existing uses of sports arena and slaughterhouses (by then redundant). The leisure department of the city council set about preparing a programme to guide the development. A wider group was then formed, drawn from the planning, leisure and real estate departments, to pursue the programme. As the project was seen to lack impetus it was thought that the inclusion of an office element would give it a greater profile and put the development on a sounder financial footing. The aim was to try to finance the sports arena from the commercial development and hence reduce the burden on local taxes. Meanwhile the planning department had been discussing the idea of decentralised sub-centres, and the bus station next to the Globe site was suggested as one possible node. There was also the desire to attract offices to the south of the city to improve the north/south imbalance of job opportunities. The Social Democrat-led municipality was keen to support new projects as there were fears of unemployment among construction workers.

This was the background to the political decision taken in 1983 to try to stimulate ideas and attract investment. However, initial interest was low. In 1985 a preliminary plan for the area was produced as a guide for a competition which was launched the same year. The purpose of the competition was to develop the design and financial aspects of the project. The plan, which contained illustrative diagrams, specified the amounts of each land use. There was the arena itself intended to house international sports and cultural activities and seat 13,000 people. Then there were the commercial activities of offices, shops and hotels amounting to between 75,000 and 90,000 square metres, but with a seven-storey height limitation.

It was intended that the development should be provided with a green setting through 2.9 hectares of park, with a garden feature acting as a focus. The final brief for the competition followed this plan with detailed specifications of land-use quantities. At the time the conservative opposition in the council tried to ensure that the preliminary plan was not a fixed condition for the competition, but they failed in their efforts. However, the foreword to the brief said that the specifications could be treated flexibly and that new ideas and suggestions were welcome: 'it is up to the participators to try out, further develop and design the brief's requirements, ideas and intentions'. Although the municipality approached about thirty big companies, only four groups submitted complete project proposals and one of these was dismissed as it was considered too expensive to realise. Because of the importance of the financial dimension of the competition the groups were made up of financial, real estate and construction companies, architects and various consultants. The winning group was announced in the spring of 1986 and provided the city with a favourable financial deal, attractive because of the city's financial limitations at the time, and also an appealing and unique design concept in the form of the Globe. This dramatic architectural statement coincided with the desire to market the attractions of Stockholm more aggressively. The Globe arena provided a facility which could be used for many large events and contributed to the tourist attraction of the city.

A process of negotiation then followed, with one politician, one official from the leisure department and one from the real estate department taking leading roles. Because the winners had dealt with concepts at the expense of detail, there was still much to discuss. An initial agreement prepared between the winners and the city was passed a few weeks later. In this it was agreed that the group would build the arena for the city and in return would receive ownership of the land with building permission for 143,000 square metres of commercial use. The Vassa Central Terminal project outlined in the last chapter set the precedent for this kind of exchange. There the developers had provided the bus terminal in exchange for the enhanced commercial opportunities. However, some people were doubtful that such an arrangement would work in the case of the Globe as the site was further out of town and the commercial returns less secure. The negotiations over the development agreement then took place in secret while the detailed plan was being prepared in parallel. It had been intended that the preliminary plan used for the competition would be taken through the necessary committees and formalised as the detailed plan for the area by 1986. In the event continual modifications had to be made to the plan as the negotiations proceeded. There was no public participation as the modifications were made. The detailed plan was finally formalised towards the end of the year taking into account the results of the negotiations. The office space was almost double that in the original plan and included four towers of 12–14 storeys – a doubling of the original height limitation. The concept of a garden focus was missing and the park area reduced from 2.9 to 1.8 hectares.

The city council at the time was run by the Social Democrats who had always seen the ownership of land by the municipality as a key tool in implementing social policy and the retention of this land as a top political objective. However, another aspect of the negotiations over the Globe Arena was that the council, for the first time, handed over the ownership of land to the developers, even though the competition brief had stated that the land would be let on leasehold. In return the city gained a world class sports and entertainment arena seating 16,000 spectators. On 19 February 1989 the Globe was opened with great ceremony, accompanied by a light and sound display from the French musician Jean-Michel Jarre. The popularity of this arena facility has possibly detracted from a full discussion of the lack of public debate in the planning process. The comprehensive plan, which is supposed to include public discussion over development guidelines, was out of date and inadequate, while the conflation of negotiations and detailed plan preparation over a short time period meant that there was no opportunity for public participation at this later stage.

The arena was used for the 1989 Ice Hockey Championships and has since become a popular venue for other attractions such as rock concerts. The rest of the development was complete by 1993 but by that time the property market had slumped and it proved difficult to gain sufficient returns from the commercial development – the company eventually going into partial liquidation. There have also been heavy running costs for the municipality as the Globe developed maintenance problems. The decision to combine financial, design and planning matters throughout the process contributed to the merging of financial negotiation and planning considerations. There have also been questions raised about whether sufficient thought had been given to the traffic generation aspects of the development, a probable consequence of the streamlined planning process. The need for a fast track planning and development process was accentuated by the requirement to produce the arena in time for the Championships. Here we see a parallel to the Olympic procedures in Barcelona mentioned in Chapter 4. It might be said that these are occasional happenings and not significant in planning terms but the increased competition between cities in Europe for such image-generating facilities is likely to maintain this pressure for speedy processes. There was of course nothing illegal about the process adopted in the Globe case. The flexibility available in the system was utilised to the full and the negotiation over the development agreement allowed to dictate the planning process. The latter was reduced to minimal proportions and broader considerations and participation avoided. This is not to say that organised opposition could not have challenged the proposal. However, the attractions of the new public facility and the image of the Globe ensured that this did not happen. One of the commentators on the Globe project, Sahlin-Andersson (1992), has suggested that the Globe was an extraordinary project and so one would expect its decision-making path to deviate from the clear rational logic of the planning system. She suggests that the actual labelling of the project as extraordinary by the actors involved gave it a certain legitimacy for avoiding the normal planning system. This of course raises the issue of what criteria are used to warrant the 'extraordinary' label – it might be suggested that this can sometimes relate to the amount of profit involved or the degree of public opposition expected. The other suggestion raised by Sahlin-Andersson is that the planning system in the Globe case provided a façade of respectability behind which the negotiation approach could hide. The need for such respectability could be said to relate to the prevailing Swedish ideology which expects consensus and openness.

THE DENNIS PACKAGE: RESTRUCTURING URBAN POLICY FOR THE 1990S

During the 1990s a number of strategic transport policies have been proposed for the Stockholm region and these have generated considerable controversy. The background to these policies will be explored with an emphasis on the decision-making process and the interests that have been involved. The case highlights the evolving relationships between central government, regional bodies, local municipalities, private interests, local and national political parties and the public. All these interests have differing attitudes to the issue and it is interesting to analyse how decisions were taken within such diversity and which interests were more powerful in shaping the outcome.

City competition and the motorway revival

The idea of an inner ring road round central Stockholm was first mooted over a hundred years ago. The western part of the ring was built during the 1960s which was a high point of such technological solutions in many European cities. However, mounting opposition prevented any further building. Environmental issues came to the fore with much direct participation opposing new roads. From then on this mood dominated the political agenda, reinforced by the shortage of public finance. The late 1980s, however, saw a resurrection of the support for roads. It was stated that the cessation of investment in infrastructure was now causing problems for the region (Stockholmsleder, 1992). If the city was to compete successfully in a European arena and attract investment something had to be done to redress the situation. National policies on regional equality meant that the capital region had been less favoured in infrastructure resources through the 1970s and 1980s. This was now said to be wrong and that in the competitive climate the Stockholm region should be forcefully promoted, as 'the Stockholm region functions as an engine for all of Sweden' (Lagerström, 1992, p.117). It was pointed out that the region contains roughly half of Sweden's research and development activities and is the home base for numerous multinational companies. It was claimed that improved accessibility across the region 'will provide the necessary precondition for favourable economic development' (Stockholmsleder, 1992, p.3).

Critics of the new policy emphasis, and its reliance on roads, are keen to point out the similarities between these economic arguments and those used in the 1960s (Eriksson, E., 1993). It was then said that if Stockholm was to survive in the contemporary world it had to modernise its city centre. City plans were drawn up to carry this forward and these were given academic support. The research of Professor Kristensson claimed to show that Greater Stockholm was the only Swedish region which could compete with the Continent's metropolitan regions, and thus it was vital for the entire country

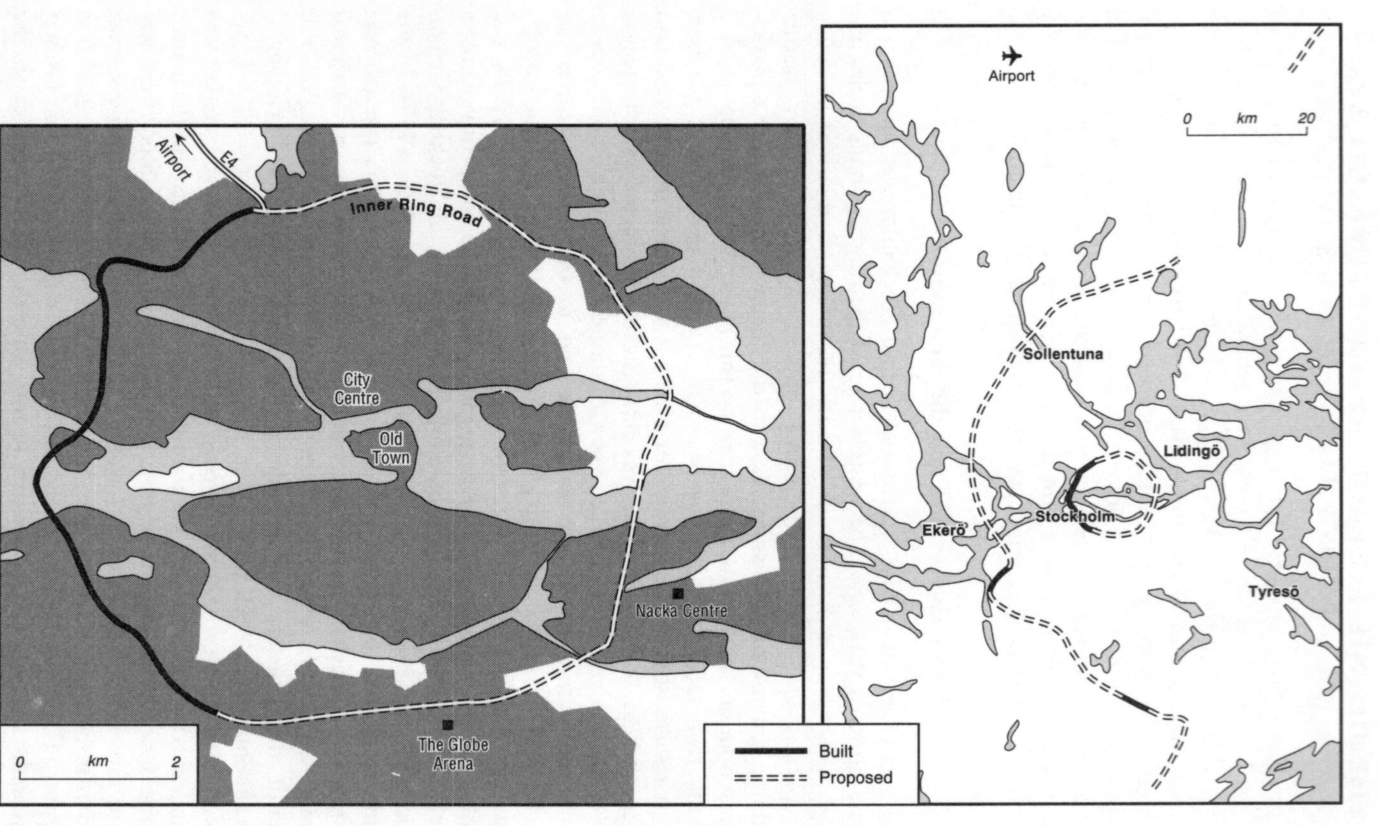

Figure 10.1 Stockholm

to take advantage of this opportunity (Hall, 1985). He believed that it was necessary to create the right kind of modern building and environment in the centre which would attract top management functions and large companies. The city plans of the time pursued these ideas and resulted in a huge redevelopment with an emphasis on car accessibility, a road tunnel, multi-storey car parks, pedestrian subways and soulless tower-blocks (for full details see Hall, 1985). Eventually, popular opposition stopped the programme but not before a large proportion of the centre had been redeveloped. These areas are now generally agreed to have destroyed the humanity of a large part of the city centre and said to be one of Stockholm's main problems (e.g. Eriksson, E, 1993: Söderlind, 1993).

This resurrection of the idea that the economic growth of the region needs better regional communications could be seen to benefit individual interests within the region, whether this be a multinational firm such as ASEA Brown Boveri with its headquarters in Västerås 100 kilometres or so to the west of Stockholm, high-tech firms in the university city of Uppsala 75 kilometres north, or Stockholm firms wanting to get to the international airport at Arlanda. After the banking crisis of 1990 and the end of the development boom, arguments arose in support of infrastructure as a means of continuing construction activity and employment. Although benefits might be seen in terms of the finances of major construction companies such as Skanska, the employment opportunities are likely to be limited because of the highly mechanised nature of this kind of work (Eriksson, E., 1993). Within this general supportive climate, there was one group of companies that took a particular interest in promoting new road building in the capital.

Private interests and Projekt Österleden

The municipality of Nacka lies to the south-east of Stockholm and although not far from the city centre communications are not good, particularly to the international airport north of Stockholm. The conservative leader of the municipality, Erik Langby, wanted to get Nacka 'on the map' and to promote its redundant manufacturing area as a major new office centre. In 1986 he joined forces with his conservative political colleagues from the city and county to resurrect and promote the idea of an eastern motorway round Stockholm which would greatly improve the accessibility of Nacka (Johansson, 1993). An eastern by-pass to the old town had been first mentioned as early as 1866 and in 1948 there was an international competition. An eastern route with motorway bridges was part of the 1960 plans but was dropped when the rate of population increase slowed down and the environmental movement raised objections. The new dimension that was introduced this time was the self-financing aspect whereby the road would pay for itself through tolls. This was particularly significant given the public expenditure difficulties.

Under Langby's leadership the municipality of Nacka organised exhibitions and seminars to promote the eastern motorway idea. Langby was well connected to the business community in Nacka who soon mobilised round the project. Four companies set up a special organisation called Projekt Österleden to advance the scheme. These companies, Arcona, Atlas Copco, Fläkt and Skanska all had major landholdings in Nacka which would increase in value if accessibility could be improved. It is interesting to note that although by 1994 no work had started on the new roads, the publicity had increased interest in the speculative Nacka Strand office centre so that by that year it was built and flourishing. However, this property dimension was not the only interest these companies had in the new road. They all had an interest in increased road building *per se*. Arcona owns Philipsons, who sell cars; Atlas Copco make machines for tunnel drilling (most of the new motorway was proposed to pass through tunnels); Fläkt make ventilation systems; and Skanska is the largest construction company in Sweden (Johansson, 1993).

Under the auspices of Projekt Österleden a publicity campaign was mounted (Tonell, 1993). This was partly oriented towards gaining public acceptance, through distributing nearly 100,000 glossy pamphlets to people's homes and issuing press releases which were taken up uncritically by the press (e.g. by a leading journalist in the influential paper *Dagens Nyheter*). At public meetings various images were deployed to gain support for the road (Tonell, 1993). It was sold as a means of 'releasing the inner city' by keeping traffic out of the medieval core. Promoting the image of Stockholm as a modern city, it was said that the only other capital city that allowed traffic in its medieval area was Tirana, capital of Albania. This particular image was dropped after a speaker at one of the meetings pointed out that Tirana had no medieval centre nor for that matter many cars. It was also repeatedly suggested that as the ring road had been started it was only logical to complete it with the 'missing link'. However, the campaign was not only oriented to general support; there was also direct political lobbying. At the time of the 1988 elections the Moderate Party (conservatives) and the Liberal Party supported the project. So in order to get a political majority the lobbying was directed mainly at the Social Democrats, with activities such as dinners and offers of talks to internal party meetings (Tonell, 1993). This lobbying was not only at the local level but also at the national level, as national legislation would be necessary to deal with the idea of road tolls.

The move to the national stage

The lobbying took place at a time when Social Democrats were concerned about their support in the major urban areas. They had set up a Commission in 1988 to look at urban problems in the three large metropolitan areas and, although this covered a wide spectrum of issues, it was the problem of traffic in city centres and the idea of by-pass roads which were taken up. Here the Österleden lobbying might have had an effect. To progress the solution to the traffic problem three influential figures were appointed, one for each city, with the responsibility for developing a transport strategy. In Gothenburg a senior conservative politician, a freelance consultant, was appointed – he had previously been leader of the Moderate Party and a mayor of Stockholm. In Malmö a Social Democrat politician who had previously been Minister of Transport and mayor of Gothenburg was appointed. The Stockholm choice was Bengt Dennis, the Director of the Bank of Sweden, who was appointed on 5 April 1990. His remit was to get a politically agreed transport strategy for the Stockholm region which would withstand the cycle of elections and overcome political fragmentation. One of the phenomena of the 1980s in Sweden was the increase in popularity of smaller parties such as the Green Party and local parties (e.g. the Stockholm Party) and this made local politics more unstable. According to a report from the director of regional planning of Stockholm county,

> the reason to choose the rather unique model of a government-appointed negotiator helping the region to reach an agreement was that there has been extensive political disagreement among local and regional officials concerning measures and their funding – which has led to paralysis in political action. Another important reason was that the region for many years had argued for more state funding.
>
> (Malmsten, 1992, p.3)

This last point relates to the previous national policies of regional balance and the move to shift more of the resources behind the Stockholm region as the economic motor for the nation.

Dennis began his work by inviting all political parties to a meeting to express their views. They came with their ideas but soon afterwards Dennis announced he could only work with the three main parties (Moderates, Liberals and Social Democrats). These three parties held between them about 70 per cent of the local votes and even more at the national level. Utilising the existing research comprising sixty-one different reports, the senior representatives of these three parties from the Stockholm municipality and Stockholm county then negotiated an agreement with Dennis. It should be pointed out that as well as the Stockholm municipality there are twenty-four other municipalities in the Stockholm county which would be affected to some extent by the transport proposals. These municipalities were not involved in the negotiations and it was assumed that the county representatives would look after their interests. However, this was not to be the case and much antagonism resulted (Gauffin, 1993). Utilising existing plans and programmes an interim agreement was reached in January 1991 and a final version in September 1992. This was a detailed document setting out all the transport projects in the region to 2006 with precise financial arrangements.

The contents of the Dennis package

The 'plan' is generally referred to as the 'Dennis agreement' or 'Dennis package'. It contained a bundle of public transport measures including new rail links, although about half the total allocation of 15.8 billion Swedish kronor (SEK) for public transport would be spent on upgrading exiting facilities, especially the subway. The elements in the road strategy were the completion of the inner ring including the eastern by-pass, which was the object of Projekt Österleden, and a western outer orbital road. The total cost was estimated at 18.2 billion SEK (although estimates have increased since), and all of this would be funded from tolls on these roads. According to Johansson (1993) all the elements in the package would have happened anyway except for the eastern by-pass, the western orbital and the tolls. The public transport measures were generally of a piecemeal nature and of the kind (e.g. upgrading of stock), which any public transport authority would be forced to undertake.

A major feature of this approach to decision-making was that the agreement had to stand as a complete package. There was no opportunity to change any individual element within the agreement. The national government did not attach its signature to the agreement which then had to be presented to Parliament for the necessary loan sanctions, financial commitments and legal requirements. The idea was that the agreement stood or fell as a complete package. The assumption was that the three parties all got something they wanted from the agreement: the Moderates and Liberals got their roads while the Social Democrats gained public transport investment and improved north/south communications through the western orbital road. The Social Democrats were keen on this last aspect because they believed that it would help to redress the imbalance between the richer north and poorer south. However, this western orbital road caused a political problem for the Liberals. The municipality most affected by the road, Ekerö, was controlled by the Liberal Party and the local politicians there were strongly against it. The Party tried to solve this dilemma through a footnote to the agreement which said that they thought that this part of the package should be delayed until 1996 for further study.

Twenty Public Boards were asked for their views on the agreement but the request came in the middle of the summer holidays and they were only given a few weeks to respond before the September deadline (Tonell, 1993). Having reached an agreement there is no further opportunity for anyone to comment upon or change the contents. The public and those bodies or parties not included in the negotiation do not get an opportunity to voice their opinion. Many public bodies, including those with transport expertise, expressed their disagreement (Gyllensten, 1993).

As the negotiations proceeded the opposition to the Dennis package increased, especially from the environmental viewpoint. For example, in

November 1993 thousands of protesters marched in a torchlight procession through the centre of Stockholm bringing back memories of the environmental opposition of the 1970s. Modifications were made to the road proposals to make them more environmentally acceptable. Most of the inner ring and some of the western orbital would be in tunnels and a lot of stress was placed on the improved environment possible in the centre. Elaborate ventilation systems would be placed in the tunnels and a special architectural competition was mounted for the design of the air vents. One idea was that these should be made of clear glass to show the cleanness of the air. However the larger environmental issue is about the increased reliance on the car that the roads will bring and concern has been expressed about the more dispersed commercial and settlement pattern that will result. Eriksson fears a spiral towards a US pattern in which cars bring increasing environmental problems – 'a car society which bites its own tail' (Erikson, E., 1993). It has also been pointed out that the method of financing the roads will encourage more and more car usage which it will be impossible to stop. Tonell believes that to meet the costs of the roads about 25–35 per cent more car traffic will be needed than that predicted (quoted in Johansson, 1993) and that this will limit the degree of support possible for public transport. E. Eriksson (1993) also pursues the 'financial trap' of the Dennis package. She points out that the money is being spent now in providing the infrastructure and this will then be paid back in the future through toll charges. Costs and interest rates are highly uncertain and society in the future will have to pick up the bill. The more cars, the more money; therefore it will be very difficult to counter car usage politically.

Fast-track implementation

In effect the Dennis package has become the official plan. As Johansson puts it,

> the Dennis package is an unusual form of regional planning, since it is not a plan but an agreement between three political parties initiated by central government. The regional plan has been adapted to fit the Dennis package; the agreement has become the regional plan.
>
> (Johansson, 1993, p.20)

The process has moved on to the implementation stage. The intention is to move amazingly fast; for example, the northern link of the ring motorway was originally scheduled to be built from 1993–7, the eastern link from 1994–8 and the southern link from 1992–7. These links and the other new road and rail proposals will need between 150–200 detailed plans to be prepared by the relevant municipalities. Most of these are in Stockholm itself – it has been said that half the time of the planning department was being spent on this task in 1994. At the same time Ekerö municipality, which was opposed to the western orbital, was doing nothing. One of the clauses in the agreement was designed to ensure that no municipality could veto the proposals; it requested that

> the government authorities, in approving this Agreement, clearly declare that the railroad and road projects contained in the Agreement are of very great national interest and therefore that none of the projects in the Agreement may be stopped by municipal veto.
>
> (Dennis Agreement, 1992, p.25)

The detailed plans contain the requirement for public participation and it will be interesting to see what effect this has. However, these plans only cover the precise design requirements for each individual section of road and do not allow for the debate over the broader issues. There is also such a head of steam behind the Dennis package, which as we have noted has to be treated as an integral whole, that politicians may ignore any opposition. There may, however, be some local difficulties where there is local Liberal opposition or where the smaller parties hold a balance of power. The Social Democrat Party also has its own internal differences. Although the party line is to support the package many of their politicians are not happy with the proposals on environmental grounds. In April 1994 the national Parliament approved the loan guarantee for the northern and southern sections of the ring road but only the planning costs of the more controversial eastern section. The Centre Party was a member of the national government coalition and had been stressing environmental issues in its party programme. It was not involved in the Dennis package as it had no representation in the Stockholm area. When the issue came to the national level it was uncomfortable with the decision to go ahead with the roads but as a minority member of the coalition it had to agree. The only other option was to break up the coalition. However the Minister for the Environment, who was a member of the Centre Party, was instrumental in setting up a special public hearing in May 1994 to discuss the package. The elections in September that year brought changes to government at national and local level and as we shall see later this brought uncertainty to the future of the package.

The agreement envisaged that the National Road Administration would set up a special organisation to co-ordinate the planning, construction and operation of the new roads. Stockholmsleder AB, a limited company, was set up to do this as a 'slim-line' agency awarding commissions to consultants and contractors. In a progress and publicity document published in December 1992 it stressed the magnitude of the operation, 'the largest road project that has ever been undertaken in Sweden'. It explained how this had to be accomplished within seven years and thus how important it was that no one rocked the boat: 'it will be necessary, particularly in the case of the decision-making bodies, to make every attempt to respect and devote the necessary efforts towards meeting the inevitably tight deadlines that the work will entail'

(Stockholmsleder, 1992, p.40). However, the streamlined moves towards implementation were set back when the National Road Administration succeeded in killing off Stockholmsleder. It wanted to keep control of the process itself, especially in the light of possible privatisation of the organisation. Responsibility for the roads has passed to its regional office and it can impose its own priorities, if needed, which may be national rather than local (Gauffin, 1993).

A restructuring of power?

One of the major issues of the case study concerns the decision-making process. There is considerable concern that the package has been put together by a small range of interests with no opportunity for public debate, and there are accusations of 'manipulation' (Gyllensten, 1993). As Eriksson puts it,

> the Dennis package is no plan. It is not founded on a thought-through notion of the future growth of Stockholm. It is a political compromise between three car-friendly political parties (the Social Democrats, the Liberals, and the Conservatives), the other parties being excluded from the beginning due to their critical assessment of the motorways.
>
> (Eriksson, E., 1993).

The package had to be discussed in the national political arena in order for it to get its financial and legal support. However, democracy at the local level was denied. In many ways this was the whole intention of the particular approach – that is, to cut through the maze of political conflicts. This intention was clearly expressed in the proposal within the agreement that national government should declare the agreement of national interest, thereby preventing a local veto and removing discussion from the local level. There is no opportunity for all local democratic interests to present their views about their particular areas – instead they just have to accept the agreement and get on with the job of implementing it, changing their plans and priorities if required. Tonell (1993) also points out that they have to carry the costs of the implementation work (such as preparing the detailed plans) from their own tax bases. Public participation in the decision-making process is limited to commenting on the specifics of the detailed plans which have to be accepted as part of the package. This does not permit the questioning of the broader aspects which has had to be channelled through the press and demonstrations. The other opportunity to react is via the electoral process, but this only happens every four years.

When an election came in the autumn of 1994, the Dennis package was an issue in the campaign in the Stockholm area. Opposition had been developing on three fronts related to environmental, democratic and economic issues. Environmental reaction increased, including squatting in a house on the route of the road. This grass-roots movement was backed up by cultural and academic figures, networks developed and articles were written. An environmental group, Nature Step, made up of respected scientists, medics and business leaders, came out against the package. A business network, including banks, was also established to challenge the value of the investment. The package was therefore beginning to face opposition from the general public, opinion setters and corporate elites. In the election in Stockholm the parties which were opposed to the package made gains; these included the Centre Party, the Green Party and the Left Party. The Social Democrats did less well in Stockholm than they did in the rest of the country. However, they were still the largest party and formed a coalition with the Greens and the Left. Meanwhile all parties that signed the Dennis agreement were looking for an escape route. The concept of a total package of policies looks doomed. The most likely outcome is that the northern and southern sections of the road will be built, with many of the public transport aspects, but that the more controversial eastern section, western orbital and the toll scheme will be thrown back into the normal political arena. As a result decisions on them will be difficult to achieve.

The lack of participation in the Dennis agreement can be contrasted with the participatory image of Swedish society. As we saw in the last chapter this consensus building has received much attention in the literature, for example through reference to the system of government commissions of inquiry which draws in the views and advice of a wide range of experts and interests (Heclo and Madsen, 1987). It was said in the 1980s that Swedish politicians and administrators placed considerable stress on 'social skills of getting along with others – not pushing advantages too far, encouraging the co-operation of others, avoiding outright confrontation, and not casting anyone in the role of permanent loser' (Heclo and Madsen, 1987, p.21, drawing upon Anton, 1980). Does this still apply in the 1990s? It might be said that the Dennis agreement is an example of this approach of compromise but it is confined to those who are party to the negotiations. This is a fairly narrow interpretation of consensus and the opposition to the agreement seems to indicate that many people consider that they will be permanent losers.

One reason that might be advanced for this lack of broader consensus building is the strategic importance of the issues. From 1991–4 the government was concerned that Sweden might become a peripheral nation and its support for membership of the European Union was part of its strategy to ensure that this did not happen. This provided a climate of opinion which was conducive to the argument that Sweden must modernise and become more efficient. More freedom should be given to economic leaders, whether individual firms or regions, which can compete internationally. In pursuing this strategy local democracy can be a luxury and handicap – sacrifices to principles might need to be made in order to achieve efficiency. The views and needs of commercial interests receive greater attention in this approach. This has been a feature of the Dennis agreement noted by many commentators (e.g. Johansson, 1993; Gyllensten, 1993). This attention has been gained not only through the influence of the Moderate Party but also through the direct advocacy of certain powerful companies.

To what extent could the Dennis agreement be said to be an example of the growth-coalition strategy discussed in Chapter 4; that is to say a concerted strategy adopted by a coalition of political elite and business interests, with the co-option of the public sector and popular support generated by the media? The early development of a campaign around the eastern by-pass in the mid-1980s and the formulation of the Projekt Österleden has many resonances of the growth-coalition approach. There are the links between Nacka politicians and local business interests with a popular campaign to form public opinion through publicity and the press. However, the need to take the strategy on to a national arena created a more complex picture. Media support weakened and it was more difficult to confine the issue to a local elite. A wider regime arose, led by central government, to confine decision-making to a small group of political leaders with a very clear strategy to restrict debate. The 'package' concept played an important role in achieving this.

Once the process had been taken up at the national level the direct involvement of business interest seemed to decline and greater reliance was placed on the traditional links between politicians and companies. This 'corporate' approach has been the other characteristic feature of Swedish decision-making – the institutionalised and organised incorporation of the major interests. This stress on the involvement of powerful organised interest can act against the philosophy of broader participation and, as we saw in the previous chapter, this was one of the concerns that led the government to set up an inquiry in 1985 into the changing power relations in Swedish society (Petersson, 1986). As Mishra points out,

> the interests of powerful organised groups are well taken care of but at the expense of those unrepresented at the bargaining table – non-producer groups and other less well-organised sectors of society. The implication is that . . . corporatist arrangements are likely to promote inequality and to curtail rather than enhance democracy. They are also likely to create a form of consensus society managed by a small elite and unresponsive to the wishes and interests of the masses.
>
> (Mashra, 1990, p.57)

How far can this be applied to the Dennis agreement? It can certainly be said that the decisions so far have been restricted to a very small elite and that an attempt has been made to impose these decisions. It is also clear that the environmental lobby, which contains very diverse interests, found it difficult to gain entry into the process. The smaller parties representing some of these interests were excluded and support within the Social Democrat party managed internally. Swedish society is changing with the dismantling of the welfare state and the weakening of Social Democrat dominance. One question which arises is whether the corporatist legacy will continue, though in a new form.

THE ÖRESUND BRIDGE: DEVELOPING A NEW EUROPEAN IMAGE

On 23 March 1991 Denmark and Sweden signed a binding treaty to build a road and rail bridge between their two countries – one of the biggest building projects ever planned in Scandinavia. The hope of its advocates is that it will create one of the strongest regions in Europe, well connected to the rest of the EU. Malmö will be transformed from a Swedish backwater into the country's European gateway. In 1993 the regional and local authorities in both countries set up a special collaborative body 'the Öresund Committee' to promote the region. The features stressed are the high-quality environment, the rich culture and the strong knowledge base, including the universities and medical schools of Roskilde, Copenhagen, Malmö and Lund. The location of the region is also emphasised – it is claimed that it is at the centre of northern Europe and the most densely populated region in Scandinavia (Ministry of Environment, 1993).

The decision to build the bridge has been fraught with controversy. One of the sources of pressure came from the Round Table group of major European industrialists who campaigned for the building of the 'missing links' in the European transport network (Harding, 1991a). This group funded a 'Scandinavian Link' organisation to lobby for the bridge. Strong advocacy came from the region's most influential economic interests such as Volvo, based in Gothenburg, the Scandinavian Airlines System (SAS) based at Copenhagen airport, and Swedish Rail which was part privatised in 1988 (Ross, 1995). The two national governments initiated a study of alternatives which included the possibility of a rail-only tunnel. Commercial and industrial interests lobbied strongly for a road link and the governments eventually chose the combined road/rail option. Although this was the most expensive alternative, the idea was that it would be self-financing through tolls. There has been considerable opposition from environmentalists objecting to the loss of good agricultural land, the increase in car usage stimulated by the need to finance the operation through increased car usage and the effects on the ecology of the Baltic. A local environmental movement, based in Malmö, arose centred around the campaign *Stoppa Bron* ('Against the Bridge') (Ross, 1995). These environmental objections have delayed the implementation of the project considerably. On the Swedish side the scheme has had to be scrutinised by the Water Court and the National Franchise Board of Environmental Protection, who in turn had to consult with about a hundred other organisations. These bodies asked for revisions to the scheme and it was the end of 1993 before the project was ready for a decision. Then a political problem arose on the Swedish side. The Centre Party, a member

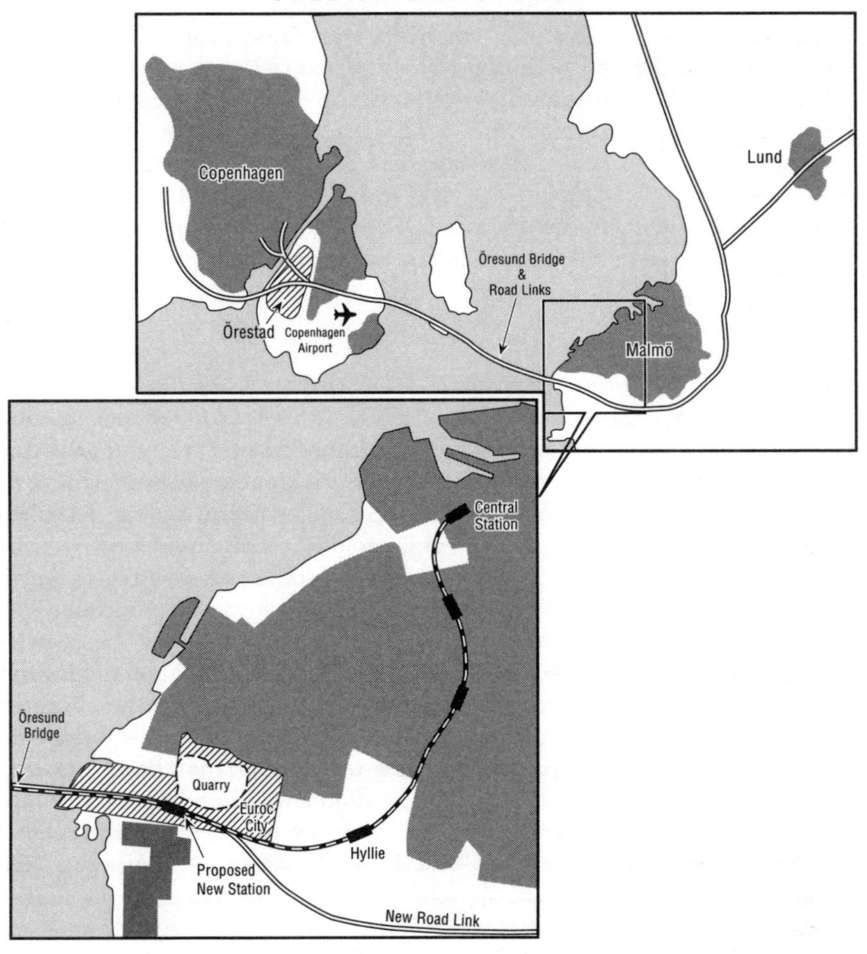

Figure 10.2 Malmö

of the coalition government, was developing a strong environmental manifesto for the forthcoming election and was unhappy to agree to the bridge. However this impasse was overcome when the Centre Party leader, who was also Minister for the Environment, resigned. The Swedish approval came at the end of 1994 and land works on the Swedish side started in spring 1995.

On the Danish side the passage of the legislation was faster and land works had started in 1993. The Danes had already spent more than 1 billion kroner on the project and demolished over 200 homes by 1994. The bridge and the marketing of the region figured prominently in the Danish national plan 'Denmark towards the year 2018'. Promotion of the region was part of the reversal, from 1989 onwards, of the Danish regional policy of seeking equality across the country. Now the emphasis was firmly one of promoting Copenhagen in the face of European competition (Jörgensen and Tonboe, 1992). While all the controversy was occurring over the bridge, plans were being formulated to exploit the development potential on each shore. Recession is affecting both Copenhagen and Malmö, and although they have been co-operating to promote the new region they naturally both want to exploit the development opportunities. We will concentrate on the plans for Malmö, but for comparison we will first outline the events in Copenhagen.

The Örestad project

On the Danish side efforts have been concentrated on Amegar near the bridge and the international airport (Hårde, 1993). This is close enough to the centre of Copenhagen to be considered an extension of the city, is near the University of Copenhagen and will have excellent communications with new roads, railway and subway. A new settlement called Örestad is proposed, with housing, research centre, high-tech park, exhibition centre and employment for about 60,000 people – its design being the subject of a proposed international competition. It is promoted as a development for the post-industrial society which will attract international investment. It is hoped that a new Baltic Exposition will take place here when the bridge opens. Thus the project is seen as an opportunity to enhance the image of Copenhagen within Europe – as the city's director of planning, Knud Rasmussen says, 'the bridge is needed more as a symbol than for speeding up transportation flows' (quoted in Hårde, 1993). The land for the new development is jointly owned by the state and the City of Copenhagen, and a new law has been passed to allow for the expropriation of existing housing for the necessary transport links. This has generated a lot of local opposition from the well-maintained middle-class neighbourhoods affected as well as from environmentalists. However, it has top level, political support. The Prime Minister commenting on Örestad in 1991, said: 'this project has created a certain turmoil in relation to the surrounding municipalities and the county. But naturally old regional plans will have to be adjusted when the fact of the matter is that we can exploit the possibilities resulting from a car–rail bridge across the Öresund' (quoted in Jörgensen and Tonboe, 1992). The Örestad project is clearly part of Copenhagen's marketing exercise to show that the city is worthy of major European status. Some think that it is totally unrealistic, comprising enough building capacity to match London Dockland and La Défense combined (Jörgensen *et al.*, 1994). The planning process has also been criticised for the way in which existing plans have been ignored and changed without participation. The Örestad scheme was the pet project of Copenhagen's chief urban planner but it only took off as an idea when it was adopted by the three-man secretariat of a commission looking at the transport problems of the region. This commission was looking for financial means to implement its plan. The Örestad scheme seemed to offer the solution – all the land in the area of the proposed development was owned by the state or municipality. Thus by promoting the scheme and selling off the land the returns could be used to pay for the infrastructure requirements of the region. This plan was then presented as a *fait accompli* to the rest of the commission and the public (Jörgensen *et al.*, 1994).

Malmö's official plans

Although the authorities on both sides of the bridge are co-operating in promoting the new region and talk about the complementary nature of their respective developments, there is clearly competition between Copenhagen and Malmö in attracting investment interest. Here Copenhagen has the edge with its better image and history as a tourist, cultural and conference centre. Once the bridge has been built considerable commuting from Malmö is expected. However Malmö has its own plans to exploit the bridge's potential.

Malmö wants the bridge to bring benefits to the city and here it is faced with a problem. The new road access to the bridge will take the form of a peripheral motorway round the city and there is the possibility that the railway will take the same route. This would mean that trains would not pass through Malmö Central Station. The council fears that the city centre could lose its vitality as a result and that the city will miss out on the advantages of the bridge. In 1990 it prepared a new comprehensive plan, with full public participation (Malmö Stadsbyggnadskontor, 1990). Among the themes of this plan are environmental sustainability, with a compact structure which allows for efficient public transport and the encouragement of greater social balance in residential areas. The plan envisages considerable urban renewal in the inner areas and new development round the central station contributing to maintaining the importance of the existing city centre. However, some of the momentum for a sustainable policy diminished when the Social Democrats lost the 1991 election and the municipality was run by the conservatives (Moderate Party), in coalition with the Liberals, who had campaigned on a platform of 'bringing the car back to the city'.

At the time of the 1990 comprehensive plan, the railway from the bridge was planned in a tunnel round the western edge of the city ending up at the central station. This would ensure that trains to the rest of Sweden would stop in central Malmö. It has proved technically very difficult, though, and so the municipality is instead promoting a new tunnel route further east which would fulfil the same purpose, and this has been the subject of a new plan (Malmö Stadsbyggnadskontor, 1992), although the finance for this has not yet been agreed. This new route has the advantage that it passes through an area of municipally owned land, Hyllie, which was identified in the comprehensive structure plan for office and high-tech development. This area lies in a break within the existing development of the southern part of the city and would bring needed employment to this largely residential area. The development of Hyllie, and additional stages to the south, has been the subject of further studies and plans (e.g. Malmö Stadsbyggnadskontor, 1991) and has been referred to as Brostaden. Although this route has the support of most authorities, has been the subject of participation and is enshrined in a new comprehensive plan, there have been some dissenting voices. One worry relates to the cut-and-cover nature of the construction which would lead to the destruction of some pleasant environments. There is also some scepticism over the costs of the tunnel (Hårde, 1993).

Euroc City

Malmö's planners have therefore been active trying to ensure that the city benefits from the advent of the bridge through preparing plans that reinforce the vitality of its city centre and bring new post-industrial development within the city fabric. However, other schemes have also been afoot. On the south-western edge of Malmö lies an area of limestone. In 1871 the company Euroc was established to exploit the stone and during the early twentieth century it developed a cement factory, concrete products, glassware, fertilisers, and other related products. In doing so it transformed the village of Limhamn into a company town and provided housing, parks, churches and other facilities (Euroc, 1991). The settlement was annexed by Malmö in 1915, and in 1977 the company sold its land to the city except for its quarry and the area around it. This land now belongs to Skanska, the largest construction company in Sweden. The quarry itself is huge, the size of eighty football pitches; it has been said that all the existing buildings of Malmö could be buried in it.

This quarry happens to be located where the bridge touches land in Sweden and Skanska are naturally keen to exploit its new potential. They paid consultants to produce a scheme which was called Euroc City. This was presented as the means by which Malmö could improve its image and move into the post-industrial future. A new settlement would be built focusing on a lake, parkland, restaurant and cable car to the bottom of the quarry. There would be 2,000 flats and from 500–600 houses, some built on the upper layers of the quarry. However, the key to the development would be 600,000 square metres of office space employing about 20,000 people, and a large international hotel. The commercial area would be clustered round a new railway station.

The scheme contravened the official plans of the city which designated the area for open space and a small amount of housing. The fear was that the Euroc development would be detrimental to enhancing the city centre and developing the Hyllie area. However, the scheme was given much publicity and Skanska was keen to attract top political support. It managed to get the leader of the municipality (a Social Democrat) to endorse the scheme with a statement of support in the publicity document. He stated that the development would contribute to Malmö's shift from manufacturing industries to high-tech activities and give new European orientation to the city: 'we are already becoming less interested in what goes on in Stockholm. In 20 years I dare suggest that we will be totally oblivious of Stockholm. What happens in Brussels will be much more important to us' (Lars Engqvist, in Euroc, 1991). Obtaining the Social Democrat leader's support for the scheme becomes less surprising when one realises that he was also on Skanska's board of directors. There is a Swedish tradition for municipal leaders to also become board members of housing companies or important local industries – in Malmö in the past this would have been shipbuilding. There is therefore a history of strong links between local political and business leaders, and this has also included Social Democrat politicians who were in continuous power in Malmö for sixty-six years.

The Euroc company also included a statement in its publicity material from the opposition leader (Moderate Party), and this proved an astute move as the municipality changed control in the election shortly afterwards. He commented that

> what I envisage is increasing diversity in the provision of services, exciting new architecture, a low-tax profile that can attract new residents, everything accented by better cultural services . . . we must have a mixed industrial structure: both manufacturing and services. At the same time I think that many knowledge-intensive activities will find the region irresistible, for example education, culture, and the mass media.
>
> (Euroc, 1991)

The slump in the property market has, at least temporarily, tempered the momentum behind the project. However, the image it presents, its potential in attracting inward investment in competition with the Danish alternatives, and its top level local political support mean that once the market picks up it may proceed, regardless of the official city plans which prefer consolidation of the existing city. The signs are that these plans are being by-passed. Outline development agreements were signed in the municipal real estate office in December 1992 which gave the go-ahead for a first stage of 200,000 square metres of development on the Skanska site, called 'Gateway to Sweden', in which the municipality will build the streets and lay the sewerage system but Euroc will pay the costs. The municipality will undertake to lobby the other authorities concerned to get the strategic road and rail connections to the development. Once these are obtained there will be further discussions on the extension of the scheme. The municipality will also pay half the costs towards a new joint company which will hire consultants to continue the search for the best image-creating use for the old quarry.

Issues raised

The competitive climate has a considerable impact on the way decisions on development are made. The bridge is seen as a catalyst for attracting international interest to the region but there is also competition between the schemes of the two municipalities. Growth coalitions and city marketing strategies arise. The literature on this shows how such strategies lead to the downplaying of local democracy and strengthen an elitist decision-making process. The same pattern can be detected on both sides of the sound. In Malmö this centralised approach is demonstrated in the way Skanska and top politicians appear to be making the running rather than the official participatory planning process. The official plans are being sidelined and concerns about a vital city centre and a sustainable environment and settlement pattern appear less important than the image of a new development. The case study also illustrates the continuing corporate nature of Swedish decision-making in which municipal leaders, quangos and private companies operate in partnership. This has been the post-war tradition which, in this case, does not seem to have changed in the more market-oriented climate.

The case also shows how different arms of the municipality can operate independently. There is the planning side with a broad range of criteria including social and environmental criteria. Planners in the municipality seem to have had some degree of autonomy in translating these criteria into the comprehensive plans which have been the subject of public and political agreement. At the same time, however, top politicians have been involved in discussions with Euroc on alternative ideas, spurred on by the desire to improve the image of the city and gain new investment. There also appears to be a touch of regional pride and desire to become more independent of Stockholm. When the time came to formalise these negotiations the real estate department drew up the development agreements. This illustrates the lack of co-ordination between departments and the way in which the development agreement can lead rather than follow the planning process. It also illustrates the way in which planning, historically based in the architectural profession, has not been integrated into economic policy issues.

The story of the Öresund bridge demonstrates the growing concern about the environmental aspects of such developments. There was significant opposition on environmental grounds at national and local levels and this created considerable modification and delay. However, this opposition was unable to fundamentally affect the decision-making process. The combination of backing from major industrialists, from municipal politicians who wanted infrastructure and development so as to promote their city in the European competition, and from government at all levels who saw financial opportunities, overcome any opposition.

CONCLUSIONS

In the last chapter it was noted that one characteristic of the Swedish decision-making system was the combination of close co-operation between elites and a culture of participation and consensus building, sometimes given the label of 'corporate pluralism'. A movement towards greater participation and decentralisation was observed since the 1970s. However, the extent of this has been questioned with the suggestion that power has remained centralised while responsibility has been devolved. One of the messages that stands out from the three case studies is the lack of participation. Thus it would appear that, as far as large projects are concerned, the corporate dimension is more important than the participatory one. This is particularly clear in the procedures of the Dennis package where, rather than any attempt at building consensus, it was explicitly accepted that a decision would need to be pushed ahead through the agreement of the leaders of the three powerful political parties, orchestrated by a government appointee. However, the downgrading of participation is also evident in the other two cases in which the development agreement is seen to lead the participatory planning process. Thus in the Globe there was very little participation, and the project was worked out by the developers with a small group of officials and politicians from the municipality. Participation in the actual decision-making process was replaced by an appeal to popular acceptance through the attraction of the arena facility and the image of the Globe. In the plans for Malmö the scheme devised by the landowners seems to be gaining acceptance through a strategy of taking on board the top politicians and widespread publicity.

In the previous chapter it was seen how municipalities have gained greater autonomy through the decentralisation process and have what is called a planning monopoly. This gives them considerable freedom in decisions over development. It was also noted that this devolved responsibility is matched by centralisation in the control of financial resources. The main impact of this centralised control has been to limit the finances available, causing municipalities to seek alternative sources. This has affected their approach to development decisions and created a willingness to prioritise the financial issues over other planning objectives. This is one reason why development agreements have dominated the process. In these agreements municipalities have traded planning leniency or land in order to achieve some financial gain. This was particularly clear in the Globe example and in other projects of the 1980s outlined in the previous chapter. However, in more recent years this approach has been given greater impetus because of the increasing competition between cities. They are all keen to develop good infrastructure and image-creating projects in order to attract the inward investment which is the prize in the competition. Again a common solution to this need is to allow developers to proceed with their schemes, such as Euroc City, even if they do not meet the social goals of the existing plan. However, the autonomy of the municipality breaks down when it comes to finance, particularly of infrastructure, and the national state is then drawn in. As we saw in the Dennis package and the Öresund bridge this often brings into play a different set of power relations. The decisive role of central government can be seen in the way it controls the financial decisions. Even the fate of the continued vitality of Malmö's city centre is likely to depend upon the central state and the decision on the financing of the rail tunnel. The central state itself is also in financial difficulties and will look for the most economic solutions such as paying for investment through tolls, or, in the Danish example, through land sales. This will of course then have a major impact on overall transport policy. Certain legal requirements may also result in national level involvement; this was the case with the environmental aspects of the bridge and the laws needed for the tolls in Stockholm.

The autonomy of municipalities allows a close relationship between developers and top politicians and the development agreement provides the vehicle allowing this elite to formulate projects. These then have to be ratified by the planning process which involves participation. However, another feature of the growth coalition is brought into play – that is the wide-scale publicity given to projects extolling their value to the city. Thus opposition is forestalled. However, the local development coalition is not completely autonomous and the decisions have to be placed in the wider national political arena if financial or legal aspects demand it.

In the future it is likely that the competitive element will become even more important. The cities and major regions are gearing up to promote themselves to a greater extent in the context of EU membership. Image, infrastructure and development opportunities will therefore remain dominant issues. The projects of the 1990s, the Öresund bridge and the Dennis package, have shown the increasing importance of environmental aspects. There is increasing concern at all levels to give greater priority to environmental issues. This could mean the resurrection of a longer-term view and the importance of comprehensive plans which emphasise such issues. The quality of the environment is also one of the 'selling factors' which is often mentioned in Scandinavian city or regional marketing material.

By concentrating on three large projects, which have been amongst the most controversial of the last decade, it might be argued that a biased picture of the Swedish planning system results. Smaller schemes probably follow more closely the processes described in the official documents (see Kalbrö and Mattsson, 1995). However, we would claim that in any country the tendencies found in the larger schemes have an influence on the operation of the system as a whole – the large schemes do not remain 'extraordinary' or 'experimental' for long. The planning system operates on different levels – there are the special concessions and procedures adopted for large schemes involving powerful actors and a rather different and stricter set of procedures for the small-scale developer or individual. Such matters would be worthy of further research.

11

CONCLUSIONS

New economic forces operating at a global scale have widened the differences in the fortunes of geographical areas. This has increased the competition between cities to attract investment and given economic objectives an urgency in urban policy. Nation-states have usually responded with deregulatory strategies and the private sector has had a more powerful role in decision-making. The fragmentation of planning decisions amongst a range of public, semi-public and private agencies follows. However, although these have been the major trends throughout Europe there has at the same time been considerable variation in the way urban planning has responded. National planning systems are differentiated along legal and administrative dimensions and national political and institutional structures create significant differences in approach to urban planning. At the urban level specific local economic and political circumstances and the relative power of the various interest groups also affect the urban planning outcome. This variety illustrates that there are opportunities to deviate from the competitive, market-oriented trend.

In recent years there has been a widespread concern across Europe to find better arrangements for urban planning and a shift from the property-led model which typified the 1980s. The problems of fragmentation have led to renewed calls for a more strategic approach. There has been a questioning of the dominance of economic objectives and the lack of 'trickle down' to deal with social needs, leading to demands for a more direct approach to social issues and greater community involvement. The strength of the environmental movement has also widened the agenda and re-emphasised the need for long-term thinking. This chapter explores the common trends and continuing differences in urban planning across contemporary Europe, focusing on what we see as four significant themes – fragmentation of government responsibility, the interaction between urban planning and economic growth, social and environmental objectives, and the continuing need for planning legitimacy.

FRAGMENTATION OF GOVERNMENT RESPONSIBILITY

Comparative studies of European planning in the past have stressed the contrast between the flexibility of the British discretionary system and the more rigid zoning approach in the rest of Europe. However, the introduction of greater flexibility has featured in the reforms in most countries in recent years. We argue that differences in the degree of decentralisation of government power from the centre is now a more important factor in determining variation in planning systems. Some countries have pursued significant decentralisation programmes, some are developing strong regional structures and some, such as Britain, are still dominated by central government. These differences present major variations in the context for urban planning.

In France there has been a continuing search for better sub-national structures. This can be seen both at the super-regional level of the Paris basin and in the *pays* of the 1994 national planning law, each covering just a few communes. The state has encouraged and required more intercommunal co-operation. The use of *Sociétés d'Economie Mixte* (SEMs) for urban development also increased following the reforms of the early 1980s. New structures were created in an already complex institutional landscape. In Sweden there has been a consistent move to decentralise responsibilities to municipalities while at the same time making increased use of the tradition of quangos. In Britain the use of quangos in urban policy was new but, as we saw in London, there was a rapid and significant increase during the late 1980s and 1990s following the abolition of metropolitan government. New agencies also appeared at subregional and city level arising from both local initiative and central government encouragement.

A common trend towards the fragmentation of responsibility can be detected. However, urban governance has not disintegrated and lines of co-ordination between levels of government remain. One reason for this is a countervailing trend towards the centralisation of power in national governments with increased control over sub-national governments in key sectors. The most significant control is that of finance. For example, in Sweden the central state has not made detailed incursions into the urban level but concentrated on identifying matters of national interest and providing broad frameworks for municipalities to work within. However, the considerable autonomy of the local level has been weakened by increasing central control of finance. In Britain both the main budgets of local councils and expenditure on urban regeneration is tightly controlled by central government. City Challenge and the SRB gave local authorities a greater role in urban regeneration than they had under the UDC regime of the 1980s. However, central government imposed detailed rules covering both bidding for funding and implementation of projects, and this resulted in a shift in emphasis from need to efficiency. The shortage of finance at the local level enhances the likelihood of central government intervention. For example, central government took control of the Heartlands project in Birmingham when local government and its partners could no longer provide financial support. In Sweden, many large projects such as the Dennis package are dependent on national finance. Financial controls have therefore served to manage the increasingly fragmented structures of governance.

As seen in Chapter 4 there are clear differences between countries in the co-ordination between levels and the degree of autonomy of the local level. In addition to financial controls the French state used contracts – for infrastructure and social investment – and redefined the offices of the *préfets* to exert influence. The integration of levels through the accumulation of political offices, as noted in the Lille case, also gave French urban planning a degree of co-ordination lacking in the conflictual central–local relationships in Britain. Financial autonomy and strong political leadership in Paris played an important role in pushing through its redevelopment proposals. A similar pattern can be detected in other cities, such as Barcelona, that are set within a Napoleonic institutional framework. Meanwhile, as a result of conflict, London lacked the institutional framework around which a coherent pro-growth strategy could develop. Those cities that were successful in European competition commanded financial and policy support from national government and this was more easily achieved with strong co-ordination between government levels.

The relative power of different levels of government has obvious impacts on approaches to urban planning. In British cities central government encouraged economic objectives and restricted social programmes. In France the position was different with cities pursuing economic goals and central government introducing a series of urban social policy initiatives. The Swedish state has defined its role in relation to issues of national interest and setting broad objectives rather than imposing urban policies on cities. The main involvement of central government has been through financial controls. However, a new dimension of control could develop as environmental issues become politically more important. Central government may become more involved in setting and monitoring standards. The approach adopted by national government and the degree of decentralised responsibility affect the particular role adopted by urban planning and the extent to which it pursues pro-growth strategies at the expense of other objectives.

URBAN PLANNING AND ECONOMIC GROWTH

Urban planning in all three countries studied in Part II, and indeed throughout Europe, has undergone constant legal adjustment and administrative reorganisation over the last decade or so. For example, in Britain planning and local government legislation has been continually modified since the early 1980s. Throughout Europe these reforms have sought to create greater flexibility in planning in relation to private sector demands and as a result public–private boundaries have become blurred. Property-led planning in Britain, 'negotiation planning' in Sweden, and the flexible approach to development plans in many French cities all confirm this common trend in shifting the balance in development decisions towards the private sector. There are, however, limits to deregulation and flexibility and distinctive national reactions. For example, in Britain NIMBYism brought a demand for strong planning protection in some areas, and the formal importance of plans was partially restored. The French state maintained a concern for the social impacts of urban change and also tried to restrain the economic development ambitions of cities. Some cities themselves sought to balance economic growth and social investment. The re-election of the Social Democrats in Sweden in 1994 indicated that right-wing ideology had not found a firm footing. The liberalisation of the planning system under consideration during the early 1990s may not therefore be pursued.

Reinforcing the trend towards greater flexibility at the national level, formal plans in cities have diminished in importance and have often been used to back up pro-growth strategies. For example, in Lille the strategic plan for the conurbation was adjusted to incorporate the development priorities set by the Communauté Urbaine de Lille (CUDL) in 1989. The Danish government changed its approach to regional planning in order to prioritise the position of the Copenhagen region. In Birmingham the Unitary Development Plan supported the international city strategy. The commercialisation of the project for Seine Rive Gauche was reflected in the revision to the regional plan for Ile-de-France. Urban planning has increasingly become associated with pro-growth strategies and the urban marketing and image promotion that accompanies them, offering a range of support including organising sites and services, co-ordinating financial and development processes and marketing development opportunities.

Variation can be detected, though, in the degree to which the formal land-use planning system is integrated into the pro-growth strategy. Sometimes, as in Birmingham, there is a close relationship in which the plan can be seen as one vehicle for city promotion. In other cases such as Malmö or Stockholm the planning system remains as a separate process and mechanisms are devised to avoid this process rather than mould it to pro-growth requirements. It is possible to consider the relationship between the pro-growth approach and the planning system as a series of stages. In the early stages of the coalition the existing plans are likely to be 'out of tune' and therefore by-passed, in the second stage some kind of synchronisation takes place, and in a third stage a reaction develops and the planning system is used as a vehicle to modify the pro-growth approach. The Swedish examples mentioned above could be said to fall into the first stage while greater synchronisation has occurred in the British examples, boosted by central government's City Pride initiative. For the relationship to move into the third stage pressure from outside the growth coalition is required. This could be expressed in popular reactions to the effects of pro-growth developments, sometimes formalised in local political opposition. The seeds of such reaction were observed in some cities such as Frankfurt and Birmingham.

In recent years there has been an increased appreciation of the need for a city-wide strategic framework. Voices within the private sector have been calling for such a framework in order to reduce uncertainty and oversupply and also to provide certain facilities such as transport. This has led to the growth of city-wide documents which offer 'visions' of urban futures. These visions have usually been stimulated by the climate of competition; for example, Barcelona developed its approach in the context of the Olympic bid and London's vision was a reaction to fears of losing its 'world city' functions to other cities. However, while urban planning may have moved beyond a project-based approach, this new concern with 'strategy' does not necessarily mark a return to comprehensive urban planning. Typically it is an extension of urban marketing.

There are differences between and within countries concerning who sets the agenda for such new city visions. The renewed interest in broad strategy can come from either the private or public sector, local initiative or national government. In France the public sector, both national and local, played a strong role in orchestrating the new strategies, whereas in Britain the situation was more varied. Sometimes the lead was taken by central government, sometimes by local government and often by the private sector. City Pride can be seen as the national state's attempt to market its most saleable cities. In Birmingham the city council took a leading role, though supporting the objectives of its private partners. In London private interests supported by central government dominated the construction of the city-wide vision.

SOCIAL AND ENVIRONMENTAL OBJECTIVES

The weakness of planning is particularly evident at the European level even though a new scale of plan-making has begun to emerge with the publication by the EC of *Europe 2000+* in 1994 and the 'trend scenario' in 1995. These initiatives were largely discussion documents and EU influence on urban planning only has real effect when linked to those aspects over which the EU has a strong mandate, such as cross-border issues, or to financial provision. For example, the EU has responded weakly to the challenge of restraining European city competition (Kunzman, 1995), reflecting the uncertainty surrounding the Commission's policy-making role. At urban, national and European levels formal planning documents have been weak and this has consequences for the integration of wider social and environmental objectives into urban planning.

Nevertheless a key factor differentiating approaches to urban planning is the degree to which they include objectives beyond the narrow pro-growth ambitions of political and business elites. In some cases attempts were made to widen the scope of urban planning, but these met many obstacles. For example, in Plaine Saint-Denis the communes struggled to get social needs on the agenda. The approach in the Greenwich Waterfront project appeared to include a wide range of interests but in the end depended on the major landowners and on the ability to attract development finance from higher-level government. Community groups unsuccessfully fought development in London Docklands throughout the 1980s, seeking the satisfaction of community needs. Similar groups in King's Cross also found it difficult to gain access to decision-making and it was only the weakness of the property market that enabled them to have some influence.

In Part II we noted several attempts by national governments to broaden urban planning objectives, but these have been heavily constrained by economic imperatives. The London Pride Prospectus incorporated social objectives, but these clashed with its economic priorities. There was a selective inclusion of community priorities in City Challenge and SRB projects, but these were small-scale programmes. In France there was continued experimentation with new urban social policies. For example, the project for the Grand Stade brought increased social spending but this only occurred because of the need to present an exemplary urban project to international visitors. There is therefore limited evidence of a softening of economic objectives to include more social concerns.

However, the ability to incorporate a wider range of objectives in urban planning is also hindered by the separation between the formal planning system and the social issues contained in other programmes of government. The divorce between formal planning systems and the financial programmes of government, for example in funding urban regeneration, is a common feature. Land-use plans have differing and often marginal relationships to state expenditure on urban policy. National urban regeneration priorities, for example in the *contrat de ville* and SRB, were developed independently of formal plans. There is also a gulf between the content of statutory plans and the priorities which cities developed in order to benefit from the financial programmes of the EU. Thus the financial means to implement social objectives occur in arenas outside the formal planning system. This formal system has a regulatory function and is reliant on the private sector to initiate development. Thus urban planning as we have defined it consists of separate, often mutually exclusive, strands.

So far attempts to include social aspects have been very restricted and new ideas are needed if the social dimension is to be fully integrated into urban planning. Policy-makers continually seek to learn lessons from other countries and cities. The detailed studies of national and urban levels in Part II allow some comparisons to be drawn. For example, one advantage of the *contrat de ville* over SRB is its three- to four-year time-span. Another is the negotiation over programmes rather than annual competition. A weakness of the

SRB projects is the reliance on new agencies to manage projects. However, these French and British initiatives are very different programmes located in differing government structures, incorporating different political priorities and drawing on different values about the role of the state in social life.

It is difficult for cities to take an independent approach to more redistributive planning because of the competitive climate and the need for certainty in public resources which is beyond the financial capacity of the city. Support of higher levels of government is required. It was noted in Chapter 3 that the EU has developed a greater interest in urban policy to tackle the problems of social exclusion. However, fundamental problems exist as there is a clash of objectives at this level. The Commission is concerned about the costs of economic change and the problem of social exclusion. However, the Commission contributes to infrastructure and development projects that reinforce competitiveness such as the Grand Palais in Lille and the Birmingham city centre development. The relationship between the EU's encouragement of competitiveness and its programmes to deal with the social impacts is not always clear. Increasing competitiveness may enlarge many social problems. There are also concerns about the equitable distribution of EU resources and further enlargement will increase the problem. It will exacerbate tensions between small and large states, poorer and richer, northern and southern, urban and rural (Gower, 1995). In such circumstances support at EU level to ameliorate the social impacts of urban change faces stiff challenges.

Can national state policy play a role in encouraging social objectives in urban planning? Countries are in competition with each other to ensure that their cities capture as much of the available investment as possible. We have seen that this leads to an attitude of 'backing winners' and a concentration of resources. However, riots in British inner cities or French peripheral estates do not help in promoting a good national image for potential investors. Thus national social policies might be developed in the interest of improving competition, though this is dependent upon national political ideology. It has been noted that there have been different attitudes towards the appropriate role of the state in a capitalist economy, ranging from the neo-liberalism of Britain in the 1980s to the social market approach of Scandinavia and Germany. The degree to which urban planning adopts stronger social objectives depends to a large degree on the position adopted by nation-states in the future.

The rise of environmentalism raises a similar set of issues for the future of urban planning. Across Europe environmental values have become increasingly important in local decision-making. Differences between national electoral systems have clearly influenced the ability of the environmental movement to gain access to formal power. In the British system the majority party can dictate the agenda and this has given the environmental movement little hold on decision-making – solutions to environmental problems have been sought within the market-oriented philosophy of the Conservative Party. However, under a proportional representative system 'green' parties have been able to influence decisions. This has been so at both the national level, as witnessed in our case studies in Sweden, and at the local level, as evidenced by the role of the Green Party in a number of German cities.

In France environmental parties gained influence at the regional level and environmentalist opposition created some difficulties for political elites in Paris and in the Ile-de-France region. However, we noted in Chapter 7 that despite some formal constraint on the growth of the region, large-scale development continued. Despite opposition within the city, the regime continued its commitment to Seine Rive Gauche. Local electoral politics nevertheless had some impact; for example, electoral losses in 1995 encouraged the city of Paris to consider seriously measures to control pollution from road traffic. In Sweden the competitive ethos behind the Dennis package and Öresund bridge led to the downgrading of environmental concerns, but there have been strong reactions from some national parties and local people. In other parts of Sweden radical experiments are underway and there is considerable activity by municipalities around Agenda 21 (Khakee *et al.*, 1995).

A major issue for the future will be the extent to which these environmental concerns are translated into urban planning policy. They would carry with them the implication that a long-term outlook is required and that there is a need to take account of the broader public interest. These implications lead to greater acceptance of public intervention and less reliance on the market. There is evidence, however, that the environment is becoming another factor in marketing the image of cities. Scandinavia is utilising its environmental quality as a selling point, for example in the Stockholm bid for the Olympics. Similarly the national plan being formulated for Switzerland is based upon the value to top decision-makers of a good housing environment and efficient public transport.

The international importance given to environmental issues allows cities to give more attention to environmental aspects and sustainability in their plans. The EU has been a major force in increasing environmental policy throughout Europe. Environmental issues are now considered a legitimate concern for EU action and attempts are being made to inform all aspects of EU policy with an environmental dimension. Again, however, greatest success has occurred where the EU has a clear mandate or can provide finance – for example, in dealing with pollution in cross-border areas or in the new *Länder*. The EU has expressed concern over the environmental costs of growth in the core area of the European economy, but because of the weaknesses of planning at this level it has little ability to generate the necessary implementation programmes to deal with such strategic issues.

At urban, national and European levels there is substantial variation in the commitment to environmental values. Unlike the issue of social planning individual cities have rather more scope to take on environmental challenges, though responsibilities are often spread between agencies, and integrated action is required by all levels of government.

LEGITIMACY AND ACCOUNTABILITY

Elected government is accountable to its citizens and can claim democratic legitimacy for its actions. Fragmentation of government dilutes that legitimacy as does the transfer of urban planning responsibilities to non-elected agencies and public–private partnerships. Issues of legitimacy and accountability occur at urban, national and European levels.

In certain parts of Europe there are fundamental problems of legitimacy for planning. There is currently a struggle taking place in eastern Europe to accept the very notion of planning as people react to their communist legacy. In parts of southern Europe there are also cultural impediments to planning regulation. For example, the spirit of untrammelled property rights is strong in Greece making it very difficult to introduce planning reforms. In several countries the disregard for the law makes planning implementation and enforcement extremely difficult. The entrenchment of corruption, most notoriously in Italy, enhances the lack of legitimacy in the planning system.

Throughout Europe planning has gained another problem of legitimacy as a result of its involvement in pro-growth strategies, which are often formulated outside a democratic context. In many cities there has been a backlash against pro-growth projects and the exclusion of communities from decision-making, for example in Birmingham, Frankfurt, and Paris. The tension between supporting the market and democratic procedures is a common European experience, but countries have adopted different mechanisms for dealing with the democratic deficit. The notion of 'partnership' has been extensively used in Britain to bind community aspirations into development projects. In France the personal authority of the mayor legitimised urban development, though there was concern about the continuing legitimacy of indirectly elected and arm's-length agencies in relation to large development projects. Eastern European countries are developing new democratic structures but tensions are often created with the desire to attract investment through the market. Hoffman (1994) argues that in the Czech Republic there may be some positive democratic outcomes from the new planning process even though the country has moved strongly in the direction of property-led planning. We noted in Chapter 4 the debate in Berlin between the promotion of prestigious development associated with world city status and the new capital location versus the desire to retain the participatory traditions of the city. In Stockholm the 'Dennis package' was a particular mechanism for achieving a strategic solution which would be acceptable to the powerful elites and avoid the problem of local democratic opposition. However, it has not proved possible to retain the legitimacy of this approach and public reaction, through both direct demonstration and the ballot box, has forced a rethink.

There are many examples of attempts to involve community associations in planning projects; for example, the Greenwich Waterfront project tried to include a broad range of interests in its decision-making structures. There is no shortage of ideas at all levels about how systems of planning and governance should be reformed. There are debates about the state of local democracy and the incorporation of community aspirations in Britain (see Stoker and Young, 1993; CLD, 1995) and France (Donzelot, 1994). At national level there are discussions about regional devolution, in practice in Spain and Belgium, and in theory on the left in Britain. In the future these ideas are increasingly likely to turn into reality, but the outcome will be greater diversity rather than conformity around a single democratic model.

Accountability is also an issue at the European level. The Commission has been criticised for its closed nature and the European Parliament for lacking the necessary power to provide democratic safeguards. The Committee of the Regions has yet to establish its role. There is likely to be considerable debate over this accountability issue in the coming years. In addition, because of its weaknesses planning lacks legitimacy at this level. A further problem of legitimacy arises from the relationship between cities and the Commission. Certain cities have greater lobbying power in the competition to gain EU resources. This restricts the debate on urban priorities and acts to the disadvantage of those cities, such as Athens (Delladetsima and Leontidou, 1995), that do not have the necessary institutional capacity to influence EU decisions. Such effects reinforce the criticism that the EU approach disadvantages southern cities. For economically successful cities a Europe of city-states may be an exciting prospect. For cities on the periphery, which are not able to adopt the competitive model of the 'successful' cities, there is the prospect of moving further behind.

Europe faces challenges of political integration, social cohesion, and environmental sustainability which will impact on urban planning. International economic forces and the resultant competitive climate will remain a major factor influencing urban planning throughout Europe. We have seen how planning at national and city level was constrained by economic competition and reshaped to assist pro-growth strategies. In the 1990s there were signs of a reaction to the property-led strategy, but urban planning can only take on social and environmental objectives with higher-level support. Although shaped by a common competitive economic context, we expect the future to continue to show great variation of experience across Europe between countries and between cities. This variety will result from different forms of interaction between economic competition, institutional structures and political demands.

REFERENCES

ABC 13 (1993a) *Seine Rive Gauche aussi* No. 60, March.

ABC 13 (1993b) *Chronique de Seine Rive Gauche* No. 61, June.

Acosta, R. (1994) 'Quelle Politique Foncière autour du Grand Stade', *Etudes Foncières* 65.

Acosta, R. and Renard, V. (1991) *Frameworks and Functioning of Urban Land and Property Markets in France,* Paris: Association des Etudes Foncières.

Aczel, G. (1993) 'Le Montage d'une opération de réhabiliation à Budapest', in V. Renard and R. Acosta (eds) *Land Tenure and property development in Eastern Europe*, Paris: Association des Etudes Foncières.

ADA 13 (1990) *Text Commun*, Association pour le Développement et l'Aménagement du 13ème arrondissement, Paris 13.

ALA (1993) *The Basics for Londoners*, London: Association of London Authorities.

Ambrose, P. (1994) *Urban Process and Power*, London: Routledge.

Andrikopoulou, E. (1992) 'Whither regional policy? Local development and the state in Greece', in M. Dunford and G. Kafkalas (eds) *Cities and Regions in the New Europe*, London: Belhaven Press.

Anton, T. (1980) *Administered Politics: Elite Political Cultures in Sweden,* Boston: Martinus Nijhoff.

Ashford, D. (1989) 'British dogmatism and French pragmatism revisited', in C. Crouch and D. Marquand (eds) *The New Centralism*, Oxford: Blackwell.

Ashworth, G. and Voogd, H. (1990) *Selling the City*, London: Belhaven.

Audit Commission (1989) *Urban Regeneration and Economic Development*, London: HMSO.

Ave, G. (ed.) (1991) *The Functioning and Framework of Urban Property Markets in Italy*, Centre for Urban Studies, COREP-Polytechnic of Turin.

Baïetto, J.-P. (1993a) Paper to Architectural Association Seminar, London, February.

Baïetto, J.-P. (1993b) 'La ville au plus près des citoyens', *Urbanisme* 261.

Bailey, N., Barker, A. and MacDonald, K. (1995) *Partnership Agencies in British Urban Policy*, London: University College Press.

Balme, R. and Le Galès, P. (1992) 'Is Europe an alternative to the Jacobin state?', Paper to EUROLO Conference, Odense, September.

Balme, R., Garraud, P., Hoffmann-Martinot, V., Le May, S. and Ritaine, E. (1994) 'Analysing territorial policies in Western Europe', *European Journal of Political Research* 25.

Bannon, M.J. (1993) 'Land use planning and development coordination in the Dublin region', Paper to The European City and its Regions Conference, Department of the Environment, Dublin, September.

Barlow, J. (1995) 'The politics of urban growth', *International Journal of Urban and Regional Research* 19, 1.

Barlow, J. and Duncan, S. (1992) 'Markets, states and housing provision: four European growth regions compared', *Progress in Planning* 38, Oxford: Pergamon.

Barlow, J. and King, A. (1992) 'The state, the market and competitive strategy: the housebuilding industry in the UK, France and Sweden', *Environment and Planning A* 24.

Barnekov, T., Boyle, R. and Rich, D. (1989) *Privatism in Urban Policy in Britain and the United States*, Oxford: Oxford University Press.

Barras, R. (1994) 'How much office building does London need?', Paper to ESRC London Seminar, London School of Economics, November.

Barry, A. (1993) 'The European Community and European government: harmonization, mobility and space', *Economy and Society* 22, 3.

Bassols, M. (1986) 'Town planning in Spain', in J.F. Garner and N.P. Gravells (eds) *Planning Law in Western Europe*, Amsterdam: Elsevier Science Publishers.

Batley, R. (1991) 'Comparisons and lessons', in R. Batley and G. Stoker (eds) *Local Government in Europe*, London: Macmillan.

Batley, R. and Stoker, G. (eds) (1991) *Local Government in Europe*, London: Macmillan.

Bayle, C. (1992) 'La Plaine-Saint-Denis', *Urbanisme* 258.

BDA (Barcelona Development Agency) (1992) 'Barcelona: Real Estate News', Fall.

Begg, H. and Pollock, S. (1991) 'Development plans in Scotland since 1975', *Scottish Geographical Magazine* 107, 1.

Begg, I. and Whyatt, A. (1994) 'Economic development in London: institutional conflict and strategic confusion', Paper to IFRESI Conference, Lille, March.

Bejrum, H., Cars, G. and Kalbro, T. (1995) 'Stockholm', in J. Berry and S. McGreal (eds) *European Cities, Planning Systems and Property Markets*, London: E & F.N. Spon.

Bell, D. (1994) *Communitarianism and its Critics*, Oxford: Oxford University Press.

Benfer, W. (1994) 'Local development policy in the United States and Germany', Paper to Cities, Enterprises and Society at the Eve of the XXIst Century Conference, Lille, March.

Bennett, R.J. (ed.) (1993) *Local Government in the New Europe*, London: Belhaven Press.

Bennington, J. (1994) *Local Democracy and the European Union*, Research Report 6, London: Commission for Local Democracy.

Bennington, J. and Harvey, J. (1994) 'Spheres or tiers?', in P. Dunleavy and J. Stanyer (eds) *Contemporary Political Studies*, Belfast: Political Studies Association.

Berggrund, L. (1994) 'Ecocycles in the Gothenburg structure plan', in B. Malbert (ed.) *Ecology-based Planning and Construction in Sweden*, Stockholm: The Swedish Council for Building Research.

Beriatos, E. (1995) 'The environmental dimension in the institutional framework of physical planning in Greece', *European Urban and Regional Studies* 2, 3.

Berry, J. and McGreal, S. (eds) (1995) *European Cities, Planning Systems and Property Markets*, London: E. & F.N. Spon.

Bertamini, F., Mutti, P. and Radaelli, A. (1994) 'Quale Milano', *Construire*, No. 133, Giugno.

BHDC (1993) *Corporate Plan*, Birmingham Heartlands Development Corporation.

Bianchini, F.., (1994) 'Milan', in A. Harding, J. Dawson, R. Evans and M. Parkinson (eds) *European Cities towards 2000*, Manchester: Manchester University Press.

Biarez, S. (1989) *Le Pouvoir Local*, Paris: Economica.

Biarez, S. (1990) 'The Metropolis debate in France: new paper and reports on the framework of decentralisation', *International Review of Administrative Sciences* 56.

Biarez, S. (1993) 'Ville, région, état, le dialogue en Europe', in J.-C. Némery and S. Wachter (eds) *Entre L'Europe et la décentralisation,* La Tour d'Aigues: DATAR/Editions de l'Aube.
Biarez, S. (1994) 'Urban policies and development strategies in France', *Local Government Studies* 19, 2.
Biggs, S. and Travers, T. (1994) 'Opportunities for city-wide government in London: the experience of the metropolitan areas', *Local Government Policy Making* 21, 2.
Birmingham City Council (1993) *Unitary Development Plan.*
Birmingham City Council (1994) *Birmingham City Pride. First Prospectus '94,* The City Pride Team.
Body-Gendrot, S. (1994) 'Immigration: la Rupture Sociale et ses Limites', *Le Débat* 80.
Bonneville, M. (1994) 'Internationalisation of non-capital cities in Europe: aspects processes and prospects', *European Planning Studies* 2, 3.
Booth, P. (1993) 'The cultural dimension in comparative research: making sense of development control in France', *European Planning Studies* 1, 2.
Booth, P. and Green, H. (1993) 'Urban policy in England and Wales and France', *Environment and Planning C: Government and Policy* 11, 4.
Borraz, O. (1994) *Mayors in France,* Paris: Centre National de la Recherche Scientifique.
Boyle, R. (1991) 'Urban economic regeneration policy in the United States of America, the United Kingdom and Germany', Paper to joint ACSP–AESOP Congress, Oxford Polytechnic, July.
Bradford City Challenge (1993) *Revised 5 Year Strategy,* Bradford: BCC Ltd.
Branegan, J. (1993) 'Model no more', *Time,* 19 July.
Bremm, H. and Ache, P. (1993) 'International changes and the single European market', *Urban Studies* 30, 6.
Brillot, F. (1993) 'Euralille: une SEM aux commandes', *Urbanisme Hors Série,* February.
Brodsky, P. (1993) Interview. Mayor, Prague, 2 November.
Brooks, M. (1993) 'City enters the bullring in defense of its inner city policy', *Municipal Journal,* 9–15 July.
Brownill, S. (1990) *Developing London's Docklands,* London: Paul Chapman.
Bruegel, I. (1993) 'Local economic development in the transformation of Berlin', *Regional Studies,* 27, 2.
Brunet, R. (1989) *Les Villes Européenes,* Rapport pour la DATAR, Montpellier: RECLUS.
Brussard, W. (1986) 'Physical planning legislation in the Netherlands', in J.F. Garner and N.P. Gravells (eds) *Planning Law in Western Europe,* Amsterdam: Elsevier Science Publishers.
Budd, L. (1994) 'Regional distributional coalitions in a global–local environment', *Current Issues in the Politics and Economics of Europe* 3, 2.
Bulpitt, J. (1989) 'Walking back to happiness? Conservative Party governments and elected local authorities in the 1980s', in C. Crouch and D. Marquand (eds) *The New Centralism,* Oxford: Blackwell.
Busquets, J. (1988) 'Barcelona 1992: new planning hypothesis from this experience', XVII Triennale di Milano Second International Conference, 9–11 November.
Calabi, D. (1984) 'Italy', in M. Wynn (ed.) *Planning and Urban Growth in Southern Europe,* London: Mansell.
Calavita, N. (1983) 'Urban planning, the state, and political regimes: the Italian case', *Environment and Planning D: Society and Space* 1.
Cappellin, R. (1989) 'Milan', in L. Klaassen, L. van den Berg and J. van der Meer (eds) *The City: Engine Behind Economic Recovery,* Aldershot: Avebury.

Carbonaro, G. and D'Arcy, E. (1994) 'Assessing property-led urban restructuring: an implementation structure approach', Paper to Cities, Enterprises and Society at the Eve of the XXIst Century Conference, Lille, March.

Carpenter, C., Chauvire, Y. and White, P. (1994) 'Marginalization, polarization and planning in Paris', *Built Environment* 20, 3.

Cars, G., Lanesjo, B., Westin, P. and Akerlund, P. (1991) 'Public/private co-operation in urban planning in Sweden – a review', Unpublished paper.

Carter, N., Oxley, M. and Gollard, G. (1994) 'Stuttgart makes adjustments to regional planning machine', *Planning* 1078.

Castells, M. (1993) 'European cities, the informational society and the global economy', *Tijdschrift voor Econ. en Soc. Geografie* 84, 4.

Castles, F. (1978) *The Social Democratic Image of Society*, London: Routledge & Kegan Paul.

Cattaneo, S. (1986) 'Planning law in Italy', in J.F. Garner and N.P. Gravells (eds) *Planning Law in Western Europe*, Amsterdam: Elsevier Science Publishers.

CBI (1991) *A London Development Agency: Optimising the Capital's Assets*, London: Conferation of British Industry.

Chapman, M. (1994) 'New policy directions in the urban environment: the role of the European Union', Paper to ESRC Urban Policy Evaluation Seminar, University of Wales, Cardiff, September.

Cheshire, P. (1990) 'Explaining the recent performance of the European Community's major regions', *Urban Studies* 27, 3.

Cheshire, P. and Hay, D. (1989) *Urban Problems in Western Europe*, London: Unwin Hyman.

Chicoye, C. (1992) 'Regional impact of the single European market in France', *Regional Studies* 26, 4.

Church, A. (1988) 'Urban regeneration in London Docklands: a five year policy review', *Environment and Planning C* 6.

Ciechocinska, M. (1994) 'Paradoxes of decentralisation: Polish lessons during the transition to a market economy', in R. Bennett, (ed.) *Local Government and Market Decentralisation,* Tokyo: United Nations University Press.

Clarke, M. and Stewart, J. (1989) *The Future for Local Government*, Local Government Training Board.

Clarke, M. and Stewart, J. (1994) 'The local authority and the new community governance', *Regional Studies* 28, 2.

Claval, P. (1994) 'La métropolisation et la nouvelle distribution des acteurs sur la scène politique mondiale', Paper to Les métropoles, la mondialisation de l'économie et la scène politique Conference, University of Paris IV, September.

CLD (Commission for Local Democracy) (1995) *Taking Charge: The Rebirth of Local Democracy*, London: Municipal Journal.

Coccossis, H. and Pyrgiotis, Y. (forthcoming) 'Greek land-use planning law', in C. Haar and J. Kayden (eds) *International Treatise on Land Use Planning Law*, Cambridge, Mass.: Lincoln Institute for Land Policy.

Cochrane, A. (1994) 'Public–private partnerships and local economic development', Paper to Conference on Cities Enterprises and Society at the Eve of the XXIst Century, Lille, March.

Cohen, N. (1993) 'Renaissance that never was', *Independent on Sunday*, 19 October.

Cole, A. and John, P. (1994) 'Beyond policy networks; research strategies for understanding local policy change in Britain and France', Department of Politics, Univeirisity of Keele.

Colenutt, B. (1994) 'Docklands after Canary Wharf', Paper to Social Justice and the City Conference, Oxford University, March.

Colenutt, B. and Ellis, G. (1993) 'The next Quangos in London', *New Statesman and Society*, 26 March.
Collier, J. (1994) 'City visions: Newcastle', Vision for London Seminar, London, April.
Collinge, C. and Hall, S. (1995) 'Hegemony and regime in urban governance', Paper to British Sociological Association Annual Conference, Leicester, April.
Condamines, E. (1993) 'L'Eurorégion', *Pouvoirs Locaux* 17.
Conseil d'Etat (1992) *L'Urbanisme: pour un droit plus éfficace*, Paris: La Documentation Française.
Coopers and Lybrand (1992) *Growing Business in the UK*, London: Business in the Community/Coopers and Lybrand.
Coopers and Lybrand Deloitte (1991) *London: World City Moving into the 21st Century*, London: HMSO.
Costa Lobo, M. (1992) 'Portugal', in A. Dal Cin and D. Lyddon (eds) *International Manual of Planning Practice* (second edition), The Hague: ISOCARP.
Coupland, A. (1992) 'Docklands: dream or disaster?', in A. Thornley (ed.) *The Crisis of London*, London: Routledge.
Cowlard, K. (1992) 'City futures', in L. Budd and S. Whimster (eds) *Global Finance and Urban Living*, London: Routledge.
Cox, K. (1991) 'Questions of abstraction in the new urban politics', *Journal of Urban Affairs* 13, 3.
Crosland, A. (1956) *The Future of Socialism*, London: Cape.
Crouch, C. and Marquand, D. (eds) (1989) *The New Centralism*, Oxford: Blackwell.
Cuñat, F., Praedilles, J.-C., and Roussier, N. (1993) 'De la conurbation urbaine aux conflits de centralité', in *Plan Urbain, Métropoles en Déséquilibre*, Paris: Economica.
Da Rosa Pires, A. (1994) 'Local economic policy, the planning system and European integration', Paper to Cities, Enterprises and Society at the Eve of the XXIst Century Conference, Lille, March.
Dahrendorf, R. (1990) *Reflections on the Revolution in Europe*, London: Chatto & Windus.
Dahrendorf, R., Greengross, A., Layfield, F., Stonefrost, M. and Swaffield, J. (1993) 'Let's rescue London before it goes under', *Observer*, 28 November.
Daily Telegraph (1994) 'Why Birmingham is Rattled', 2 February.
Dal Cin, A. and Lyddon, D. (eds) (1992) *International Manual of Planning Practice* second edition, The Hague: ISOCARP.
DATAR (1993) *Débat National pour l'Aménagement du Territoire. Document Introductif*, Paris: La Documentation Française.
DATAR (1994) *Débat National pour l'Aménagement du Territoire. Document d'Etape*, Paris: La Documentation Française.
DATAR (1995) *Schéma de développement de l'espace communautaire*, Paris: DATAR.
DATAR/CPPR (Conférence Permanante des Présidents des Régions) (1994) *Charte du Bassin Parisien*, Paris: La Documentation Française.
David, R. and Brierley, J. (1985) *Major Legal Systems in the World Today*, London: Stevens & Sons.
Davies, C. (1991) 'Eastern promise', *Architects Journal*, 11–18 December.
Davies, H.W.E. (1989) 'The Netherlands', in H.W.E. Davies, D. Edwards, A.J. Hooper and J.V. Punter, *Planning Control in Western Europe*, London: HMSO.
Davies, H.W.E. (1993) 'Europe and the future of planning', *Town Planning Review* 64, 3.
Davies, H.W.E. (1994) 'Towards a European planning system', *Planning Practice and Research* 9,1.
Davies, H.W.E., Edwards, D., Hooper, A.J. and Punter, J.V. (1989) *Planning Control in Western Europe*, London: HMSO.

Davies, H.W.E and Gosling, J. (1994) *The Impact of the European Community on Land Use Planning in the United Kingdom*, London: Royal Town Planning Institute.

Davies, R.L. (ed.) (1979) *Retail Planning in the European Community*, Hants: Saxon House.

DCC (Docklands Consultative Committee) (1990) *The Dockland Experiment: A Critical Review of Eight Years of the LDDC,* London: DCC.

DCC (Docklands Consultative Committee) (1994) *Community Empowerment in Urban Regeneration*, London: DCC.

De Forn I Foxà, M. (1993) 'Barcelona: Strategies for urban economic transformation', Stategic Planning Department, May.

De Vos, E. (1993) 'The Sector of Roissy', M.Phil Thesis, University of Reading.

Deacon B. (ed.) (1992) *Social Policy, Social Justice and Citizenship in Eastern Europe*, Aldershot: Avebury.

Deacon, B., Castle-Kanerova, M., Manning, N., Millard, F., Orosz, E., Szalai, J. and Vidinova, A. (1992) *The New Eastern Europe: Social Policy, Past, Present and Future*, London: Sage.

Debenham Thorpe Zadelhoff (1992) *The Property Market in Prague Czech Republic*, Prague: DTZ.

Delladetsima, P. and Leontidou, L. (1995) 'Athens', in J. Berry and S. McGreal (eds) *European Cities, Planning Systems and Property Markets*, London: E. & F.N. Spon.

Delluc, M. (1994) 'Pandore et sa boîte', *L'Architecture d'Aujourd'hui* 295.

Delmartino, F. (1988) 'Regionalisation in Belgium', *European Journal of Political Research* 16.

Demangeau, A. (1993) *Lille, Métropole Européene*, Supplement to *Bulletin d'Institut Français d'Architecture* 171.

Dennis Agreement (1992) *Agreement on Enhancing the Traffic Infrastructure in the Stockholm Region*, Social Democratic Party, Moderate Party, Liberal Party, 29 September.

Devès, C. and Bizet, J.-F. (1991) *Les Sociétés d'Economie Mixte Locales*, Paris: Economica.

Dickens, P., Duncan, S., Goodwin, M. and Gray, F. (1985) *Housing, States and Localities*, London: Methuen.

Dielman, M. and Hamnett, C. (1994) 'Globalisation, regulation and the urban system', *Urban Studies* 31, 3.

Dieterich, H. and Dransfeld, E. (1995) 'Düsseldorf', in J. Berry and S. McGreal (eds) *European Cities, Planning Systems and Property Markets*, London: E. & F.N. Spon.

Dieterich, H., Dransfeld, E. and Voss, W. (1993) *Urban Land and Property Markets in Germany*, London: UCL Press.

DiGaetano, A. and Klemanski, J. (1993a) 'Urban regimes in comparative perspective', *Urban Affairs Quarterly* 29, 1.

DiGaetano, A. and Klemanski, J. (1993b) 'Urban regime capacity', *Journal of Urban Affairs* 15, 4.

Dion, S. (1986) *La Politicisation des Mairies,* Paris: Economica.

DoE (Department of Environment) (1990) *This Common Inheritance*, London: HMSO.

DoE (Department of Environment) (1991) *Local Government Review*, London: HMSO.

DoE (Department of Environment) (1992) Planning Policy and Guidance Note, *General Policy and Principles*, PPG1, London: HMSO.

DoE (Department of Environment) (1993a), Planning Policy Guidance Note, *Town Centres and retail developments,* PPG6, London: HMSO.

DoE (Department of Environment) (1993b), *London: Making the Best Better*, London: DoE.

DoE (Department of Environment) (1993c), *Partnership Opportunities*, London: Private Finance Unit, DoE.

DoE (Department of Environment) (1994), Planning Policy Guidance Note, *Transport*, PPG13, London: HMSO.

Donzelot, J. (1994) *L'Etat animateur*, Paris: Editions Esprit.

Dostál, P. and Kára J. (1992) 'Territorial administration in Czechoslovakia: an overview', in P. Dostál, M. Illner, J. Kára and M. Barlow (eds) *Changing Territorial Administration in Czechoslovakia*, Amsterdam: Instituut voor Social Geografie, Universiteit van Amsterdam.

DREIF (1994) 'Ile-de-France 2015. Schéma Directeur', Direction Régionale de l'Equipement d'Ile-de-France, Paris: DREIF.

Drouet, D. (1994) 'Les services urbains', in J-C. Némery and S. Wachter (eds) *Gouverner les territoires*, La Tour d'Aigues: DATAR/ Editions de l'Aube.

Dumont, M-J. (1994) 'Du Vieux Paris au Paris Nouveaux', *Le Débat* 80, May–August.

Duncan, S. (1985) 'Land policy in Sweden: separating ownership from development', in S. Barrett and P. Healey (eds) *Land Policy: Problems and Alternatives*, Aldershot: Avebury.

Duncan, S. and Goodwin, M. (1988) *The Local State and Uneven Development*, Cambridge: Polity Press.

Dunford, M. (1994) 'Winners and losers: the new map of economic inequality in the European Union', *European Urban and Regional Studies* 1, 2.

Dunford, M. and Kafkalas, G. (eds) (1992) *Cities and Regions in the New Europe*, London: Belhaven Press.

Dutt, A.K. and Costa, F.J. (eds) (1985) *Public Planning in the Netherlands*, Oxford: Oxford University Press.

East 8 (1993) *Real Estate Developments in Eastern Europe: The Czech & Slovak Republics*, London: Interforum Publications Limited.

EC (1990) *Green Paper on the Urban Environment*, Brussels: Commission of the European Communities.

EC (1991) *Europe 2000: Outlook for the Development of the Community's Territory*, Brussels: Commission of the European Communities.

EC (1992) *Towards Sustainability*, Brussels: Commission of the European Communities.

EC (1993a) *Community Structural Funds 1994–1999. Regulations and Commentary*, Brussels: Commission of the European Communities.

EC (1993b) *Growth Competitiveness and Employment*, Brussels: Commission of the European Communities.

EC (1994a) *Community Initiatives Concerning Urban Areas*, Brussels: Commission of the European Communities.

EC (1994b) *Europe 2000+ Cooperation for European Territorial Development*, Luxembourg: Office for Official Publications of the European Communities.

Edwards, D. (1989) 'Denmark' in H.W.E. Davies, D. Edwards, A.J. Hooper and J.V. Punter, *Planning Control in Western Europe*, London: HMSO.

Edwards, M. (1992) 'A microcosm. Redevelopment proposals at Kings Cross', in A. Thornley (ed.) *The Crisis of London*, London: Routledge.

El Guedj, F. (1993a) 'Le Grand Lyon I', *Pouvoirs Locaux* 16.

El Guedj, F. (1993b) 'Le Grand Lyon II', *Pouvoirs Locaux* 17.

Elander, I. (1991) 'Good dwelling for all: the case of social rented housing in Sweden', *Housing Studies* 6, 1.

Elander, I. (1994) 'Paradise lost? Desubsidization and social housing in Sweden', in B. Danermark and I. Elander (eds) *Social Rented Housing in Europe: Policy, Tenure and Design*, Delft: Delft University Press.

Elander, I. and Gustafsson, M. (1993) 'The re-emergence of local self-government in Central Europe', *European Journal of Political Research* 23.

Elander, I., Gustafsson, M., Sandell, K. and Lidskog, R. (1995) 'Environmentalism, sustainability and urban reality', in A. Khakee, I. Elander and S. Sunesson (eds) *Remaking the Welfare State*, Aldershot: Avebury.

Elander, I. and Montin, S. (1990) 'Decentralisation and control: central–local government relations in Sweden', *Policy and Politics* 18, 3.

Elander, I. and Montin, S. (1995) 'Citizenship, consumerism and local government in Sweden', *Scandinavian Political Studies*, forthcoming.

Ellger, C. (1992) 'Berlin: legacies of division and problems of unification', *The Geographical Journal* 158, 1.

Engrand, G. (1992) 'Euralille. Relazione B1', Paper to Marketing Urbino Conference, Turin, September.

Engström, C.-J. (1991) 'Föfhandlingsplanering', *Plan* 1.

Eriksson, E. (1993) 'Vi får en bilism som biter sig själv i svanson' (We will get a car society that bites its own tail), *Dagens Nyheter*, 13 December.

Eriksson, K. (1993) Sollentuna Municipality, Interview, November.

Esping-Andersen, G. (1990) *The Three Worlds of Welfare Capitalism*, Cambridge: Polity Press.

Estates Europe (1993/4) 'Milan', *Estates Europe*, December 1993/January 1994.

Etzioni, A. (1993) *Spirit of Community*, New York: Simon & Schuster.

Euralille (1994) 'Dossier de Presse', Euralille, Pavillon Souham, Lille.

Euroc (1991) *Euroc City*, Malmö: Euroc.

European (1993) '300 towns that just can't find a mayor', 5–11 November.

Europroperty (1993) 'Milan on the move', *Europroperty*, March.

Fainstein, N. and Fainstein, S. (1982) 'Restoration and struggle: urban policy and social forces', in N. Fainstein and S. Fainstein (eds) *Urban Policy under Capitalism*, Newbury Park, Calif.: Sage.

Fainstein, S. (1994) *The City Builders. Property, Politics and Planning in London and New York,* Oxford: Blackwell.

Fainstein, S. Gordon, I. and Harloe, M. (1992) *Divided Cities*, Oxford: Blackwell.

Falkanger, T. (1986) 'Planning law in Norway', in J.F. Garner and N.P. Gravells (eds) *Planning Law in Western Europe*, Amsterdam: Elsevier Science Publishers.

Faludi, A. and Valk, A. (1994) *Rule and Order: Dutch Planning Doctrine in the Twentieth Century,* Dordrecht: Kluwer Academic Publishers.

Fareri, P. (1991) 'Milano', in *La Costruzione della Città Europea Negli Anni '80*, Roma: Credito Fondiario SPA.

Fehily, J. and Grist B. (1992) 'Ireland', in A. Dal Cin and D. Lyddon (eds) *International Manual of Planning Practice* (second edition), The Hague: ISOCARP.

Fiddeman, P. (1994) 'Single regeneration budget', in M. Edwards and J. Ryser (eds) *Bidding for the Single Regeneration Budget. Seminar Proceedings,* University College London.

Fog, H. (1991) 'The beauty of a city', in A. Fredlund (ed.) *Swedish Planning in Time of Transition,* Gävle: Swedish Society for Town and Country Planning.

Fog, H., Bröchner, J., Törnqvist, A. and Åström, K. (1993) *Mark, politik och rätt,* Stockholm.

Foley, D. (1960) 'British town planning: one ideology or three?', *British Journal of Sociology* 11.

Fournier, S and Axelsson, S. (1993) 'The shift from manufacturing to services in Sweden', *Urban Studies* 30, 2.

Frick, D. (1991) 'City development and planning in the Berlin conurbation', *Town Planning Review*, 62, 1.

Friedman, J. (1986) 'The world city hypothesis', *Development and Change* 17,1.
Gamble, A. (1988) *The Free Economy and the Strong State*, London: Macmillan.
Garcia, S. (1991) *Urbanisation and the functions of Cities in the European Community. City Case Study: Barcelona*, Liverpool: John Moores University.
García-Bellido, J. (1991) 'European viewpoint: a (r)evolutionary framework for Spanish town planning', *Town Planning Review* 62, 4.
Garner, J.F. and Gravells, N.P. (eds) (1986) *Planning Law in Western Europe*, Amsterdam: Elsevier Science Publishers.
Gauffin, C. (Stockholm City Council) (1993) Interview, 9 November.
Getimis, P. (1992) 'Social conflicts and the limits of urban policies in Greece', in M. Dunford and G. Kafkalas (eds) *Cities and Regions in the New Europe*, London: Belhaven Press.
Glasgow Action (1987) *Glasgow Action. The First Steps*, Glasgow: Glasgow Action.
GOL (1995) 'Strategic Planning Guidance for London Planning Authorities. Consultation Draft', Government Office for London.
Gold, V. and Ward, S. (1994) *Place and Promotion*, Chichester: Wiley & Sons.
Goldsmith, M. (1993) 'The Europeanisation of local government', *Urban Studies* 30, 4/5.
Goldsmith, M. and Newton, K. (1988) 'Centralisation and decentralisation: changing patterns of intergovernmental relations in advanced western societies', *European Journal of Political Research* 16.
Goodwin, M., Duncan, S. and Halford, S. (1993) 'Regulation theory, the local state and the transition of urban politics', *Environment and Planning D: Society and Space* 11, 1.
Gordon, I. (1995) ' "London World City": organisational and political constraints on territorial competition', in P. Cheshire and I. Gordon (eds) *Territorial Competition in an Integrating Europe*, Aldershot: Avebury.
Gourevitch, P. (1986) *Politics in Hard Times*, Ithaca, N.Y.: Cornell University Press.
Gower, J. (1995) 'Can the European Union be expanded to include the countries of central and eastern Europe?', in J. Lovenduski and J. Stanyer (eds) *Contemporary Political Studies*, Belfast: Political Studies Association.
Grant, M. (1991) 'Recent developments', *Encyclopedia of Planning Law and Practice Monthly Bulletin*, July.
Guichard, O. (1986) *Propositions pour l'aménagement du territoire*, Paris: La Documentation Française.
Gustafsson, A. (1988) *Local Government in Sweden*, Stockholm: Swedish Institute.
Gustafsson, A. (1991) 'The changing local government and politics of Sweden', in R. Batley and G. Stoker (eds) *Local Government in Europe*, London: Macmillan.
Gyllensten, L. (1993) 'Storföretagens intressen styr Dennispaketet' (The interests of big business rule the Dennis package), *Dagens Nyheter*, 4 November.
Haar, C. and Kayden, J. (eds) (forthcoming) *International Treatise on Land Use Planning Law*, Cambridge, Mass.: Lincoln Institute for Land Policy.
Haegel, F. (1994) *Un Maire à Paris*, Paris: Presses de la Fondation Nationale des Sciences Politiques.
Häggroth, S. (1993) *From Corporation to Political Enterprise*, Stockholm: Civildepartementet, Ds 1993:6.
Hague, C. (1993) 'The restructuring of the image of Edinburgh: the politics of urban planning in a European arena', Ninth Urban Change and Conflict Conference, Sheffield, September.
Haila, A. (1990) 'The end of hierarchical planning in Finland?', Paper presented to the AESOP Congress, Reggio Calabria, November.
Hall, P. (1992) 'Agenda for a new government', *The Planner* 78, 21.
Hall, P. (1993) 'Forces shaping urban Europe', *Urban Studies* 30, 6.

Hall, P. (1994) 'London 1994: retrospect and prospect', in J. Simmie (ed.) *Planning London*, London: UCL Press.

Hall, S. (1988) *The Hard Road to Renewal*, London: Verso.

Hall, T. (1985) 'The central business district: planning in Stockholm, 1928-1978', in T. Hall *'I nationell skala . . .': Studier kring cityplaneringen i Stockholm*, Stockholm: Svensk stadsmiljö.

Hall, T. (1991a) 'Urban Planning in Sweden', in T. Hall (ed.) *Planning and Urban Growth in the Nordic Countries*, London: E & F.N. Spon.

Hall, T. (ed.) (1991b) *Planning and Urban Growth in the Nordic Countries*, London: E. & F.N. Spon.

Hambleton, R. (1994) 'Lessons from America', *Planning Week* 2, 24.

Hamnett, C. (1994) 'Socio-economic change in London', Paper to ESRC London Seminar, London School of Economics, June.

Hamnett, S. (1982) 'The Netherlands: planning and the politics of accommodation', in D.H. McKay (ed.) *Planning and Politics in Western Europe*, London: Macmillan.

Hårde, U. (1993) 'Bron som till himlen bär', *Arkitektur* 4.

Harding, A. (1991a) 'Copenhagen: comfortable capital or growing Eurocity', *Urbanisation and the Functions of Cities in the European Community*, A Study for the EC, John Moores University, Liverpool.

Harding, A. (1991b) 'The rise of urban growth coalitions, UK-style?', *Environment and Planning C* 9, 3.

Harding, A. (1994) 'Urban regimes and growth machines: towards a cross-national research agenda', *Urban Affairs Quarterly* 29, 3.

Harding, A., Dawson, J., Evans, R. and Parkinson, M. (eds) (1994) *European Cities. Towards 2000*, Manchester: Manchester University Press.

Harding, A. and Garside, P. (1995) 'Urban economic development', in J. Stewart and G. Stoker, (eds) *Local Government in the 1990s*, London: Macmillan.

Harding, A. and Le Galès, P. (1994) 'Globalization, urban change and urban policy', Paper to Post-Fordism and Local Governance Conference, Lancaster University, February.

Harloe, M. and Fainstein, S. (1992) 'Conclusion: the divided cities', in S. Fainstein, I. Gordon and M. Harloe (eds) *Divided Cities*, Oxford: Blackwell.

Häussermann, H. (1992) 'Perspectives on regional development in East Germany and urban change in Berlin', Paper for the ISA Conference, Los Angeles, 24 April.

Häussermann, H. and Strom, E. (1994) 'Berlin: the once and future capital', *International Journal of Urban and Regional Research*, 18, 2.

Hay, C. and Jessop, B. (1993) 'The post-Fordist local state: some sceptical remarks', 9th Urban Change and Conflict Conference, University of Sheffield, September.

Healey, P. (ed.) (1994) *Trends in development plan-making in European Planning Systems*, Working Paper No. 42, Department of Town and Country Planning, University of Newcastle upon Tyne.

Healey, P., Davoudi, S., O'Toole, M., Tavsanoglu, S. and Usher, D. (eds) (1992) *Rebuilding the City*, London: Spon.

Healey, P. and Williams, R.H. (1993) 'European urban planning systems: diversity and convergence', *Urban Studies* 30, 4/5.

Hebbert, M. and Travers, T. (1988) *The London Government Handbook*, London: Cassell.

Heclo, H. and Madsen, H. (1987) *Policy and Politics in Sweden: Principled Pragmatism*, Philadelphia, Pa.: Temple University Press.

Heinz, W. (1994) *Partenariat public–privé dans l'aménagement urbain*, Paris: L'Harmattan.

Hellsten, P. (1993) Tyresö Municipality, Interview, November.

Hennings, G. and Kunzman, K. (1993) 'Local economic development in a traditional industrial area; the case of the Ruhrgebiet', in P. Meyer (ed.) *Comparative Studies in Local Economic Development*, Westport, Conn.: Greenwood Press.
Hesse, J.J. (1987) 'The Federal Republic of Germany: from co-operative federalism to joint policy-making', in R.A.W. Rhodes and V. Wright (eds) *Tensions in the Territorial Politics of Western Europe*, London: Frank Cass.
Hoffman, L. (1994) 'After the fall: crisis and renewal in urban planning in the Czech Republic', *International Journal of Urban and Regional Research* 18, 4.
Holm, P. and Fredlund, A. (1991) 'The idea of community planning', in A. Fredlund (ed.) *Swedish Planning in Times of Transition*, Gävle: Swedish Society for Town and Country Planning.
Holmberg, S. (1984) *Väljare i förändring*, Stockholm: Liber.
Holt-Jensen, A (1990) 'Planning in Norway 1965-1990: changing philosophies, planning laws and practical results', Paper to AESOP Congress, Reggio Calabria, November.
Holt-Jensen, A. (1994) 'Key issues in Norwegian development planning', in P. Healey (ed.) *Trends in Development Plan-making in European Planning Systems*, Working Paper No. 42, Department of Town and Country Planning, University of Newcastle upon Tyne.
Hooper, A.J. (1989) 'Germany', in H.W.E. Davies, D. Edwards, A.J. Hooper and J.V. Punter, *Planning Control in Western Europe*, London: HMSO.
Hooper, J. (1994) 'An unnatural disaster', *Guardian*, 16 November.
Hudson, P. (1993) 'Urban renewal case study: Szczecin, Poland', Unpublished paper, University of Westminister.
Hupe, P. (1990) 'Implementing a meta-policy: the case of decentralisation in the Netherlands', *Policy and Politics* 18, 3.
Hutton, W. (1995) *The State We're In*, London: Jonathan Cape.
Illner, M. (1992) 'Municipalities and industrial paternalism in a "real socialist" society', in P. Dostál, M. Illner, J. Kára and M. Barlow (eds) *Changing Territorial Administration in Czechoslovakia*, Amsterdam: Instituut voor Social Geografie, Universiteit van Amsterdam.
Imrie, R. and Thomas, H. (1993) 'The limits of property-led regeneration', *Environment and Planning C* 11.
INSEE (1993) *La France et ses Régions*, Paris: Institut National de la Statistique et des Etudes Economiques.
Jessop, B., Bonnett, K., Bromley, S. and Ling, T. (1988) *Thatcherism*, Cambridge: Polity Press.
Johansson, J. (1993) 'Infrastruktur till Varje Pris' (Infrastructure at any price), *Arkitektur* 2.
John, P. (1994) 'UK sub-national offices in Brussels: diversification or regionalisation?', *Regional Studies* 28, 7.
Johnston, B. (1993) 'London shapes vision into next century', *Planning* 1026.
Jones Lang Wooton (1992) *City Report Berlin 1992*, Frankfurt: Jones Lang Wooton.
Jones Lang Wootton (1993a) *The French Property Market*, London: Jones Lang Wootton.
Jones Lang Wootton (1993b), *Prague City Review*, London: Jones Lang Wootton.
Jörgensen, I. and Tonboe, J. (1992) *Space and Welfare - the EC and the Eclipse of the Scandinavian Model*, Aalborg University Working Paper, March.
Jörgensen, I., Kjaersdam, F. and Nielsen, J. (1994) 'The State of Development Plan-making in Denmark', in P. Healey (ed.) *Trends in Development Plan-making in European Planning Systems*, Working Paper No. 42, Department of Town and Country Planning, University of Newcastle upon Tyne.
Judd, D. and Parkinson, M. (eds) (1990) *Leadership in Urban Regeneration*, Newbury Park, Calif.: Sage.

Judge, E. (1994) 'Poles feeling good on the road to market', *Planning*, 21 January.

Kalbro, T. and Mattsson, H. (1995) *Urban Land and Property Markets in Sweden*, London : UCL Press.

Kantor, P. and Savitch, H.V. (1993) 'Can politicians bargain with business?', *Urban Affairs Quarterly* 29, 2.

Kearns, G. and Philo, C. (1993) *Selling Places: The City as Cultural Capital, Past and Present*, Oxford: Pergamon Press.

Keating, M. (1991) *Comparative Urban Politics*, Aldershot: Edward Elgar.

Keating, M. (1993a) 'The continental meso: regions in the European community', in L.J. Sharpe (ed.) *The Rise of Meso Government in Europe*, Newbury Park, Calif.: Sage.

Keating, M. (1993b) 'The politics of economic development', *Urban Affairs Quarterly* 28, 3.

Keating, M. and Midwinter, A. (1994) 'The politics of central–local grants in Britain and France', *Environment and Planning C: Government and Policy* 12, 2.

Keil, R. and Lieser, P. (1992) 'Frankfurt; global city - local politics', *Comparative Urban and Community Research* 4.

Kerswell, H. (1994) 'Getting the partners together', in M. Edwards and J. Ryser (eds) *Bidding for the Single Regeneration Budget, Seminar Proceedings*, University College London.

Keyes, J., Munt, I. and Riera, P. (1991) *Land Use Planning and the Control of Development in Spain*, Centre for European Property Research, University of Reading.

Keyes, J., Munt, I. and Riera, P. (1993) 'The control of development in Spain', *Town Planning Review* 64, 1.

Khakee, A., Elander, I. and Sunesson, S. (eds) (1995) *Remaking the Welfare State*, Aldershot: Avebury.

King, A.D. (1990) *Global Cities. Post Imperialism and the Internationalisation of London*, London: Routledge.

King, D.S. (1987) *The New Right: Politics, Markets and Citizenship*, London: Macmillan.

Kitchen, T. (1993) 'The Manchester Olympic bid and urban regeneration', *The Planner TCPSS Proceedings*.

Kjaersdam, F. (1992) 'Denmark', in A. Dal Cin and D. Lyddon (eds) *International Manual of Planning Practice* (second edition), The Hague: ISOCARP.

Knapp, A. and Le Galès, P. (1993) 'Top-down to bottom-up? Centre–periphery relations and power structures in France's Gaullist party', *West European Politics* 16, 3.

Krätke, S. (1992) 'Berlin: the rise of a new metropolis in a post-Fordist landscape', in M. Dunford and G. Kafkalas, *Cities and Regions in the New Europe*, London: Belhaven Press.

Kühne, J. and Weber, G. (1986) 'Planning law in Austria', in J.F. Garner and N.P. Gravells (eds) *Planning Law in Western Europe*, Amsterdam: Elsevier Science Publishers.

Kukawka, P. (1993) 'L'avenir de la régionalisation', in J.-C. Némery and S. Wachter (eds) *Entre L'Europe et la Décentralisation*, La Tour d' Aigues: DATAR/Editions de l'Aube.

Kunzman, K. (1995) 'Europe 2000+. A critical analysis', Paper to Regional Studies Association Conference, Gothenburg, May.

La Lettre de Matignon (1994) No. 447, Paris: Palais de Matignon.

La Plaine Renaissance (1992) 'Le Projet Urbain: La Synthèse. Conférence de Presse', 25 May.

Labour Party (1994) *Empowering Urban Communities*, London: Labour Party Environment Group.

Lagerström, P. (Chairman, Stockholm County Council) (1992) 'Stockholm: foreword', in J. Chevrot (ed.) *Magazine of the European Regions No. 5*, Neuilly sur Seine: EPI.

Lalenis, K.S. (1993) 'Public participation strategies in urban planning in Greece', Unpublished Ph.D. thesis, University of Westminster.

Lambert, A. (1992) 'Fast forward', *Geographical Magazine* 643.

Lamorlette, B. and Demoureaux, J.-P. (1994) 'L'Erreur Manifeste d'Urbanisme', *Etudes Foncières* 62.

Läpple, D. and Krüger, T. (1993) 'Boom Immobilier à Hambourg', *Les Annales de la Recherche Urbaine* 55–6.

Lawless, P. (1994) 'Partnership in urban regeneration in the UK: The Sheffield Central Area Study', *Urban Studies* 31, 8.

Le Débat (1994) 'Entretien avec Jacques Chirac. Vu de l'Hôtel de Ville', No. 82, November/December

Le Galès, P. (1992) 'New directions in decentralisation and urban policy in France; the search for a postdecentralisation state', *Environment and Planning C: Government and Policy* 10, 1.

Le Galès, P. (1993) *Politique Urbaine et Développement Local*, Paris: L'Harmattan.

Le Galès, P. and Mawson, J. (1994) *Management Innovations in Urban Policy: Lessons from France*, Luton: Local Government Management Board.

Le Journal de Saint Denis (1990) 'Unanimité pour La Plaine', No. 48, November.

Le Journal de Saint Denis (1994) '35 milliards de regards braques sur saint denis', No. 139, May.

Le Monde (1992) 'Les maires et la construction', 22 October.

Le Monde (1993) 'L'économie mixte au service du développement local', 5 December.

Le Monde (1994a) 'Le groupement des communes est un succès', 23/24 January.

Le Monde (1994b) 'Le nouveau rôle de la DDE', 28 February/1 March.

Le Monde (1994c) 'Les associations prennent le contre-pouvoir', 27 March.

Le Monde (1994d) 'Jean Pierre Raffarin aménage son territoire', 15/16 May.

Le Monde (1994e) 'Les élus communistes jugent la charte du Bassin Parisien contraire à la décentralisation', 27 May.

Le Monde (1994f) 'L'adoption du project de loi sur l'aménagement du territoire', 14 July.

Le Monde (1994g) 'Signature du contrat de plan entre l'état et la région', 22 July.

Le Monde (1994h) 'Des habitants de Saint-Denis dénoncent des réculades', 30 July.

Le Monde (1994i) 'Contrats de ville entre l'état et quatorze communes', 9 August.

Le Moniteur (1994) 'Saint-Denis avant le Grand Stade', 29 July.

Le Nord (1994) No. 81, Lille, Conseil Général du Nord.

Le Point (1994a) 'La révolte des maires des grandes villes', No. 1130, 14 May.

Le Point (1994b) 'Grandes Villes – Nice', No. 1134, 11 June.

Leemans, A.F. (1970) *Changing Patterns of Local Government*, The Hague: International Union of Local Authorities.

Lendi, M. (1986) 'Planning law in Switzerland', in J.F. Garner and N.P. Gravells (eds) *Planning Law in Western Europe*, Amsterdam: Elsevier Science Publishers.

Lenfant-Valerie, C. (1993) 'Montpellier', *Urbanisme* 261.

Leonardi, R., Nanetti, R.Y. and Putman, R.D. (1987) 'Italy - territorial politics in the post-war years: the case of regional reform', in R.A.W. Rhodes and V. Wright (eds) *Tensions in the Territorial Politics of Western Europe*, London: Frank Cass.

Leontidou, L. (1995) 'Repolarization of the Mediterranean: Spanish and Greek Cities in Neo-liberal Europe', *European Planning Studies* 3, 2.

Letruelle, E. (1993) 'Euralille: le pari du cercle de qualité urbaine', *Urbanisme* 266.

Lever, W.F. (1993) 'Competition within the European urban system', *Urban Studies* 30, 6.

Levine, M.A. (1994) 'The transformation of urban politics in France', *Urban Affairs Quarterly* 29, 3.

Levitas, R. (ed.) (1986) *The Ideology of the New Right*, Cambridge: Polity Press.

Liberal Democratic Party (1994) '*Reclaiming the City*', London: Liberal Democratic Party.

Libération (1993) 'Urbanisme: Une loi bien betonnée', 22 December.

Lim, H. (1993) 'Cultural strategies for revitalising the city: a review and evaluation', *Regional Studies* 27, 6.

Limonov, L. (1993) 'Emergence des marchés foncier et immobilier en Russie', in V. Renard and R. Acosta (eds) *Land Tenure and Property Development in Eastern Europe*, Paris: Association des Etudes Foncières.

Lizieri, C. (1991) 'City office markets: global finance and local constraints', Paper to 8th Urban Change and Conflict Conference, University of Lancaster, September.

Lloyd, G. and Black, S. (1995) 'Edinburgh', in J. Berry and S. McGreal (eds) *European Cities, Planning Systems and Property Markets*, London: E. & F.N. Spon.

Lloyd, G. and Newlands, D. (1990) 'Business interests and planning initiatives: a case study of Aberdeen', in J. Montgomery and A. Thornley (eds) *Radical Planning Initiatives,* Aldershot: Gower.

Llusa, R. (1994) Gabinet de Programació, Barcelona City Council, Personal communication.

Lodden, P. (1991) 'The "Free Local Government" experiment in Norway', in R. Batley and G. Stoker (eds) *Local Government in Europe*, London: Macmillan.

Loftman, P. and Nevin, B. (1992) *Urban Regeneration and Social Equity*, Research Paper 8, Faculty of the Built Environment, University of Central England.

Loftman, P. and Nevin, B. (1993) 'Prestige projects or urban burdens?', *Municipal Journal* 21-27 May.

Loftman, P. and Nevin, B. (1994) 'Planning for whom?', *Town and Country Planning*, April.

Logan, J. and Molotch, H. (1987) *Urban Fortunes*, Berkeley, Calif.: University of California Press.

Logan, J.R. and Swanstrom, T. (1990) 'Urban restructuring: a critical view', in J.R. Logan and T. Swanstrom (eds) *Beyond the City Limits*, Philadelphia, Pa.: Temple University Press.

London Borough of Camden (1993) 'A revised community planning brief. Kings Cross railway lands'.

London Borough of Greenwich (1991) 'Greenwich Waterfront Strategy Summary'.

London Borough of Greenwich (1992) 'The Greenwich Waterfront Development Partnership'.

London First (1992) *London First: A World Class Capital*, London First.

London Pride Partnership (1995) *London Pride Prospectus.*

Lorange, E. and Myhre, J.E. (1991) 'Urban planning in Norway', in T. Hall (ed.) *Planning and Urban Growth in the Nordic Countries*, London: E. & F.N. Spon.

Lorrain, D. (1991) 'Public goods and private operators in France', in R. Batley and G. Stoker (eds) *Local Government in Europe*, London: Macmillan.

Lorrain, D. (1992) 'Le Modèle Ensemblier en France', in E. Campagnac (ed.), *Les Grands Groupes de la Construction: de Nouveaux Acteurs Urbains?*, Paris: L'Harmattan.

Lorrain, D. (1994) 'La Production Urbaine Après la Décentralisation', *Cahiers Techniques Territoires et Sociétés* 26, Paris-La Défense, Ministère de l'Equipement DRAST, January.

LPAC (1994) 'Advice on strategic planning guidance for London', London Planning Advisory Committee.

McGuirk, P. (1994) 'Economic restructuring and the realignment of the urban planning system: the case of Dublin', *Urban Studies* 31, 2.

Mackintosh, M. (1992) 'Partnership: issues of policy and negotiation', *Local Economy* 7, 3.
McKay, D.H. (1982) *Planning and Politics in Western Europe*, London: Macmillan.
Mairie de Paris (1993) *Paris: No Greater City for Working and Living*, Direction de la Communication.
Makridou-Papadaki, E. (1992) 'Greece', in A. Dal Cin and D. Lyddon (eds) *International Manual of Planning Practice* (second edition), The Hague: ISOCARP.
Malbert, B. (1994) 'Green structure in Lidingö', in B. Malbert (ed.) *Ecology-based Planning and Construction in Sweden*, Stockholm: The Swedish Council for Building Research.
Malmö Stadsbyggnadskontor (1990) *Översiktsplan för Malmö 1990*, Malmö: Malmö Stadsbyggnadskontor.
Malmö Stadsbyggnadskontor (1991) *Brozonen, Porten till Sweden, Program till fördjupadöversiktsplan*, Malmö: Malmö Stadsbyggnadskontor.
Malmö Stadsbyggnadskontor (1992) *Översiktsplan för Citytunneln i Malmö*, Malmö: Malmö Stadsbyggnadskontor.
Malmsten, B. (1992) *Traffic and Environment in the Stockholm Region: The Greater Stockholm Negotiation*, Stockholm: Stockholm Council Office of Regional Planning and Urban Transportation.
Malusardi, F. and Talia, M. (1992) 'Italy', in A. Dal Cin and D. Lyddon (eds) *International Manual of Planning Practice* (second edition), The Hague: ISOCARP.
Manceau, B. (1989) *1982–1988: Histoire d'une legislation*, Supplement to *Cahiers Français*, No. 239.
Mansikka, M. and Rautsi, J. (1992) 'Finland', in A. Dal Cin and D. Lyddon (eds) *International Manual of Planning Practice* (second edition), The Hague: ISOCARP.
Marcou, G. (1993) 'New tendencies of local government development in Europe', in R. Bennett (ed.) *Local Government in the New Europe*, London: Belhaven Press.
Marcou, G. (1994) 'Governance and developments in Paris and Ile-de-France', in R.J. Bennett (ed.) *Local Government and Market Decentralisation*, Tokyo: United Nations University Press.
Marcou, G. and Verebelyi, I. (eds) (1993) *New Trends in Local Government in Western and Eastern Europe*, Brussels: International Institute of Administrative Sciences.
Markowski, T. and Kot, J. (1993) 'Planning for strategic economic development of Łódź - concepts, problems and future vision of a city', in T. Marszal and W. Michalski (eds) *Planning and Environment in the Lódz Region*, Lódz: Zarzad Miasta Łodzi.
Marsh, D. and Rhodes, R.A.W. (1992) *Policy Networks in British Government*, Oxford: Clarendon Press.
Marshall, T. (1990) 'Letter from Barcelona', *Planning Practice and Research* 5, 3.
Marshall, T. (forthcoming) 'Barcelona – fast forward? City entrepreneurialism in the 1980s and 1990s', *European Planning Studies*.
Martin, S. and Pearce, G. (1992) 'The internationalisation of local authority economic development strategies: Birmingham in the 1980s', *Regional Studies* 26, 5.
Martinez Cearra, A. (1994) 'Revitalization strategies in metropolitan Bilbao', Paper to Cities, Enterprises and Society at the Eve of the XXIst Century Conference, Lille, March.
Massey, D. (1991) *Docklands – A Microcosm of Broader Social and Economic Trends*, London: Docklands Forum.
Matheou, D. (1992) 'A who's who of rebuilding Berlin', *Architectural Journal*, 24 June.
Mattsson, H., Kalbro, T. and Miller, T. (1989) 'Development agreements for residential development in Sweden', in D.R. Porter and L.L. March (eds) *Development Agreements: Practice, Policy and Prospects*, Washington: Urban Land Institute.

Mauroy, P. (1992) 'The Lille metropolis and city networks: stakes and perspectives', *Ekistics* 352/353.
Mauroy, P. (1994) *Parole de Lillois*, Paris: LieuCommun.
Mayer, M. (1994) 'Post-Fordist city politics', in A. Amin (ed.) *Post-Fordism: A Reader*, Oxford: Blackwell.
Mazey, S. (1993) 'Developments at the French meso level', in L.J. Sharpe (ed.) *The Rise of Meso Government in Europe*, London: Sage.
Mazza, L. (1991) 'European viewpoint: a new status for Italian metropolitan areas', *Town Planning Review* 62, 2.
Mény, Y. (1988) 'France: the construction and reconstruction of the centre', in R.A.W. Rhodes and V. Wright (eds) *Tensions in the Territorial Politics of Western Europe*, London: Frank Cass.
Mény, Y. (1992) 'La République des fiefs: L'état de la décentralisation', *Cahiers Français* 256.
Mény, Y. (1994) 'Leçons Françaises du cas Italien', *Le Monde*, 30 September.
MHLG (Ministry of Housing and Local Government) (1962) *Town Centres: Approach to Renewal,* London: HMSO.
Ministry of the Environment (1993) *The Öresund Region - A Europole*, Copenhagen: Spatial Planning Department.
Ministry of the Environment and Natural Resources (1994) 'Towards Sustainable Development in Sweden', Swedish Government Bill 1993/1994: 111, Stockholm.
Mishra, R. (1990) *The Welfare State in Capitalist Society*, Hemel Hempstead: Harvester Wheatsheaf.
Modeen, T. (1986) 'Town and country planning law in Finland', in J.F. Garner and N.P. Gravells (eds) *Planning Law in Western Europe*, Amsterdam: Elsevier Science Publishers.
Mollenkopf, J.H. and Castells, M. (eds) (1991) *Dual City: Restructuring New York*, New York: Russell Sage Foundation.
Molotch, H. (1976) 'The city as a growth machine', *American Journal of Sociology* 82, 2.
Montin, S. (1992) 'Recent trends in the relationship between politics and administration in local government: the case of Sweden', *Local Government Studies* 18, 1.
Montin, S. (1993) *Swedish Local Government in Transition*, Örebro Studies No. 8, University of Örebro.
Moran, F. (1960) 'The Republic of Ireland', *The Law Quarterly Review* 76.
Moulaert, F. and Demazière, C. (1994) 'Local development in western Europe', Paper to Cities, Enterprises and Society at the Eve of the XXIst Century Conference, Lille, March.
Mouritzen, P.E. (ed.) (1992) *Managing Cities in Austerity*, London: Sage.
Moylan, M. (1995) 'Developments in member states and their preparation for the 1996 IGC', Paper to Practical Implications of Europe 2000+ Guidelines Conference, Association of District Councils, London, March.
Muccini, P. (1993) 'Northern exposure', *Europroperty*, September.
Müller (1993) *Property Report Eastern Europe*, Frankfurt/Vienna: Müller International Immobilien GmbH.
NCVO (1993) *Community Involvement in City Challenge: A Policy Report and Good Practice Guide*, London: National Council for Voluntary Organisations.
Needham, B., Koenders, P. and Kruijt, B. (1993) *Urban Land and Property Markets in the Netherlands*, London: UCL Press.
Némery, J.-C. (1993) 'Les Institutions Territoriales Françaises à l'Epreuve de l'Europe', in J.-C. Némery and S. Wachter (eds) *Entre L'Europe et la Décentralisation,* La Tour d'Aigues: DATAR/Editions de l'Aube.
Netherlands Scientific Council for Government Policy (1990) *37. Institutions and Cities. The Dutch Experience*, The Hague: NSC.

Nevin, B. and Shiner, P. (1993) 'Britain's urban problems: communities hold the key', *Local Work* 50.

Newman, P. (1990) 'Master planning', Paper to Planning Theory in the 1990s Conference, Oxford Polytechnic, March.

Newman, P. (1995) 'London Pride', *Local Economy*, 10, 2.

Newman, P. and Thornley, A. (1992) 'Londoners lose out in World City stakes', *Planning* 961.

Newman, P. and Thornley, A. (1995) 'Euralille. Boosterism at the centre of Europe', *European Urban and Regional Studies* 2, 3.

Nijkamp, P. (1993) 'Towards a network of regions: the United States of Europe', *European Planning Studies* 1, 2.

Norell, P.O. (1993) 'Facing fiscal stress and demand for renewal: the Swedish municipal case', Paper presented to seminar on Local Government Finance, University of Neuchâtel, 25–28 June.

Norton, A. (1991) 'Western European local government in comparative perspective', in R. Batley and G. Stoker (eds) *Local Government in Europe*, London: Macmillan.

Novarina, G. (1994) 'Concurrence entre villes et transformation des politiques urbaines. L'example de Grenoble (1960–1990)', Paper to Cities, Enterprises and Society at the Eve of the XXIst Century Conference, IFRESI, Lille, March.

NPPA (1991) *Perspectieven in Europa*, The Hague: National Physical Planning Agency.

Ödmann, E. (1987) 'Vague legal norms in a tough reality - some remarks on the new Swedish Planning Act, *Scandinavian Housing and Planning Research* 4.

Östergård, N. (1994) *Spatial Planning in Denmark*, Copenhagen: Ministry of the Environment.

Ozanne, J. (1991) '21st century bonanza', *Estates Times*, 25 October.

Padioleau, J.-G. and Demesteere, R. (1992) 'Les Démarches Stratégiques de Planification des Villes', *Les Annales de la Recherche Urbaine* 51.

Page, E.C. and Goldsmith, M.J. (eds) (1987) *Central and Local Government Relations. A Comparative Analysis of West European Unitary States*, London: Sage.

Parkinson, M. (1992) 'Strategic responses to economic restructuring: the rise of the "entrepreneurial" European city', Paper to Marketing Urbino Conference, Turin, June.

Parkinson, M., Bianchini, F., Dawson, J., Evans, R. and Harding, A. (1992) *Urbanisation and the Functions of Cities in the European Community*, Liverpool: European Institute of Urban Affairs, John Moores University.

Parkinson, M. and Le Galès, P. (1994) *Politiques de la Ville en France et en Grande-Bretagne: Etude Comparative*, Rapport au Conseil Franco-Britannique, Paris.

Pearce, D., Markandya, A. and Barbier, E. (1989) *Blueprint for a Green Economy*, London: Earthscan Publications Ltd.

Peck, J. and Tickell, A. (1992) 'Local modes of social regulation?', *Geoforum* 23.

Peck, J. and Tickell, A. (1993) 'Business goes local: dissecting the business agenda in post-democratic Manchester', Paper to the Ninth Urban Change and Conflict Conference, Sheffield, September.

Petersson, O. (1986) *Study of Power and Democracy in Sweden: Outline of the Final report*, English Series Report No. 3, Stockholm: SOU.

Petersson, O. (1987) *Democracy - Ideal and Reality, The Study of Power and Democracy in Sweden*, English Series Report No. 7.

Petersson, O. (1991) 'Shifts in power', in A. Fredlund (ed.) *Swedish Planning in Times of Transition*, Gävle: Swedish Society for Town and Country Planning.

PFO (1993) 'Public finance initiative', Public Finance Office, Treasury.

Piven, F.F. (1991) *Labor Parties in post industrial societies*, Cambridge: Polity Press.

Power, A. (1993) *From Hovels to High Rise*, London: Routledge.

Poznan Municipal Town Planning Office (n.d.) *Proposals of Hotel Sites in Poznan*, Poznan: Department of Information and Development.
Poznan Municipal Town Planning Office (1992) *The Guidelines for the General Physical Plan of Poznan City*, Poznan: Municipal Town Planning Office.
Prague City Architect's Office (1992) *Prague: Metropolitan Area Report*, Prague: City Architect's Office.
Prague City Council (1993) *Handbook for Real Estate Investors in Prague*, Prague, Prague City Council.
Preteceille, E. (1988) 'Decentralisation in France: new citizenship or restructuring hegemony?', *European Journal of Political Research* 16.
Preteceille, E. (1990) 'Political paradoxes of urban restructuring', in J.R. Logan and T. Swanstrom (eds) *Beyond the City Limits*, Philadelphia, Pa.: Temple University Press.
Preteceille, E. (1991) *Capital Cities: The Contradictions of Success*, Document du Travail, Centre de Sociologie Urbaine, Paris.
Provan, B. (1994) 'Problem housing estates in Britain and France', Ph.D. thesis, London School of Economics.
Querrien, A. (1992) 'La Planification Urbaine et ses Doubles', *Les Annales de la Recherche Urbaine* 51.
Railway Lands Group (1993) 'Interim uses initiative', London: Railway Lands Group.
Reade, E. (ed.) (1989) *Britain and Sweden: Current Issues in Local Government*, Gävle: National Swedish Institute for Building Research.
Regulski, J. (1989) 'Polish local government in transition', *Environment and Planning C: Government and Policy* 7.
Regulski, J. and Kocan, W. (1994) 'From communism towards democracy: local government reform in Poland', in R. Bennett (ed.) *Local Government and Market Decentralisation*, Tokyo: United Nations University Press.
Renard, V. (1994) 'Le Sol et La Ville. Pathologies Françaises', *Le Débat* 80.
Rhodes, R.A.W. and Wright, V. (eds) (1987) *Tensions in the Territorial Politics of Western Europe*, London: Frank Cass.
Richard Ellis (1990) *The Berlin Office Market*, Berlin: Richard Ellis.
Ridley, N. (1988) *The Local Right: Enabling not Providing*, London: Centre for Policy Studies.
Riera, P. and Keogh, G. (1995) 'Barcelona', in J. Berry and S. McGreal (eds) *European Cities, Planning Systems and Property Markets*, London: Spon.
Ringli, H. (1992) 'Switzerland', in A. Dal Cin and D. Lyddon (eds) *International Manual of Planning Practice* (second edition), The Hague: ISOCARP.
Ringli, H. (1994) 'Tendencies in development plan-making in Switzerland', in P. Healey (ed.) *Trends in Development Plan-making in European Planning Systems*, Working Paper No. 42, Department of Town and Country Planning, University of Newcastle upon Tyne.
Robert, J.-P. (1994) 'Paris brûle-t-il? Un bilan critique', *L'Architecture d'Aujourd'hui* 295.
Robinson, S. (1990) 'City under stress', *The Planner* 76, 10.
Robson, B., Bradford, M., Deas, I., Ham, E. and Harrison, E. (1994) *Assessing the Impact of Urban Policy*, London: HMSO.
Ronneburger, K. and Keil, R. (1992) 'Riding the tiger of modernisation', Paper to International Sociological Association Conference, Los Angeles, April.
Rose, R. (1991) 'Comparing forms of comparative analysis', *Political Studies* XXXIX, 3.
Ross, J. (1995) 'When co-operation divides: Öresund, the Channel Tunnel and the new politics of European transport', *Journal of European Public Policy* 2,1.
Rouzeau, B. (1993) Interview. La Plaine Développement, April

Rudberg, E. (1991) 'From model town plans to municipal planning guidelines', in A. Fredlund (ed.) *Swedish Planning in Times of Transition*, Gävle: Swedish Society for Town and Country Planning.
Ryan, L. (1995), 'Bureaupolitics, the SRB and urban policy in Britain', in J. Lovenduski and J. Stanyer (eds) *Contemporary Political Studies*, Belfast: Political Studies Association.
Sahlin-Andersson, K. (1992) 'The social construction of projects', *Scandinavian Housing and Planning Research* 9, 2.
Sassen, S. (1991) *The Global City: New York, London and Tokyo*, Princeton: Princeton University Press.
Sassen, S. (1994) 'La Ville Globale. Eléments pour une lecture de Paris', *Le Débat* 80, May–August.
Savitch, H.V. (1988) *Post-Industrial Cities*, Princeton, N.J.: Princeton University Press.
SCET (1994) 'La Ville, La Vie', Presentation at MIPIM Conference, Nice.
Schulman, H. and Verwijnen, J. (1993) *External Trends and Local Strategies in the Development of the Helsinki Region*, Helsinki: Centre for Urban and Regional Studies.
SEMAPA (1993) Interview with M.G. de Montmarin, April.
Shachar, A. (1994) 'Randstad Holland: A "World City"?', *Urban Studies* 31, 3.
Simmie, J. (1993) 'Technopole planning in Britain, France, Japan and the USA', *Planning Practice and Research* 9, 4.
Skelcher, C. and Stewart, J. (1993) *The Appointed Government of London*, London: Association of London Authorities.
Smith, M. and Feagin, J. (eds) (1987) *The Capitalist City*, Oxford: Blackwell.
Snickars, F. (1991) 'Negotiations and urban planning', in A. Fredlund (ed.) *Swedish Planning in Times of Transition*, Gävle: Swedish Society for Town and Country Planning.
Snickars, F. and Cars, G. (1991) 'Negotiated decisionmaking in urban planning', Paper presented to AESOP/ACP Congress, Oxford.
Söderlind, J. (1993) 'Stockholm Stad' (Stockholm City), *Arkitektur* 2.
Sotarauta, M. (1994) 'Finnish municipalities and planning in transition: empowerment, impulses and strategies', *European Planning Studies* 2, 3.
Soulignac, F. (1993) *La Banlieue Parisienne*, Paris: La Documentation Française.
South, G. (1992) 'Divisions still run deep in troubled city of Berlin', *Chartered Surveyor*, 28 May.
Stevens, J. (1993) 'La Metropolisation en Région Urbaine: L'Expérience Lilloise', The European City and its Region Conference, Department of the Environment, Ireland, September.
Stevenson, D. (1994) 'Partnerships and the new sub regions needed for resource bidding', in M. Edwards and J. Ryser (eds) *Bidding for the Single Regeneration Budget. Seminar Proceedings,* University College London.
Stewart, J. and Stoker, G. (1989a) *From Local Administration to Community Government*, Research Series 351, London: Fabian Society.
Stewart, J. and Stoker, G. (1989b) 'The "Free Local Government" experiments and the programme of public service reform in Scandinavia', in C. Crouch and D. Marquand (eds) *The New Centralism*, Oxford: Blackwell.
Stockholm County Council (1994) *Stockholm Läns Landsting och Europa*, Stockholm: Stockholm County Council.
Stockholmsleder AB (1992) *Ringen – The Stockholm Ring Road: Implementation Plan*, Stockholm: Stockholmsleder AB.
Stoker, G. (1990) 'Regulation theory, local government and the transition from Fordism', in D. King and J. Pierre (eds) *Challenges to Local Government*, Newbury Park, Calif.: Sage.
Stoker, G. (1991a) *The Politics of Local Government*, London: Macmillan.

Stoker, G. (1991b) 'Trends in western European local government', in R. Batley and G. Stoker (eds) *Local Government in Europe*, London: Macmillan.

Stoker, G. (1992) *The Local Governance Initiative*, Local Government Management Board, Research Link 2.

Stoker, G. (1993) *Urban Regime Theory in Comparative Perspective*, Department of Government, University of Strathclyde.

Stoker, G. and Mossberger, K. (1994) 'Urban regime theory in comparative perspective', *Environment and Planning C: Government and Policy* 12, 2.

Stoker, G. and Young, S. (1993) *Cities in the 1990s*, Harlow: Longman.

Stone, C. (1991) 'The hedgehog, the fox and the new urban politics', *Journal of Urban Affairs* 13, 3.

Stone, C., Orr, M. and Imbroscio, D. (1991) 'The reshaping of urban leadership in US cities: a regime analysis', in M. Gottdiener and C. Pickvance (eds) *Urban Life in Transition*, Newbury Park, Calif.: Sage.

Stopford, J. and Strange, S. (1991) *Rival States, Rival Firms: Competition for World Market Shares*, Cambridge: Cambridge University Press.

Suetens, L.P. (1986) 'Town and country planning law in Belgium', in J.F. Garner and N.P. Gravells (eds) *Planning Law in Western Europe*, Amsterdam: Elsevier Science Publishers.

Swianiewicz, P. (1992) 'The Polish experience of local democracy: is progress being made?', *Policy and Politics* 20, 2.

Sýkora, L. (1994) 'Local urban restructuring as a mirror of globalisation processes: Prague in the 1990s', *Urban Studies* 31, 7.

Sýkora, L. (1995), 'Prague', in J. Berry and S. McGreal (eds) *European Cities, Planning Systems and Property Markets*, London, E. & F.N. Spon.

Taylor, R. (1991) *The Economic Policies of Sweden's Political Parties*, Stockholm: Swedish Institute.

Technopolis International (1993) 'Le Nouveau Quartier Latin', Interview with Jacques Toubon, April.

Thomas, D., Minett, J., Hopkins, S., Hamnett, S., Faludi, A. and Barrell, D. (1983) *Flexibility and Commitment in Planning*, The Hague: Martinus Nijhoff.

Thornley, A. (1992a) 'Ideology and the by-passing of the planning system', *European Planning Studies* 1, 2.

Thornley, A. (ed.) (1992b) *The Crisis of London*, London: Routledge.

Thornley, A. (1993a) *Urban Planning under Thatcherism: The Challenge of the Market*, London: Routledge.

Thornley, A. (1993b) 'Letter from Sofia: building the foundations for a market-oriented planning system in Bulgaria', *Planning Practice and Research* 8, 4.

Thrift, N. (1994) 'Globalisation, regulation, urbanisation: the case of the Netherlands', *Urban Studies* 31, 3.

Times (1994) 'Geordies seek home rule to revive region', 30 September.

Times (1995a) 'Festival fires gas works site with life', 14 February.

Times (1995b) 'Canary Wharf expansion to go ahead', 6 April.

Tonell, L. (1991) '1980-talet–förhandlingsplaneringens och de stora projektens decennium', in T. Hall (ed.) *Perspektiv på planering*, Uppsala: HSFR.

Tonell, L. (1993) Interview. Department of Geography, Stockholm University, November.

Toonen, T.A.J. (1987) 'A decentralised, unitary state in a welfare society', in R.A.W. Rhodes and V. Wright (eds) *Tensions in the Territorial Politics of Western Europe*, London: Frank Cass.

Toonen, T.A.J. (1993) 'Dutch provinces and the struggle for the meso', in L.J. Sharpe (ed.) *The Rise of Meso Government in Europe*, London: Sage.

Travers, T., Jones, G., Hebbert, M. and Burnham, J. (1992) *The Government of London*, York: Joseph Rowntree Foundation.

Turok, I. (1992) 'Property-led urban regeneration: panacea or placebo', *Environment and Planning A* 24.

Unal, M. (1994) 'Dossier: Quel Avenir pour le Bassin Parisien', *MagaVille* 4, Sciences Po-Urbain.

UNCED (1992) *Agenda 21,* Conches, Switzerland: United Nations Conference on Environment and Development.

Valles, J.M. and Foix, M.C. (1988) 'Decentralisation in Spain: a review', *European Journal of Political Research* 16.

Van den Berg, l., Van Klink, H.A. and Van der Meer, J. (1993) *Governing Metropolitan Regions*, Aldershot: Avebury.

Van den Bos, J. (1992) 'Reborn Berlin', *Europroperty*, July/August.

Van Wunnick, P.M. (1992) 'Belgium', in A. Dal Cin and D. Lyddon (eds) *International Manual of Planning Practice* (second edition), The Hague: ISOCARP.

Vasconcelos, L. and Reis, C. (1994) 'The Portuguese planning process and local development plans', in P. Healey (ed.) *Trends in Development Plan-making in European Planning Systems*, Working Paper No. 42, Department of Town and Country Planning, University of Newcastle upon Tyne.

Vedung, E. (1993) *Statens markpolitik Kommunerna och historiens ironi*, Stockholm: SNS Förlag.

Vegara, J. (1992) *Barcelona Olympic Games: A Successful Example of Project Management*, Barcelona City Council.

Veraldo, R. (1979) 'Retail planning in Italy', in R.L. Davies (ed.) *Retail Planning in the European Community*, Hants: Saxon House.

Verpraet, G. (1992) 'Le Dispositif Partenarial des Projets Intégrés', *Les Annales de la Recherche Urbaine* 51.

Vicari, S. and Molotch, H. (1990) 'Building Milan: alternative machines of growth', *International Journal of Urban and Regional Research*, 14, 4.

Ville de Montpellier (1990) *Plan d'Occupation des Sols. Rapport de Présentation*, Direction de l'aménagement et de la programmation.

Virtanen, P. (1994) 'A review of development plan-making in Finland', in P. Healey (ed.) *Trends in Development Plan-making in European Planning Systems*, Working Paper No. 42, Department of Town and Country Planning, University of Newcastle upon Tyne.

VOSA (1992) *La Vila Olimpica*, June.

Wassenhoven, L. (1984) 'Greece', in M. Wynn (ed.) *Planning and Urban Growth in Southern Europe*, London: Mansell.

Webster, J. (1992) 'Serious games', *New Statesman and Society*, 31 July.

Westerståhl, J. (1987) *Staten, kommunerna och den statliga styrningen* (The state, the municipalities and state control), Committee on Local Democracy, Ministry of Public Administration.

White, P. (1995) 'Public transport in London', Paper to ESRC London Seminar, London School of Economics, April.

Williams, A.M. (1984) 'Portugal' in M. Wynn (ed.) *Planning and Urban Growth in Southern Europe*, London: Mansell.

Wise, C. and Amnå, E. (1992) *Reform in Swedish Local Government*, Stockholm: Swedish Institute.

Wolman, H. (1990) 'Decentralisation: what it is and why we should care', in R. Bennett (ed.) *Decentralisation, Local Governments and Markets: Towards a Post-welfare Agenda,* Oxford: Clarendon Press.

Wood, B. and Williams, R. (1992) *Industrial Property Markets in Western Europe*, London: Spon.

Wood, C. (1994) 'Local urban regeneration initiatives', *Cities* 11, 1.

Wretblad, L. and Lindgren, E. (1993) Interview, Lidingö Municipality, November.
Wynn, M. (ed.) (1984) *Planning and Urban Growth in Southern Europe*, London: Mansell.
Yarranton, L. (1993) 'Rising star, fading fast', *Europroperty*, June.
Zarecky, P. (1994) 'Central–local relations under the new Czechoslovak constitutional law', in R. Bennett (ed.) *Local Government and Market Decentralisation*, Tokyo: United Nations University Press.
Zweigert, K. and Kötz, H. (1987) *Introduction to Comparative Law*, Oxford: Clarendon Press.

INDEX

ABC 13 (1993) 181
accountability, legitimacy and 253–4
Ache, P. 14, 15, 16, 17
Acosta, R. 46, 169, 178, 188
Aczel, G. 14, 80
ADA 13 (1990) 181
Adjustment Areas (Berlin) 101
administrative systems 27–38
Against the Bridge campaign 236
Agence de Dévéloppement et d'Urbanisme 191
Agences d'Urbanisme 155, 162
agency model 31
Agenda 21 programme 18, 120, 220, 252
Aims of Industry group 24
Alpine Arc region 20
aménagement de territoire 154, 196; DATAR 16, 45, 169–60, 167, 171–2, 174, 185, 198
Amnå, E. 206
Andrikopoulou, E. 58, 59
Anti-Poverty programmes 19
Anton, T 234
Architectural Competitions 115
Areas of New Centrality 92, 105
Areas of Outstanding Natural Beauty 115
arrondissements 162, 181, 182
Ashford, D. 32
Ashworth, G. 16
Association of European Planning Schools 3
Association of London Authorities 149, 152–3
Association of Metropolitan Interests 90
Atelier Parisien d'Urbanisme 177–8, 179, 182, 196, 197–8
Athens 60, 105, 254
Atlantic Arc regions 20, 21, 159
Audit Commission 112, 122, 123, 124
Austria 62
authoritarian decentralism 112, 115, 116
Autonomous Communities 41
autonomy: local level 79, 88, 121, 124–5, 242–4, 247; national framework 28, 31–2, 40–1, 48, 56, 61, 67, 71–2
Ave, G. 50–1, 52
Axelsson, S. 200

Baïetto, J.P. 192, 193, 194
Bailey, N. 127
Balludur government 168
Balme, R. 40, 50, 62, 190
Baltic Regional Plan 21
banking sector (Sweden) 200–1, 219
Bannon, M.J. 76
Barcelona (urban governance/planning) 76, 91–5, 104–6, 247, 249
Barcelona Development Agency 93
Barcelona Economic and Social Plan 2000 94
Barlow, J. 13, 82, 84, 164, 209
Barnekov, T. 83
Barras, R. 138
Barry, A. 23
Basic Laws (Germany) 34, 60

Bassin Parisien 159–60, 173
Bassols, M. 56
Batley, R. 31, 38, 40, 76
Bayle, C. 185
Bebauungsplan 61
Begg, H. 43, 150, 151
Bejrum, H. 213
Belgium 41, 54–5
Bell, D. 127
Benfer, W. 82
Bennett, R.J. 10, 23, 31, 32, 33, 38, 41
Bennington, J. 22
Berggrund, L. 211
Beriatos, E. 60
Berlin (urban governance/planning) 76, 98–104, 105, 106, 253
Berlin Wall 98, 99, 102
Berry, J. 3, 103
Bertamini, F. 89
Bestemmingsplan 48, 50
Bianchini, F. 89, 90
Biarez, S. 155, 157, 165, 174
Biggs, S. 147
'Bilbao-Metropoli-30' 79
Birmingham: case study 128–35, 249; Chamber of Commerce 130–1; Heartlands scheme 128, 131–4, 151, 152, 247; Marketing Partnership 133; 2000 (membership) 130; Unitary Development Plan 129, 132, 134, 151, 248
Bizet, J.F. 163
Black, S. 43
Body-Gendrot, S. 167
Bonneville, M. 15, 128, 134
boosterism in cities 22, 116–18
Booth, P. 6, 168
Borraz, O. 162, 164–5
Bousquet, Mayor (Nîmes) 164
Bouygues 163, 188
Boyle, R. 79
Bradford Breakthrough 117–18
Brandenburg 100
Branegan, J. 201
Bremm, H. 14, 15, 16, 17
Brierley, J. 28
Brillot, F. 191
Britain: English case studies 128–53; legal/administrative systems 29–31, 39, 71; planning systems 42–5, 71–4, 246; Thatcherism legacy 111–27; *see also individual cities*
British Gas 143, 144
British Rail 139–42
British Steel 117
Brodsky, P. 96
Brooks, M. 131
Brownhill, S. 116
Bruegel, I. 98, 101
Brunet, R. 16
Brussard, W. 47, 48
Building and Planning Act (Sweden, 1987) 210–12, 216
Bulgaria 70–1
Bulpitt, J. 31, 112
Business-Centred Activist regime 82
Business in Cities Forum 117
Business in the Community 78, 117, 149, 150, 152
Business Leadership Teams 150
Busquets, J. 91
Butlers Wharf Company 138

Caisse des Dépôts et Consignations 80, 163, 187
Calabi, D. 52
Calavita, N. 52
Camden 140, 141
Canary Wharf 7, 135, 136–9, 145, 148
cantons (in Switzerland) 62–3
Cappellin, R. 87, 88
Carbonaro, G. 78
'Caretaker and Exclusionary' regime 82, 86
Carignon, Alain 164–5
Carpenter, C. 178
Cars, G. 213–14, 215, 221
Carter, N. 76–7
case studies: England 128–53; France 176–99; Sweden 221–44
Castells, M. 9, 10, 13, 26
Castles, F. 202
Catholic Church (France) 32, 39
Cattaneo, S. 50

CBI 117, 149
Central Area Study 117
central government 5; administrative systems 27–38
centralisation 33, 36–7, 49, 68; case study (London) 135–45
Centre Capitals region 20
Centre Party (Sweden) 204, 210, 217, 232, 234, 236–7
Chambre de Commerce et d'Industrie 164, 165, 187
Channel Tunnel High Speed Link 126
Chapman, M. 18
Charte d'Objectif 159, 161, 174
Charte du Bassin Parisien 160
Charter for Capital City of Prague 96
Checkpoint Charlie 100
Cheshire, P. 5, 16
Chicoye, C. 159
Chirac, Jacques 164, 177, 182, 194
Christian Democrats 86, 87, 102, 104
Church, A. 116
Ciechocinska, M. 36, 37
Cité Europe (at Coquelles) 194
cities: boosterism 22, 116–18; competition 225–7; entrepreneurial 78–80; global 12–14, 26; hierarchy 14–17; international context 12–17, 21–3; regimes 176–83
Citizen's Charter 40, 126, 207
city-states 26
City Challenge 22, 117, 118, 122–6, 132, 144, 146, 152, 246, 250
City Grant 132
City Pride initiative 124, 133–5, 149, 151, 152, 248–9
'City 2020' inquiry 127
Civil Code (France) 31–2, 39
CLAQ 181, 182
Clarke, M. 31, 114
Claval, P. 12
client-performer model 206
Clinton administration 127
Coccossis, H. 57, 59, 60
Code d'Urbanisme 45, 170
Cohen, N. 131
Cole, A. 83
Colenutt, B. 77, 137, 139, 149
Collier, J. 117
Collinge, C. 151
Comité Interministériel pour l'Aménagement du Territoire 159
Comités d'Expansion 164
command and control functions 12, 14
Commission for Local Democracy 114, 254
Committee of the Regions 18, 22, 130, 254
Common Law 30
communal public sector (France) 162–7
Communauté Urbaine de Lille (CUDL) 190–5, 197, 248
Communauté Urbaine de Lyon 165
communautés de communes 163
communautés de villes 163
communautés urbaines 155, 162–3, 197
communes: Eastern Europe 38, 70; France 32–3, 45–7, 155–6, 161–2, 174; Germany 34, 61; Italy 50, 51; Switzerland 63
communication systems 14, 19, 20
communism 4, 6, 9, 23–5, 35–7, 253
Communist Party: France 160, 183–5; Milan 87; Poland 25, 36–7
community empowerment 127
Community Initiatives 19
Community Support Frameworks 18
Community Trusts 127, 142
Compagnie Générale des Eaux 163
Compendium of European Planning Systems 3
competition: city (Stockholm) 225–7; European (Euralille project) 189–94; international 176–83
Condamines, E. 157
Conférénce de la Région Urbaine de Lyon 165
Conseil d'Etat 170, 181
Conseil National de l'Aménagement et du Développement du Territoire 172
Conseils Généraux 160, 172
consensus 114; culture (Sweden) 201–3, 207, 210, 235, 243
consensus-building 83; Barcelona 91–5, 105; Netherlands 40

Conservation Areas 115
Conservative Party: Britain 3, 120–1, 147, 252; Sweden 233
Constitutional Court (Spain) 56
Constitutions: Austria 62; Belgium 54; Britain 30; Germany 34, 60; Greece 58; Netherlands 47, 48; Spain 55–6; Sweden 211
Continental Diagonal region 20
contrats de plan 157, 159, 169, 173, 174, 187
contrats de ville 22, 168–9, 174, 186, 250
Coopers and Lybrand 78, 79; Deloitte 12, 149
Coordination et Liaison des Associations de Quartier 181, 182
Copenhagen 248; Öresbund Bridge case study 221, 236–42, 244, 252
corporate pluralism 202, 243
corporatism 205, 209, 235–6
corruption scandals: France 170–1; Italy 89–90, 253
Costa, F.J. 49
Costa Lobo, M. 53
Council of State (Luxembourg) 46–7
County Administrative Board (Sweden) 66
Coupland, A. 137
Cour des Comptes 46
Cowlard, K. 137
Cox, K. 85
Craxi, Bettino 87, 89
Crosland Anthony 202
Cross Millennia Art, Technology and Ecology project 142
Crouch, C. 40, 205
culture/cultural difference 6, 12, 16
Cuñat, F. 190
Czechoslovakia 38

Da Rosa Pires, A. 78
Dahrendorf, R. 24, 152, 201
D/Arcy, E. 78
DATAR 16, 45, 159–60, 167, 171–2, 174, 185, 198
David, R. 28
Davies, C. 139
Davies, H.W.E. 17, 47–50, 71, 73–4
de Forn i Foxà, M. 92
de Vos, E. 163
Deacon, B. 24
Debenham Thorpe Zadelhoff 95
decentralisation 76, 109; authoritarian 112, 115, 116; fragmentation of government responsibility 246–7; national framework 36, 40–1, 48–9, 59, 63, 69; post-decentralisation planning 169–71; power 204–8, 243
Défense, La 173, 179, 180, 185
Defferre, Gaston 155–6, 174
deindustrialisation 10, 23, 200, 213
Délégation à l'Aménagement du Territoire et à l'Action Régionale 16, 45, 159–60, 167, 171–2, 174, 185, 198
Délégation Intérministerielle à la Ville 168, 184–5
Delladetsima, P. 58, 59, 105, 254
Delluc, M. 178
Delmartino, F. 41
Delors, Jacques 19, 23
Delta project 94
Demazière, C. 9
Demesteere, R. 165
Demoureaux, J.-P. 170
Democratic Left Alliance (Poland) 25
Denmark 248; Öresund Bridge case study 221, 236–42, 244, 252; planning system 63–5
Dennis, Bengt 229
Bennis package (Stockholm) 221, 225–36, 243–4, 247, 252, 253
Département 45–6, 155, 157, 160, 173
Départements d'Outre-Mer 20
deregulation 9–10, 40, 148, 216–17
Detailed Zoning Plan (Belgium) 55
Development Areas (Berlin) 101
development agreements (Sweden) 214–15
Development Corporations model 151
development projects 109; Britain (Thatcher reforms) 111–27; English case studies 128–53; France (state reorganisation) 154–75; French case studies 176–99; Sweden (market)

200–220; Swedish case studies 221–44
Développement Social des Quartiers 167, 168–9
Devès, C. 163
Dickens, P. 208
Dielman, M. 11, 13
Dieterich, H. 3, 60, 61
DiGaetano, A. 82, 83–4, 85, 130
Dion, S. 164
Direction Départementale de l'Equipement (DDE) 155, 161, 184
Direction Régionale de l'Equipement (DRE) 184
Docklands Consultative Committee 116, 127
Donau-City project 15
Donzelot, J. 254
dorsale 16, 19
Dostál, P. 38
Dransfield, E. 61
DREIF 178
Drouet, D. 13, 163
dual polity 31, 40
Dumont, M.-J. 178
Duncan, S. 13, 31, 40, 112, 209
Dunford, M. 3, 14
Dutt, A.K. 49

East 8 (1993) 95
East Thames Corridor 139, 144
Eastern Europe 77, 253; legal and administrative systems 29, 35–8; market creation 23–6; planning systems 69–71
Ecological City (OECD project) 211
economic: crisis (Sweden) 200–1; growth and urban planning 247–9; liberalism (UK) 112, 115; relations 9–12, 14–17
Economic Development Corporation 86
Economic and Social Committees 156
Edinburgh Vision 117
Education Acts 113
Edwards, D. 65
Edwards, M. 140
Ekerö, 230, 231–2
El Guedj, F. 165
Elander, I. 37, 38, 41, 206–7, 209, 216, 220
Electricité de France 185, 187
Ellger, C. 98, 99
Ellis, G. 149
empowerment zones 127
Emscher Park (Ruhr) 79
English case studies 152–3; Birmingham 128–35; London 135–45; London (fragmentation of planning) 145–51
English Partnership 123
Engqvist, Lars 241
Engrand, G. 193
Engström, C.-J. 214
Englightenment 32, 33
Enterprise Communities 127
Enterprise Zones 115, 137, 140
entrepreneurial cities 78–80
environment: Agenda 21, 18, 120, 220, 252; greening of Thatcherism 119–20; impact analysis 17, 119, 218
Environmental Assessment regulations (Denmark) 64
Environmental Impact Assessment 17, 119
environmental objectives 249–53
Eriksson, E. 225, 227, 231, 233
Eriksson, K. 219
Estate Action Funding 132
Estates Europe 90
Etzioni, A. 127
Euralille Project 8, 176, 189–94, 195, 196, 197–9, 247, 248
Euroc City scheme 240–1, 242, 243
Eurocities 15, 17, 130
Europe: approaches/definitions 3–8; legal/administrative systems 28–38; nation-states and cities 21–3; Single European Act 3, 17, 90, 148; urban/regional policy 17–21; urban hierarchy 14–17
Europe 2000 18, 19–20
Europe 2000+ 20, 21, 22, 23, 73–4, 249
European City of Culture festival 16

European Council of Town Planners 3
Europeanisation 3, 22
Europoles 15
Europroperty 90

Fainstein, N. 114
Fainstein, S. 5, 13, 23, 77, 83, 114
Falkanger, T. 67
Faludi, A. 47
Fareri 88
fast-track implementation (Dennis package) 231–3
Feagin, J. 12
Federal Building Law (Germany) 61
Federal Comprehensive Regional Planning Law (Germany) 61
Federal Law of Spatial Planning (Switzerland) 63
Fehily, J. 45
Feldt, Kjell-Olof 203
Fiddeman, P. 150
finance policies 81–2
Finland (planning system) 68–9
Flächennutzungsplan 61, 86, 102
Fog, H. 211, 213
Foix, M.C. 41
Foley, D. 114
Fordism 11, 15
Forza Italia 90
Fournier, S. 200
fragmentation (of government responsibility) 246–7
France: case studies 176–99; legal and administrative systems 31–3, 41; planning system 45–6, 246, 247; state reorganisation 154–75
Franco, Francisco 55–6
Frankfurt (urban governance/planning) 76–7, 85–7, 104–6, 249
Fredlund, A. 65, 210
Friedman, J. 12
French case studies: Euralille project 8, 176, 189–94, 195, 196, 197–9, 247, 248; La Plaine Saint-Denis 8, 176, 183–8, 194–5, 196–9, 250; Seine Rive Gauche 8, 176–83, 194–5, 197–8, 248, 252; urban planning and development (approaches) 194–9
French Revolution 31, 35, 45
Frick, D. 98, 101
functional urban regions 5

Gamble, A. 112
Garcia, S. 91, 92
García-Bellido, J. 55
Garside, P. 132
gateway cities 14–15, 16
GATT rounds 9
Gauffin, C. 229, 233
Gaz de France 185, 187
General Zoning Plan (Belgium) 55
Générale des Eaux 188
Germanic family: legal/administrative systems 29, 33–4, 39–40; planning systems 60–3, 72
Germany 79, 81; legal/administrative system 33–4, 41; planning system 60–2; *see also individual cities*
Getimis, P. 57, 59
Glasgow Action 78, 81, 117
global cities 12–14, 26
globalisation 9–12
Globe Arena project 8, 215, 221–4, 243
Gold, V. 16
Goldsmith, M. 3, 17, 28, 41
Goodwin, M. 11, 31, 40, 112
Gorbachev, Mikhail 202
Gordon, I. 147
Gosling, J. 17
Gothenburg 211, 229, 236
Gourevitch, P. 4
governance and planning, urban 76–7; approaches to 80–5; Barcelona 91–5; Berlin 98–104; Frankfurt 85–7; Milan 87–91; partnerships 78–80; Prague 95–8
Government Office for London 147–8, 150
government responsibility (fragmented) 246–7
governmental approaches (trends) 38–42
Gower, J. 251
Grand Palais (Lille) 191, 192, 193, 251
Grand Stade 186, 188, 194–5, 196, 197, 250

grands projets 45, 178, 187
Grant, M. 121
Greater London Council 122, 128, 140, 145, 146, 148
Greece (planning system) 57–60
Greek Civil Code 34
Green, H. 168
green belts 86, 115, 120
Green Paper on the Urban Environment (EC) 19, 119
Green Party 120, 121; France 156, 191; Germany 86, 252; Sweden 229, 234
greening of Thatcherism 119–20
Greenwich Waterfront 7, 135, 136, 142–5, 148, 152, 250, 254
Greenwich Waterfront Development Partnership 143–4
Grist, B. 45
growth coalities 81–3, 86, 88–9, 94, 104–6, 151–2, 242, 244
Guichard Report (1986) 159
Guidelines for Swiss Spatial Development 63
Gummer, Selwyn 126
Gustafsson, A. 206
Gustafsson, M. 37, 38
Gyllensten, L. 230, 233, 235

Haegel, F. 178, 182
Häggroth, S. 206
Hague, C. 117
Haila, A. 69
Hall, Peter 4, 16, 123
Hall, S. 112, 151
Hall, T. 73, 211, 213, 216, 227
Hambleton, R. 127
Hamburg 79, 81
Hamburgische Gesellschaft für Wirtschaftföderung 79–80
Hamnett, C. 11, 13
Hamnett, S. 48
Hårde, U. 238, 240
Harding, A. 3, 10, 81–2, 85, 132, 236
Harloe, M. 13, 23, 77
harmonisation process 23, 73
Harvey, J. 22
Hassemer, Senator 103
Häussermann, H. 98, 100, 101
Hay, C. 11
Hay, D. 5
Healey, P. 27, 72, 73, 122
Heartlands project 128, 131–3, 134, 151–2, 247
Hebbert, M. 145
Heclo, H. 202, 234
Heinz, W. 78
Hellsten, P. 218
Hennings, G. 79, 81
Heseltine, Michael 124
Hesse, J.J. 40, 41
hierarchy of cities 14–17
Hoffman, L. 96, 253
Holm, P. 65, 210
Holmberg, S. 204
Holt-Jensen, A. 67–8
Hooper, A.J. 60, 61
Hooper, J. 52
housing programme (Sweden) 208–10, 212–14, 216, 218–20
Howard, Michael 119
Hudson, P. 14, 80
Hupe, P. 48
Hutton, W. 24
Hyatt Hotel 129

ICE (in Germany) 14
ideological shift (Sweden) 203–4
Ile-de-France 173, 178-9, 184, 198, 248
Illner, M. 38
IMF 25, 37
Imrie, R. 82, 122
information society 9
'Initiatives beyond Charity' 117
INSEE 161, 162
institutional innovation 78–80
institutional responses to European competition (Euralille) 189–94
Intergovernmental Conference 26
intergovernmental rivalry 176, 183–8
Integrated Operation 130
Integrated Regional Offices 124
international competition (case study) 176–83
international context 9–26

International Convention Centre 129, 131
International Monetary Fund 25, 37
INTERREG programme 19
Ireland (planning system) 44–5
Italy: planning system 50–3, 253; urban governance (Milan) 76, 87–91, 105, 106

Jessop, B. 11, 112
Johansson, J. 227–8, 230, 231, 235
John, P. 22, 83
Johnston, B. 147
Jones Lang Wootton 95, 100, 137–8, 180
Jörgensen, I. 238–9
Jubilee Line extension 138, 144
Judd, D. 78, 84
Judge, E. 70

Kaderlaw 49
Kafkalas, G. 3
Kalbro, T. 65–6, 201, 209, 210, 212, 244
Kantor, P. 82, 84, 86
Kára, J. 38
Kearns, G. 16
Keating, M. 22, 26, 81, 85, 168
Keil, R. 15, 77, 86
Kentish Properties 138
Keogh, G. 92, 93
Kerswell, H. 144
Keyes, J. 56, 57
Khakee, A. 252
King, A. 12, 209
King, D.S. 112
King's Cross railway lands 7, 135, 136, 139–42, 145, 152, 250
Kitchen, T. 93
Kjaersdam, F. 65
Klemanski, J. 82, 83–4, 85, 130
Knapp, A. 164, 171
Kocan, W. 36, 37
kommuneplan 64, 67
KONVER programme 144
Koolhaas, Rem 192
Kot, J. 70
Kötz, H. 28, 32
Krätke, S. 103–4
Krüger, T. 79
Kühne, J. 62
Kukawa, P. 157
Kunzman, K. 21, 79, 81, 249

La Plaine Saint-Denis (case study) 176, 183–8, 194–5, 196–9, 250
Labour Party 112, 124, 126-7, 129, 131, 147, 149–50
Lagerström, P. 225
Lalenis, K.S. 58, 59
Lambert, A. 14
Lamorlette, B. 170
land-ownership (Sweden) 209–10, 212, 213–14, 216
Länder 20, 34, 60–2, 76, 98, 252
Langby, Erik 227–8
Läpple, D. 79
law/legal systems 27, 28–38
Lawless, P. 117
Le Galès, P. 6, 10, 156, 159, 161, 164, 168–71, 174, 190
leadership 84
Leeds City Development Company 118
Leemans, A.F. 31, 33
Left Party (Sweden) 234
Lega Nord (Milan) 89–90
legal systems 27, 28–38
legislative framework 5
legitimacy, accountability and 253–4
Lendi, M. 63
Leonardi, R. 41
Leontidou, L. 58, 59, 78, 105, 254
Letruelle, E. 192, 193
Lever, W.F. 16
Levine, M.A. 81, 82, 85
Levitas, R. 112
Liberal Democratic Congress (Poland) 25
Liberal Democratic Party 127
Liberal Party (Sweden) 203–4, 207, 210, 228–30, 232, 233, 239
Lidingö 211
Lieser, P. 15, 86
Light Railway (Canary Wharf) 138
Lille: Euralille project 8, 176, 189–94, 195, 196, 197–9, 247, 248; Grand Palais 191, 192, 193, 251

Lim, H. 16
Limehouse Link 138
Limonov, L. 15
Lindgren, E. 219
Livre Blanc du Bassin Parisien, Le 159–60
Lizieri, C. 13
Lloyd, G. 43, 80
Llusa, R. 94–5
local choice strategy 121–2
local democracy 114, 152–3, 171, 242, 254
local government: administrative system 27–38; Thatcher reforms 112–14; urban governance 76–106
Local Government (Planning and Development) Act (1963) 44
Local Government Act (1985) 41
Local Government Act (1991) 44, 206
local network theory 83
local politics 5
Lodden, P. 67
Łódź 70
Loftman, P. 129, 130, 134
Logan, J. 6, 11, 81
Loi d'Orientation Foncière 46
Loi d'Orientation sur la Ville 168
Loi Defferre 156, 174
London 77, 116, 247, 249; case studies 128–51; fragmentation of planning 145–51; projects/partnerships 135–45; Unitary Development Plan 140, 145, 147, 151
London Borough of Camden 140, 141
London Borough of Greenwich 143
London Boroughs Association 149
London Chamber of Commerce 149
London Docklands 13, 77, 139, 143, 147, 250
London First 149, 150, 151
London Forum 149
London Partnership 149
London Planning Advisory Committee 146–7, 149
London Pride Partnership 149–50, 151, 152–3, 250
London Regeneration Consortium 140
London Voluntary Services Council 149
Lorange, E. 67
Lorrain, D. 162, 163, 164, 171, 181
Luxembourg (planning system) 46–7
Lyon 2010 strategy 165
Lyonnaise des Eaux 163, 188

Maastricht Treaty 18, 19–20, 21, 22, 160
McGreal, S. 3, 103
Mackinstosh, M. 78
Madrid 211
Madsen, H. 202, 234
Major, John 40, 207
Makridou-Papadaki, E. 58, 59
Malbert, B. 211
Malmö 215–16, 229; Öresund Bridge project 221, 236–42, 244, 252
Malmsten, B. 229
Malusardi, F. 52
Marceau, B. 156
Mansikka, M. 68–9
Marcou, G. 28, 38, 40, 77, 173, 195
market: Eastern Europe 23–6; orientation (urban regeneration) 126; ideology (Sweden) 216–20
Markowski, T. 70
Marquand, D. 40, 205
Marsh, D. 83
Marshall, T. 91, 94
Martin, S. 130
Martinez Cearra, A. 79–80
Massey, D. 138
master plans: Italy 50–3; Luxembourg 47; Milan 87–8, 105; Prague 97; Sweden 208
Matheou, D. 102, 103
Mattsson, H. 65–6, 201, 209–10, 212, 214, 244
Mauroy, Pierre 190–1, 193–4
Mawson, J. 159, 168, 190
Mayer, M. 77
mayors (role in France) 164–5, 172, 173, 174, 182, 194, 196
Mazey, S. 160
Mazza, L. 51
Mény, Y. 41, 170, 171
Metropoles 15
Midwinter, A. 168

Milan (urban governance/planning) 76, 87–91, 105, 106
Millennium Commission 133, 142
Millennium project 144–5, 150
Million Homes Programme 208
Mishra, R. 235
Modeen, T. 69
Moderate Party (Sweden) 201, 203–4, 207–8, 210, 217–19, 228–30, 235, 239, 241
Mollenkopf, J.H. 13
Molotch, H. 81, 82, 87–9
Montijn Commission 49
Montin, S. 41, 206–7
Montpellier 157
Moran, F. 30
Mossberger, K. 82, 83–4
motorway revival (Stockholm) 225–7
Moulaert, F. 9
Mouritzen, P.E. 41, 76
Moylan, M. 21
MSI (in Italy) 50
Muccini, P. 89
Müller 95
Myrhe, J.E. 67

Nacka 227–8, 235
Napoleonic family: legal/administrative systems 29, 31–3, 39–40; planning systems 45–60, 72
nation–states 6, 10–11, 19, 21–3, 26, 34, 46–7, 245
national: economic planning 9–10; framework 27–75; governmental approaches (trends) 38–42; planning re-emergence (France) 171–3; planning systems 42–71, 245
National Association of Municipal Companies 209
National Council for Voluntary Organisations 146
National Development Plan 44
National Exhibition Centre 128
National Federation of Tenants' Associations 209
National Franchise Board of Environmental Protection 236
National Freight 140
National Indoor Arena 129
National Parks 115
National Physical Planning Agency (Netherlands) 15, 47
national planning approaches 109; Britain (Thatcher reforms) 111–27; English case studies 128–53; France (state reorganisation) 154–75; French case studies 176–99; Sweden (planning model) 200–20; Swedish case studies 221–44
National Planning Guidelines 43
National Planning Policy Guidelines 43
National Road Administration 232–3
Natural Resources Act (Sweden, 1987) 65, 211–12, 218
Nature Step 234
Needham, B. 3, 49
negotiation planning 212–16, 219–24, 248
neighbourhood councils 206
Némery, J.C. 159
Netherlands 15, 40; planning system 47–50; Scientific Council 78
network theory 83, 85
Nevin, B. 127, 129, 130, 134
New Democrats (Sweden) 203–4
New Economic Approach 126
New Right 111
Newcastle Initiative 117
Newlands, D. 80
Newman, P. 82, 140, 147, 151, 193
Newton, K. 41
Nijkamp, P. 16, 190
NIMBYism 96, 120, 126, 210, 218, 248
No Turning Back Group 113
Noir, Michel 164, 165
Norell, P.O. 207
North Sea Regions 20
North West Business Leadership Team 118
North West Regional Alliance 118
Northern Ireland Office 43
Northern League 41, 50
Norton, A. 30, 33
Norway (planning system) 67–8
Novarina, G. 77, 165

Objective 2 designation 18, 150
Ödmann, E. 210, 211
OECD 211
Office for Modern Architecture 192
Olympia and York 137–9
Olympic Games (Barcelona) 91–5, 105
Olympic Holdings (HOLSA) 93
Operation for Urban Reconstruction (Greece) 59
Ordinamento della Autonomie Locali 51
Örestad project 238–9
Öresund Bridge project (case study) 221, 236–42, 244, 252
Östergård, N. 63–4
Österleden Projekt 227–8, 229, 230, 235
Ozanne, J. 99

Padioleau, J.-G. 165
Page, E.C. 28
Paris (case studies): La Plaine Saint-Denis 8, 176, 183–8, 194–5, 196–9, 250; Seine Rive Gauche 8, 176–83, 194, 195, 197–8, 248, 252
Parkinson, M. 3, 15, 18, 78–9, 84, 174
partnerships 5, 253; in cities 78–80; in London (case studies) 135–45; public/private 4, 77, 78–80, 84–5
Patten, Chris 121
Pearce, David 119
Pearce, G. 130
Peck, J. 11, 118
Pelli, Cesar 137
Petersson, O. 204–5, 235
Philo, C. 16
Physical Planning Act (Netherlands) 47
Pierre, Abbé 181
Piven, F.F. 4, 23
Plaine Développement 185, 196
Plaine Saint-Denis, La (case study) 8, 176, 183–8, 194–5, 196–9, 250
plan-led system (Thatcher reforms) 120–2
Plan d'Aménagement du Zone 178–9, 180–1, 182
Plan d'Occupation des Sols 46, 155, 161, 163, 166, 169, 178, 198
Plan Programme de l'Est de Paris 177–8
Planning Appeals Board 45
planning approaches, national 109; Britain (Thatcher reforms) 111–27; English case studies 128–53; France (state reorganisation) 154–75; French case studies 176–99; Sweden (planning model) 200–20; Swedish case studies 221–44
Planning and Building Acts 65–7, 69
Planning Commission (Luxembourg) 47
Planning and Compensation Act 43, 120
planning and governance *see* governance and planning, urban
planning model (Sweden) 208–10
Planning Policy Guidance Notes 42–3, 44, 71, 121, 126
planning system and urban policy (Britain 1990s) 119–27
planning systems: British family 42–5; Eastern Europe 69–71; Germanic family 60–3; Napoleonic family 45–60; Scandinavian family 63–9
Poblenou site (Barcelona) 92, 93
Poland 25, 36–8, 70
polarisation (global cities) 12–13
political influence (Milan) 87–91
Pollock, S. 43
Port Greenwich scheme 144
Port of London Authority 116
Portugal (planning system) 53–4
post-decentralisation problems (France) 169–71
post-Fordism 15
Power, A. 93, 167
power: decentralisation (Sweden) 204–8; relations 5; restructuring (Sweden) 233–6
Poznan 70
Prague (urban governance/planning) 76, 95–8, 104–5
préfets 33, 46, 155–7, 159, 161, 163, 168–9, 172, 174, 177, 247
Presidential Decrees (Greece) 58, 59

Preteceille, E. 11, 26, 41
private/public partnership 4, 77, 77–80, 84–5
private/public strategic planning (London) 147–50
Private Finance Initiative 126
private interests (Projekt Österleden) 227–8
privatisation policies (Sweden) 206–8, 216
pro–growth strategies 81–3, 86, 88–9, 94, 104–6, 129, 133, 151–2, 165, 242, 244, 247–9, 253, 254
Progetto Passante 88, 89, 105
Programme of Urban Action (Spain) 57
Progressive regimes 82
Project and Development Plan (Berlin) 61–2, 103
property: development 13–14; restitution (Eastern Europe) 36, 96
Provan, B. 167
public/private partnership 4, 77, 77–80, 84–5
public/private strategic planning (London) 147–50
public finance 81–2, 138
Public Finance Office 138
public sector, communal 162–7
Pyrgiotis, Y. 57, 59, 60

quality circles 192
quangos 114, 246

Railway Lands Group 140–2
Rasmussen, Knud 238
Rautsi, J. 68–9
Read, E. 205
Reformation 32
regime theory 82–4, 86–7
Regional Administrative Tribunal 89
Regional Challenge 124
Regional Co-ordination Commission 53
regional councils 159–60, 172, 182
Regional Development Plans 58
Regional National Committees 38
regional offices initiative 123, 124
regional planning 68, 156–61
regional policy 17–21
Regional and Strategic Guidance 43
regionalism 40, 71, 72, 148
Regulski, J. 36, 37
regulation theory 11
Reis, C. 53, 54
Renard, V. 46, 169, 170, 178
representative democracy (Sweden) 204–5
Réseau Express Regional (RER) 179
Rhodes, R.A.W. 83
Ridley, Nicholas 113
Riera, P. 92, 93
Ringli, H. 63
Rio Summit 120
Robert, J.-P. 178
Robinson, S. 149
Robson, B. 112, 122
Rocard government 186
Roissy 180
Roman Law 30, 33
Ronneburger, K. 77, 86
Rose, R. 6
Ross, J. 236
Rouzeau, B. 185
RPR 171, 182
Rudberg, E. 208

Sahlin-Anderson, K. 221, 224
St Pancras station 141, 142
Sassen, S. 12, 178
Savitch, H.V. 82, 84, 86
Scandinavia: legal/administrative systems 29, 34–5, 39; planning systems 63–9, 72–3
Scandinavian Airlines System 236
Scandinavian Link 236
Schèmas Directeurs 46, 173
Schulman, H. 15
Scottish Development Agency 81
Scottish Office 43
Seine Rive Gauche (case study) 8, 176–83, 194, 195, 197–8, 248, 252
self–government 38, 63, 69, 96, 205
Shachar, A. 12
Sheffield Economic Regeneration Committee (SERC) 117
Shiner, P. 127

Simplified Planning Zones 115, 131
Single European Act 3, 17, 90, 148
Single Market 19–20, 22–3, 159, 171, 190
Single Regeneration Budget 123–4, 126, 135, 144, 146, 148, 151–2, 246, 250
Skanska 227, 240–1, 242
Skelcher, C. 113, 145–6
smaller cities 15
Smith, N. 12
SNCF 157, 180, 185, 191–2, 195
Snickars, F. 213
Social Democrats: Berlin 102, 104; Sweden 201, 203–10, 216, 218–19, 223, 228–30, 232–3, 235, 239, 241
social objectives 249–53
social regulation 11
social responsibility/cohesion 152
Socialist Party: France 162, 167, 191, 194; Frankfurt 86; Milan 87, 89
socialist reformism (Sweden) 202
Société Centrale pour l'Equipment du Territoire (SCET) 191–2
Société d'Economie Mixte (SEM) 163–4, 174, 179–80, 185, 191–7, 246
Société d'Economie Mixte d'Aménagement de Paris (SEMAPA) 179–82, 196–7
Société Générale 180
Société National des Chemins de Fer 157, 180, 185, 191–2, 195
Société Nationale d'Economie Mixte (SANEM) 187, 188, 197
society/social change (Sweden) 205
Söderlind, J. 227
Sotarauta, N. 68, 69
South, G. 100
Spain 41; Barcelona (urban governance and planning) 76, 91–8, 104–6, 247, 249; planning system 55–7
Special Area Regulations 66, 212
Special Development Orders 115
Stadtforum (in Berlin) 103
Stevens, J. 190
Stevenson, D. 150, 152
Stewart, J. 31, 113–14, 145–6, 205–6
Stockholm 213, 220; Dennis package 221, 225–36, 243–4, 247, 252, 253; Globe project 8, 215, 221–4, 243
Stockholmsleder 232–3
Stoker, G. 11, 31, 38, 76, 82–4, 112, 113, 206, 254
Stone, C. 82–3, 85
Stopford, J. 10
Strange, S. 10
strategic planning (London) 147–50
Strom, E. 98, 100, 101
Structural Funds 17, 18, 19, 21, 23
subregional alliances (London) 150–1
subsidiarity principle 30, 39, 61
Suetens, L.P. 55
Swanstrom, T. 6, 11
Sweden 41; case studies 221–44; planning model 200–20; planning system 65–6, 246, 247
Swedish case studies: Dennis package 225–36; Globe Arena project 221–4; Öresund Bridge 236–42
Swedish Rail 236
Swianiewicz, P. 37
Switzerland (planning system) 62–3
Sýkora, L. 24, 95–6, 97
Syndicats à Vocation Multiple 162
Syndicats à Vocation Unique 162

Talia, N. 52
Tam-Tam 181
tax base: Paris 195–6; Sweden 200, 203
taxe professionnelle 161–2, 163, 170, 173
Taylor, R. 203, 205
Technopolis International 178
Territorial Administration of the Republic 159, 161
TGV 14, 19, 45, 173, 186, 189–91, 193, 194–5, 198
Thames Gateway Framework 150
Thatcher, M. 23, 24, 30, 37
Thatcherism (legacy of) 111; planning system/urban policy 120–7; project (themes) 112–18
This Common Inheritance 119–20
Thomas, D. 49

Thomas, H. 82, 122
Thornley, A. 71, 82, 112, 137, 147, 193
Thrift, N. 9, 10
Tickell, A. 11, 118
Tobacco Dock retail development 138
Tonboe, J. 238–9
Tonell, L. 213, 216, 228, 230–1, 233
Toonen, T.A.J. 40, 47
Toubon, Jacques 182
Towards Sustainability (EC) 17–18
Training and Enterprise Councils 117, 118, 123–4, 143, 145–6
Travers, T. 145, 147, 148
Treaty of European Union 18, 19–20, 21, 22, 160
Treaty of Rome 17, 18
Treuhandanstalt 100
Triangle project 215–16
Tribunal Administratif de Paris 181
trickle–down effect 114, 122, 138–9, 245
Turok, I. 122
TYBO (in Tyresö) 218–19

ultra peripheral regions 20
Unal, M. 160
UNCED Agenda 21 programmes 18, 120, 220, 252
Unitary Development Plan 43; Birmingham 129, 132, 134, 151, 248; London 140, 145, 147, 151
Urban Development Act (Greece) 58, 59
Urban Development Board 219
Urban Development Corporations 115–16, 122, 123, 135, 137, 151, 246
urban governance and planning 76–7; approaches 80–5; Barcelona 91–5; Berlin 98–104; Frankfurt 85–7; Milan 87–91; partnerships 78–80; Prague 95–8
urban hierarchy 14–17
urban planning: approaches/definitions 3–8; conclusions 245–54; development approaches (France) 194–9; and economic growth 247–9; issues (in France) 174–5; styles 165–7
urban policy: Britain 119–27; Europe 17–21; France 167–9
urban regeneration 116–18, 122–7, 146, 148, 150
urban regimes 82–4, 86–7
urban villages 127

Valk, A. 47
Valles, J.M. 41
Van den Berg, I. 190
Van den Bos, J. 100, 103
Van Wunnick, P.M. 55
Vasa Terminal 215, 223
Vasconcelos, L. 53, 54
Vedung, E. 209
Vegara, J. 93
Veraldo, R. 51
Verebelyi, I. 38, 39, 40, 77
Verpraet, G. 193
Verts du 13ème 181
Verwijnen, J. 15
Vicari, S. 82, 87–9
Vila Olimpica (VOSA) 93
Virtanen, P. 68, 69
voivodships (in Poland) 38
Volvo 236
Voogd, H. 16
Vorhaben-und-Erlschliessungsplan 61–2, 103

Ward, S. 16
Wassenhoven, L. 58, 59, 60
Water Court 236
Waterlinks business village 132
Weber, G. 62
welfare state 10–11, 23, 24; Britain 111, 114; Sweden 200–1, 203, 205, 208, 235
Weltstadt 12
West Midlands County Council 130, 135
Westerståhl, J. 204
White, P. 138
Whyatt, A. 150, 151
Williams, A.M. 53, 54
Williams, R. 13
Williams, R.H. 27, 72, 73–4
Wolman, H. 41

Wood, B. 13, 131–2
Woolwich Building Society 143
Woolwich Revival project 144
World Bank 37
World Cup (1998) 186, 187
World Ice Hockey Championship (1989) 222, 224
World Trade Centre 215
Wretblad, L. 219
Wynn, M. 55

Yarranton, L. 101
Young, S. 254

Zarecky, P. 38
Zones d'Aménagement Concerté 46, 59, 178, 179–80, 182, 185
Zones a Urbaniser en Priorité 45
Zoning and Planning Act (Belgium) 54–5
Zweigert, K. 28, 32